New Directions f

CHEMICAL ENGINEERING

Committee on Chemical Engineering in the 21st Century:
Challenges and Opportunities

Board on Chemical Sciences and Technology

Division on Earth and Life Studies

National Academy of Engineering

A Consensus Study Report of
The National Academies of
SCIENCES · ENGINEERING · MEDICINE

THE NATIONAL ACADEMIES PRESS
Washington, DC
www.nap.edu

THE NATIONAL ACADEMIES PRESS 500 Fifth Street, NW Washington, DC 20001

This material is based upon work supported by the U.S. Department of Energy, Office of Science, Biological and Environmental Research Program under Award Number DE-SC0019159; the U.S. Department of Energy, Office of Energy Efficiency & Renewable Energy, Advanced Manufacturing Office under Award Number DE-EP0000026/89243420FEE400139; and the U.S. Department of Energy, Office of Fossil Energy and Carbon Management under Award Number DE–EP0000026/89303018 FFE400005. This report was prepared as an account of work sponsored by an agency of the United States Government. Neither the United States Government nor any agency thereof, nor any of their employees, makes any warranty, express or implied; or assumes any legal liability or responsibility for the accuracy, completeness, or usefulness of any information, apparatus, product, or process disclosed; or represents that its use would not infringe privately owned rights. Reference herein to any specific commercial product, process, or service by trade name, trademark, manufacturer, or otherwise does not necessarily constitute or imply its endorsement, recommendation, or favoring by the United States Government or any agency thereof. The views and opinions of authors expressed herein do not necessarily state or reflect those of the United States Government or any agency thereof. The activity was supported by the National Science Foundation under Award Number CHE - 1926880, as well as private contributions from universities, industry, and professional organizations (Appendix D). Any opinions, findings, conclusions, or recommendations expressed in this publication do not necessarily reflect the views of the National Science Foundation or any organization or agency that provided support for the project.

International Standard Book Number-13: 978-0-309-26842-4
International Standard Book Number-10: 0-309-26842-7
Library of Congress Control Number: 2022937743
Digital Object Identifier: https://doi.org/10.17226/26342

Additional copies of this publication are available from the National Academies Press, 500 Fifth Street, NW, Keck 360, Washington, DC 20001; (800) 624-6242 or (202) 334-3313; http://www.nap.edu.

Printed in the United States of America

Suggested citation: National Academies of Sciences, Engineering, and Medicine. 2022. *New Directions for Chemical Engineering*. Washington, DC: The National Academies Press. https://doi.org/10.17226/26342.

The National Academies of

SCIENCES • ENGINEERING • MEDICINE

The **National Academy of Sciences** was established in 1863 by an Act of Congress, signed by President Lincoln, as a private, nongovernmental institution to advise the nation on issues related to science and technology. Members are elected by their peers for outstanding contributions to research. Dr. Marcia McNutt is president.

The **National Academy of Engineering** was established in 1964 under the charter of the National Academy of Sciences to bring the practices of engineering to advising the nation. Members are elected by their peers for extraordinary contributions to engineering. Dr. John L. Anderson is president.

The **National Academy of Medicine** (formerly the Institute of Medicine) was established in 1970 under the charter of the National Academy of Sciences to advise the nation on medical and health issues. Members are elected by their peers for distinguished contributions to medicine and health. Dr. Victor J. Dzau is president.

The three Academies work together as the **National Academies of Sciences, Engineering, and Medicine** to provide independent, objective analysis and advice to the nation and conduct other activities to solve complex problems and inform public policy decisions. The National Academies also encourage education and research, recognize outstanding contributions to knowledge, and increase public understanding in matters of science, engineering, and medicine.

Learn more about the National Academies of Sciences, Engineering, and Medicine at **www.nationalacademies.org**.

COMMITTEE ON CHEMICAL ENGINEERING IN THE 21st CENTURY: CHALLENGES AND OPPORTUNITIES

Members

ERIC W. KALER, NAE (*Chair*), Case Western Reserve University
MONTY M. ALGER, NAE, The Pennsylvania State University
GILDA A. BARABINO, NAE, NAM, Olin College of Engineering
GREGG T. BECKHAM, National Renewable Energy Laboratory
DIMITRIS I. COLLIAS, The Procter & Gamble Co.
JUAN J. DE PABLO, NAE, University of Chicago
SHARON C. GLOTZER, NAS, NAE, University of Michigan
PAULA T. HAMMOND, NAS, NAE, NAM, Massachusetts Institute of Technology
ENRIQUE IGLESIA, NAE, University of California, Berkeley
SANGTAE KIM, NAE, Purdue University
SAMIR MITRAGOTRI, NAE, NAM, Harvard University
BABATUNDE A. OGUNNAIKE,[1] NAE, University of Delaware
ANNE S. ROBINSON, Carnegie Mellon University
JOSÉ G. SANTIESTEBAN, NAE, ExxonMobil Research and Engineering Company, *retired*
RACHEL A. SEGALMAN, NAE, University of California, Santa Barbara
DAVID S. SHOLL, Oak Ridge National Laboratory
KATHLEEN J. STEBE, NAE, University of Pennsylvania
CHERYL TEICH, Teich Process Development, LLC (*until September 2020*)

Consultants

PHILIP B. HENDERSON, EMD Electronics
REINALDO M. MACHADO, EMD Electronics
LAURA MATZ, EMD Electronics

Staff

MAGGIE L. WALSER, Study Director
BRENNA ALBIN, Program Assistant
BRITTANY BISHOP, Christine Mirzayan Science Policy Fellow
KESIAH CLEMENT, Research Assistant
ANNE MARIE HOUPPERT, Senior Librarian
GURU MADHAVAN, NAE Senior Director of Programs
REBECCA MORGAN, Senior Librarian
NICHOLAS ROGERS, Deputy Director, Program Finance
LIANA VACCARI, Program Officer
JESSICA WOLFMAN, Research Associate
ELISE ZAIDI, Communications Associate

---

[1] Deceased, February 20, 2022

Reviewers

This Consensus Study Report was reviewed in draft form by individuals chosen for their diverse perspectives and technical expertise. The purpose of this independent review is to provide candid and critical comments that will assist the National Academies of Sciences, Engineering, and Medicine in making each published report as sound as possible and to ensure that it meets the institutional standards for quality, objectivity, evidence, and responsiveness to the study charge. The review comments and draft manuscript remain confidential to protect the integrity of the deliberative process.

We thank the following individuals for their review of this report:

Nicholas Abbott, NAE, Cornell University
Aristos Aristidou, NAE, Cargill, Inc.
Gretchen Baier, The Dow Chemical Company
Donna Blackmond, NAS/NAE, The Scripps Research Institute
Joan Brennecke, NAE, The University of Texas at Austin
Prodromos Daoutidis, University of Minnesota
Alice Gast, NAE, Imperial College London
Julia Kornfield, NAE, California Institute of Technology
Cato Laurencin, NAS/NAE/NAM, University of Connecticut Health Center
Jodie Lutkenhaus, Texas A&M University
Phillip Westmoreland, North Carolina State University

Although the reviewers listed above provided many constructive comments and suggestions, they were not asked to endorse the conclusions or recommendations of this report nor did they see the final draft before its release. The review of this report was overseen by **Thomas Connolly Jr.**, American Chemical Society, and **Elsa Reichmanis**, Lehigh University. They were responsible for making certain that an independent examination of this report was carried out in accordance with the standards of the National Academies and that all review comments were carefully considered. Responsibility for the final content rests entirely with the authoring committee and the National Academies.

Acknowledgments

This study would not have been completed successfully without the contributions of many individuals and organizations. The committee would especially like to thank the individuals who participated in our town hall at the American Institute of Chemical Engineers (AIChE) 2019 Annual Meeting and the AIChE Virtual Local Section meeting in spring 2020. We are grateful as well for the insights provided by respondents to our community questionnaire in spring 2021 (Appendix C), as well as by the numerous individuals who spoke to the committee during an open information-gathering session or otherwise provided input, and we thank the organizations that contributed financial support for this study (Appendix D). We also are grateful to Elsevier for providing access to its SciVal tool.

In Memory of Babatunde A. Ogunnaike

Professor Babatunde (Tunde) Ogunnaike was a valuable member of the report committee who passed away just after the report was released in 2022. Tunde's contributions can be seen in every part of the report. He had a broad and deep knowledge of our field, and his perspective and clear thinking both empowered forward thinking and constrained the growth of bad ideas. His kind spirit and easy style of collaboration made him a true joy to work with; his fluid and precise writing style, tireless energy, and ability to meet a deadline made him an ideal committee member. Tunde was a warm and engaged scholar with valuable insights and a broad vision that spanned many areas of chemical engineering. Beyond his engineering contributions, Tunde was incredibly generous with his time, teaching and mentoring countless early career scientists and engineers and leading his College. He was a true friend to many, and those of us who were lucky to know him carry with us a bit of Tunde in the example he leaves for us. Our community has lost a giant.

Eric W. Kaler, *Chair*
On Behalf of the Committee and Staff

Preface

"It is hard to make predictions, especially about the future."

Attributed to many. And true.

Yet we as a group accepted the challenge of developing a report designed to articulate the status of and challenges and promising opportunities for the field of chemical engineering in the United States, as well as benchmark its international stature, for the next 10 to 30 years. A committee comprising 17 chemical engineers with diverse backgrounds, expertise, and life experiences explored a question not investigated by the National Academies since the 1980s: What is the future of chemical engineering?

As the only engineering field with molecules and molecular transformations at its core, chemical engineering represents an area of intellectual inquiry and commercial applications that is profoundly important for society's future advances in such vital areas as energy, food, water, medicine, and manufacturing. Chemical engineering is also the natural door through which the implications and applications of molecular biology—writ large in its current incarnations, including genetic engineering, personalized medicine, organs-on-chips, and even artificial intelligence—enter the realm of practice and application. The future of this field has crucial implications for what the future looks like for everyone.

Despite remarkable advances and contributions, the legacy of chemical engineering is complicated. As a profession and a discipline, chemical engineering has enabled the cost-effective production of materials and chemicals. On the other hand, the durability of some of these products, such as plastics and fluorinated chemicals, continues to have unintended consequences for the environment. At the same time, energy transformations have generated greenhouse gases that threaten Earth's climate. It is essential, therefore, that any future advances in the field address the history of the advances of the past—an emphasis throughout this report.

The report describes how chemical engineering is well positioned to serve as the enabling discipline in advancing the decarbonization of energy systems and materials without compromising reliability and cost, and while remaining cognizant of the existential threat of global climate change. In the foreseeable future, no single energy carrier will be able to meet the energy demands of all sectors, and the work of chemical engineers will play a vital role in informing the selection of options for the scale-up, delivery, systems integration, and optimization of the mix of energy carriers that will address the world's energy needs with lower carbon emissions and costs across all regions and sectors of society. At the same time, global pressures associated with climate change, energy demand, and population growth will change, in unprecedented ways, the ways in which humans meet their needs for food and water. As in the past, chemical engineers will confront these challenges through such enabling technologies as precision agriculture, the development of protein alternatives, and the reduction or elimination of food waste.

Chemical engineers also will be leaders in the engineering of targeted and accessible solutions for human health. Their domain of influence will range from personalized medicine to the application of systems engineering to biology and health. This work will include strategic modification of the molecular pathways and genomic networks involved in the regulation of both normal physiology and disease states. It will also include the application of systems-level thinking to the production of and end-of-life considerations for useful materials, including polymers and a variety of other hard and soft materials, in a circular economy. Chemical engineers will lead the way as well in the application of new tools—such as machine learning and artificial intelligence—to solve complex problems.

As for the U.S. position in chemical engineering, it is critical to note that China is making large investments in technologies that are either central or highly relevant to chemical engineering. These investments, combined with China's accelerating productivity and scholarly output, makes investment in the U.S. research enterprise imperative. Failure to make these investments will cede global leadership not only of chemical engineering, but of technology more broadly.

I commend the committee members for their enthusiastic engagement and hard work. We all found our ways to collaborate and communicate while constrained by the COVID-19 pandemic, but I know we also all missed the synergies and spontaneous insights that in-person conversations would have generated. While we engaged in virtual meetings and chats instead of face-to-face meetings, at the end of the day, the creative engagement and critical thinking of the group made it possible to crystallize important ideas. Finally, but of crucial importance, the expert guidance, gifted diplomacy, and detailed engagement of the National Academies staff, led by Dr. Maggie Walser and including Kesiah Clement, Dr. Liana Vaccari, and Jessica Wolfman, made this report possible.

Eric W. Kaler, *Chair*
Committee on Chemical Engineering in the 21st Century:
Challenges and Opportunities

Contents

APPENDIXES

Summary

Chemical engineering is the engineering of systems—at scales ranging from the molecular to the macroscopic—that integrate chemical, physical, and biological elements to design processes and produce materials and products for the benefit of society. Chemical transformations are at the heart of the technologies that enable modern society, and the work of chemical engineers has affected societies and individual lives around the world. Without synthetic fertilizers made with chemical engineering processes, for example, a Green Revolution to feed the world would not have been possible. Without the invention of Ziegler-Natta catalysts, polyolefins would not exist, and the myriad benefits of plastics would not have been realized. Without the invention of tough, stable polymers such as Teflon® and Kevlar®, the commercial and medical devices made from those polymers would not have emerged. Without the contributions of many chemical engineers, the silicon chips, glass materials, and plastics that make up today's ubiquitous electronic devices would not have been developed. And without an army of chemical engineers, there would be no oil and gas industry to power the world and no pharmaceutical industry to discover and produce the medicines, therapeutics, and vaccines needed for a long and healthy life. More recently, chemical engineers have contributed to the tools of directed evolution, which has allowed for the engineering of improved function in proteins, metabolic pathways, and genomes.

Unfortunately, the discoveries of chemical engineers have also been responsible for the production of chemicals that will persist in the environment indefinitely, greenhouse gas emissions that contribute to climate change, plastic materials that accumulate in landfills and the oceans, and the chemicals of war that have inflicted long-term or permanent damage on humans and the environment. Thus the field of chemical engineering today faces opportunities and challenges not only to innovate for the future, but also to innovate in ways that repair the unintended consequences of the past.

Chemical engineering is a discipline and a profession that evolved from the roots of industrial and applied chemistry, which in turn emerged from such ancient chemical processes as fermentation and leather tanning. Its academic legacy traces back to the late 1880s, when steam engines still powered the world, and internal combustion was a nascent idea. The world has of course changed since then, and continues to do so at a rapid rate, as illustrated by the pace at which technology is disrupting established practices and organizations. Yet the core chemical engineering curriculum has evolved more slowly over the preceding decades, even as the challenges facing engineers have expanded and become more difficult.

At its most fundamental level, engineering is about solving problems, and it is natural to expect that as one problem is solved, another will emerge or grow in importance. The last time the National Academies of Sciences, Engineering, and Medicine surveyed the challenges and opportunities for chemical engineering—in the 1988 report *Frontiers in Chemical Engineering: Research Needs and Opportunities*, better known as the "Amundson Report"—the conclusions of that study suggested an approach to developing new technologies and maintaining leadership in established ones. Given that the pace of change has only increased since the 1980s, and in the face of a rapidly evolving landscape for higher education in general, a fresh look at what new challenges and opportunities lie ahead for both the discipline and the profession of chemical engineering could not be more timely.

Challenges faced today include not only addressing climate change and the energy transition, but also reducing raw material usage and increasing recycling to move from a linear to a circular economy, generating and distributing food worldwide while conserving water and other resources, and creating and scaling the manufacture and distribution of new medicines and therapies. Across all these applications, chemical engineers have opportunities to address today's most important problems by collaborating with multiple disciplines and engaging systems-level thinking. To leverage these opportunities, now and in the coming decades, chemical engineering will need to define and pursue new directions. To this end, this report details a vision for the future of chemical engineering research, innovation, and education.

To provide a framework for discussion in this report, the study committee examined the role of chemical engineering in addressing key challenges that face society. While several organizations have outlined grand challenges in various areas, this report focuses on the areas of energy and the energy transition; water, food, and air; health and medicine; manufacturing and the circular economy; and materials. Also included is a discussion of tools and techniques with the potential to enable future advances across all of these areas. In addition, the report examines the current state of chemical engineering education and the need for innovation to ensure that the next generation of chemical engineers is equipped to address the challenges that lie ahead. Observations on U.S. international leadership in chemical engineering are provided as well.

DECARBONIZATION OF ENERGY SYSTEMS

Mitigation of climate change is one of, if not the most, pressing problems facing humankind and the planet today. Addressing this problem will require decarbonization of current energy systems, a challenge rendered all the more difficult by the complexity and magnitude of the energy landscape and the resultant inability of any single energy carrier to meet the energy demands of all sectors in the foreseeable future. The field of chemical engineering continues to make important contributions to the scalability, delivery, systems integration, and optimization of the mix of energy carriers that will meet energy needs across different regions and sectors of society with lower carbon emissions and costs. Chemical engineers will enable technological advances at every point in the energy value chain, from sources to end uses, and bring to bear the systems-level thinking necessary to balance the economic and environmental trade-offs that will be necessary to transition to a low-carbon energy system.

The increasing market penetration of electric vehicles for personal transportation, for example, calls for a reimagining of petroleum refineries that were designed to produce gasoline and diesel fuel as their main products. The transition to a low-carbon energy system will require a bridging strategy that relies on a hybrid system consisting of a mix of energy carriers. Chemical engineering is rooted in the transformation of stored energy carriers into forms that are more convenient and into chemicals and materials, and chemical engineers have an important opportunity to continue applying their skillsets to non-fossil-based energy sources and carriers.

In the long term, achieving net-zero carbon emissions will require significant advances in photochemistry, electrochemistry, and engineering to enable efficient use of the predominant source of energy for Earth—the solar flux. To this end, novel systems will be required to improve the efficiency of photon capture and conversion to electrons; improve the storage of electrons; and advance the direct and/or sequential conversion of photons to energy carriers via reactions with water, nitrogen, and CO_2 to produce hydrogen, ammonia, and liquid fuels, respectively.

Successful mitigation of climate change and the transition to a low-carbon energy system will also require chemical engineers to collaborate with other disciplines, including chemistry, biology, economics, social science, and others. In the energy sector, coordination between academic researchers and industrial practitioners, as well as international collaboration, will be crucial to ensuring that solutions are economically competitive and deployable at scale.

Recommendation 3-1:[1] Across the energy value chain, federal research funding should be directed to advancing technologies that shift the energy mix to lower-carbon-intensity sources; developing novel low- or zero-carbon energy technologies; advancing the field of photochemistry; minimizing water use associated with energy systems; and developing cost-effective and secure carbon capture, use, and storage methods.

Recommendation 3-2: Researchers in academic and government laboratories and industry practitioners should form interdisciplinary, cross-sector collaborations focused on pilot- and demonstration-scale projects and modeling and analysis for low-carbon energy technologies.

SUSTAINABLE ENGINEERING SOLUTIONS FOR ENVIRONMENTAL SYSTEMS

Chemical engineers have historically played a central role in the energy sector, but their contributions have been more modest in solving problems in the interconnected space of water, food, and air quality. Yet while water, food, and air have historically been the focus of other disciplines, chemical engineers bring both molecular- and systems-level thinking to pioneering efforts in this highly interconnected space. The positive impact of chemical engineers will be magnified as they adapt to thinking beyond the traditional unit operation scale to focus at a global scale. A continued increase in the world's population will lead to increased resource demands, a challenge that is key to defining the future of chemical engineering.

---

[1] The committee's recommendations are numbered according to the chapter of the main report in which they appear.

Chemical engineers can support water conservation by both designing higher-efficiency processes and developing methods for using alternative fluids to freshwater. Specific research opportunities range from better understanding the fundamentals of water structure and dynamics to developing membranes and other separation methods. In the domain of water use and purification, U.S. chemical engineers would benefit from collaborations with civil engineers and other scientists and engineers in arid regions that have more experience with desalination.

Global pressures associated with climate change and population growth will require substantial changes in the world's food sources, a need that chemical engineers can help address through enabling technologies. Specific opportunities for chemical engineers include precision agriculture, non–animal-based food and low-carbon-intensity food production, and reduction or elimination of food waste. Advanced agricultural practices designed to improve yield while reducing demand for both energy and water will require collaboration with other disciplines, as well as systems-level approaches such as life-cycle assessments. A particularly valuable opportunity for collaboration is with researchers who are pioneering initial demonstrations of "lab-grown" foods on small scales.

The Earth's atmosphere, with its large range of spatial and temporal scales, presents intriguing challenges for chemical engineers. Chemical engineers have contributed to fundamental understanding in this area, and their work will continue to contribute to improving air quality, including through the removal of CO_2 and other heat-trapping gases. Chemical engineers have contributed to current understanding of aerosol particles in particular, and will have an opportunity to aid in improving air quality by advancing understanding of the nature and physics of aerosol particles and applying separation technologies, as well as the molecular- and systems-level thinking that will be necessary to address this global challenge. Atmospheric science is already an interdisciplinary field that includes chemistry, physics, meteorology, and climatology, making it a promising area in which chemical engineering can contribute through increased collaboration.

Recommendation 4-1: Federal research funding should be directed to both basic and applied research to advance fundamental understanding of the structure and dynamics of water and develop the advanced separation technologies necessary to remove and recover increasingly challenging contaminants.

Recommendation 4-2: To minimize the land, water, and nutrient demands of agriculture and food production, researchers in academic and government laboratories and industry practitioners should form interdisciplinary, cross-sector collaborations focused on the scale-up of innovations in metabolic engineering, bioprocess development, precision agriculture, and lab-grown foods, as well as the development of sustainable technologies for improved food preservation, storage, and packaging.

ENGINEERING TARGETED AND ACCESSIBLE MEDICINE

There are few areas of science and engineering in which the rate of progress has been, and continues to be, more rapid than advances in biology and biochemistry aimed at treatments and cures for human illness. Specific contributions of chemical engineers include reactor design and separations, and more recently cell engineering, formulations, and other aspects of drug manufacturing. Since the first attempts to isolate small molecules from biological organisms and control and reengineer cell behavior, the development of biologically derived products has increased, with major advances resulting from recombinant DNA technology, the sequencing of genomes, the development of polymerase chain reaction, the discovery of induced pluripotent stem cells, and the discovery and implementation of gene editing.

All of these challenges present opportunities for chemical engineers to apply systems-level approaches at scales ranging from molecules to manufacturing facilities, and to coordinate and collaborate across disciplines. Opportunities to apply quantitative chemical engineering skills to immunology include cancer immunotherapies, vaccine design, and therapeutic treatments for infectious diseases and autoimmune disorders. The development of completely noninvasive methods for drug delivery represents an exciting frontier of device- and materials-based strategies. Chemical engineers are also well positioned to advance work with sustained-release depots and targeted delivery of therapeutics.

In addition, the demand for monoclonal antibodies, therapeutic proteins, and messenger RNA (mRNA) therapeutics will continue to grow, in part in response to the aging U.S. population. At the same time, the cost to produce biologics and the subsequent cost to the consumer create pressure to improve flexibility and reduce costs so as to increase health care equity while maintaining reliability and stability during manufacturing and distribution. This challenge provides an opportunity for chemical engineers to develop novel bioprocess and cell-based improvements through collaborations with biologists and biochemists.

Recommendation 5-1: Federal research investments in biomolecular engineering should be directed to fundamental research to

- **advance personalized medicine and the engineering of biological molecules, including proteins, nucleic acids, and other entities such as viruses and cells;**
- **bridge the interface between materials and devices and health;**
- **improve the use of tools from systems and synthetic biology to understand biological networks and the intersections with data science and computational approaches; and**
- **develop engineering approaches to reduce costs and improve equity and access to health care.**

Recommendation 5-2: Researchers in academic and government laboratories and industry practitioners should form interdisciplinary, cross-sector collaborations to develop pilot- and demonstration-scale projects in advanced pharmaceutical manufacturing processes.

FLEXIBLE MANUFACTURING AND THE CIRCULAR ECONOMY

Chemical engineering as a discipline was founded in the need to deal with heterogeneous raw materials, especially petroleum, and this need will be amplified in the transition to more sustainable feedstocks. The production and manufacturing of useful materials and molecules enabled by chemical engineers are now creating previously unforeseen problems that must be solved at scale. Chemical engineers play a critical role in manufacturing and can thus contribute to more sustainable manufacturing through efficiency, nimbleness, and process intensification.

A sustainable future will require a shift to a circular economy in which the end of life of products is accounted for, utilizing new developments and advances in green chemistry and engineering. This shift represents another opportunity for chemical engineers to innovate from the molecular to manufacturing scales. The continued drive toward more efficient, environmentally friendly, and cost-effective manufacturing processes will benefit from a wider range of feedstocks for the production of chemicals and materials. The challenge of feedstock flexibility offers chemical engineers an opportunity to develop advances in reductive chemistry and processes that will allow the use of oxygenated feedstocks such as lignocellulosic biomass. Chemical engineers also have substantial opportunities to develop scaled-out, distributed manufacturing systems and innovative, large-scale processes that can compete with the conversion of fossil resources.

Current challenges in process design include the need for improvements in distributed manufacturing and process intensification—areas in which the chemical engineering research community can provide intellectual leadership. Collaborations between academic researchers and industrial practitioners will be important for demonstration at process scale. In the transition from a linear to a circular economy, specific opportunities for chemical engineers include redesigning processes and products to reduce or eliminate pollution, developing new ways to reduce and utilize waste, designing products to be used longer and to be recyclable, and designing processes and products using sustainable feedstocks.

Recommendation 6-1: Federal research funding should be directed to both basic and applied research to advance distributed manufacturing and process intensification, as well as the innovative technologies, including improved product designs and recycling processes, necessary to transition to a circular economy.

Recommendation 6-2: Researchers in academic and government laboratories and industry practitioners should form interdisciplinary, cross-sector collaborations focused on pilot- and demonstration-scale projects in advanced manufacturing, including scaled-down and scaled-out processes; process intensification; and the transition from fossil-based organic feedstocks and virgin-extracted inorganic feedstocks to new, more sustainable feedstocks for chemical and materials manufacturing.

NOVEL AND IMPROVED MATERIALS FOR THE 21st CENTURY

Chemical engineers have a critical role to play in the development of new materials and materials processes from the molecular to macroscopic scales. Their integration of theory, modeling, simulation, experiment, and machine learning is accelerating the discovery, design, and innovation of new materials and new materials processes.

Chemical engineers can contribute to materials development across a range of material types and applications. The combination of molecular-level understanding and thermodynamic and transport concepts yields important insights and enables advances. In particular, chemical engineers have a unique role to play in the continued development of polymer science and engineering because of their understanding of chemical synthesis and catalysis, thermodynamics, transport and rheology, and process and systems design. Chemical engineering is also the logical home for research and development of complex fluids and soft matter. The science and application of nanoparticles by chemical engineers in both industry and biomedicine are rapidly accelerating, offering the opportunity for breakthroughs. Chemical engineers play an essential role in advancing the development of biomaterials for both regenerative engineering and organ-on-a-chip technology, and chemical engineering principles are at the heart of understanding and improving targeted drug delivery both spatially and temporally. Chemical engineering expertise around reactor design, separations, and process intensification is critical to the success and growth of the electronic materials industry.

Recommendation 7-1: Federal and industry research investments in materials should be directed to

- **polymer science and engineering, with a focus on life-cycle considerations, multiscale simulation, artificial intelligence, and structure/property/processing approaches;**
- **basic research to build new knowledge in complex fluids and soft matter;**
- **nanoparticle synthesis and assembly, with the goal of creating new materials by self- or directed assembly, as well as improvements in the safety and efficacy of nanoparticle therapies; and**
- **discovery and design of new reaction schemes and purification processes, with a steady focus on process intensification, especially for applications in electronic materials.**

TOOLS TO ENABLE THE FUTURE OF CHEMICAL ENGINEERING

Current and future chemical engineers will need to navigate the interface between the natural world and the data that describe it, as well as use the tools that turn data into useful information, knowledge, and understanding. Some emerging and future tools will be developed in other fields but will have a significant impact on the work of chemical engineers; others will be developed directly by chemical engineers and have an impact in science and engineering more broadly. Some tools and capabilities will be evolutionary, with gradual and predictable development and applications, while others will be revolutionary and will change chemical engineering research and practice in ways that may be difficult to predict or anticipate today. While the list of tools and capabilities—many of which will drive innovation when used in combination—is virtually endless, this report focuses on data science and computational tools, modeling and simulation, novel instruments, and sensors.

Developing tools that synthesize available data in real time and frameworks or models that transform data into information and actionable knowledge could become one of chemical engineers' key contributions to society over the next decades. It is easy to imagine a not-too-distant future characterized by data-on-demand—where data on anything, at any level of granularity, will be readily and instantly accessible. Such a future suggests profound and exciting opportunities for chemical engineers, who are trained in process integration and systems-level thinking—skills that will be required to synthesize disparate data streams into information and knowledge.

The systems thinking, analytical approaches, and creative problem-solving skills of today's chemical engineering graduates give them a distinct advantage in using artificial intelligence in real-world contexts. The evolution of artificial intelligence in the next decade will have enormous implications not only for the types of problems chemical engineers will be able to solve but also for how they will do so. Chemical engineers are poised to contribute significantly to the development of modeling and simulation tools that will influence education, research, and industry. They will continue developing and disseminating methods, algorithms, techniques, and open-source codes, making it easier for nonexperts to use computing tools for scientific research.

The increasing operational complexities and decreasing capital investments and economic margins in the petrochemical industry, coupled with stringent environmental and quality demands on the manufacture of specialty chemicals and polymers, will continue to drive increased use of modeling and simulation to run scenarios and test hypotheses. While the pharmaceutical industry currently lags behind the chemical industry in its use of simulation tools, fundamental changes in regulatory requirements are motivating greater use of mathematical models and simulation, especially in the rapidly growing biomanufacturing sector.

Recommendation 8-1: Federal and industry research investments should be directed to advancing the use of artificial intelligence, machine learning, and other data science tools; improving modeling and simulation and life-cycle assessment capabilities; and developing novel instruments and sensors. Such investments should focus on applications in basic chemical engineering research and materials development, as well as on accelerating the transition to a low-carbon energy system; improving the sustainability of food production, water management, and manufacturing; and increasing the accessibility of health care.

TRAINING AND FOSTERING THE NEXT GENERATION OF CHEMICAL ENGINEERS

Chemical engineers are in high demand across most professions and job levels, and chemical engineering provides an excellent foundation for many career paths. The undergraduate chemical engineering curriculum has served the discipline well and has continued to evolve, slowly, in response to scientific discoveries, technological advances, and societal needs. The undergraduate curriculum provides a mathematical framework for designing and describing (electro-/photo-/bio-) chemical and physical processes across diverse spatial and temporal scales. Data science and statistics may be delivered most effectively in a separate course embedded within the core curriculum and taught with specific emphasis on matters of chemistry and engineering. In addition, experiential learning is important, and the majority of industrial and academic chemical engineers interviewed by the committee discussed the importance of internships and other practical experiences. However, there are far fewer internships available than the number of students who would benefit from them, and the density of the core undergraduate curriculum leaves few openings for incorporating an additional hands-on laboratory course earlier in the curriculum.

The current chemical engineering curriculum is well suited to preparing students for a wide variety of industrial roles. Graduate research increasingly encompasses a diverse range of topics that do not all require the same level of traditionally curated knowledge currently delivered in graduate chemical engineering curricula, and so graduate curricula may need to be adjusted. Internships for graduate students are currently rare, and new models will need to address issues of equity and inclusion, suitable compensation, intellectual property considerations, and adequate intern mentoring.

Women and members of historically excluded groups are underrepresented in chemical engineering relative to their numbers in the general population, even by comparison with chemical and biological sciences and related fields. Diversifying the profession will bring valuable new perspectives, and is therefore essential to the field's survival and potential for impact. At all points along their academic path, chemical engineering students need role models and effective, inclusive mentors, including those who reflect the diversity of backgrounds that the field needs. Leveraging of professional societies and associated affinity groups could provide valuable support for people of diverse backgrounds entering the field, and strong university support for student chapters of professional organizations will improve access and success.

Additionally, the general affordability of community colleges is a major attraction for a diverse body of students, ranging from budget-minded high school seniors to non-traditional students. Increased engagement of transfer students therefore represents an untapped opportunity to broaden participation in and access to the chemical engineering profession.

Both students from 2-year colleges and those who change their major to chemical engineering would benefit from a redesign of the curriculum that would allow them to complete the degree in less time. Better academic and social support structures are needed to enable successful pathways for these students. New methods that would make it possible to offer portions of the curriculum in a distributed manner, as well as more general restructuring, may require flexibility in curriculum design and changes in university policies and graduation and accreditation requirements.

Recommendation 9-1: Chemical engineering departments should consider revisions to their undergraduate curricula that would

- **help students understand how individual core concepts merge into the practice of chemical engineering,**
- **include earlier and more frequent experiential learning through physical laboratories and virtual simulations, and**
- **bring mathematics and statistics into the core curriculum in a more structured manner by either complementing or replacing some of the education that currently occurs outside the core curriculum.**

Recommendation 9-2: To provide graduate students with experiential learning opportunities, universities, industry, funding agencies, and the American Institute of Chemical Engineers should coordinate to revise graduate training programs and funding structures to provide opportunities for and remove barriers to systematic placement of graduate students in internships.

Recommendation 9-3: To increase recruitment and retention of women and Black, Indigenous, and People of Color (BIPOC) individuals in undergraduate programs, chemical engineering departments should emphasize opportunities for chemical engineers to make positive societal impacts, and should build effective mentoring and support structures for students who are members of such historically excluded groups. To provide more opportunities for BIPOC students, departments should consider redesigning their undergraduate curricula to allow students from 2-year colleges and those who change their major to chemical engineering to complete their degree without extending their time to degree, and provide the support structures necessary to ensure the retention and success of transfer students.

Recommendation 9-4: To increase the recruitment of students from historically excluded communities into graduate programs, chemical engineering departments should consider revising their admissions criteria to remove barriers faced by, for example, students who attended less prestigious universities or did not participate in undergraduate research. To provide more opportunities for women and Black, Indigenous, and People of Color (BIPOC) individuals, departments should welcome students with degrees in related disciplines and consider additions to their graduate curricula that present the core components of the undergraduate curriculum tailored for postgraduate scientists and engineers.

Recommendation 9-5: A consortium of universities, together with the American Institute of Chemical Engineers, should create incentives and practices for building and sharing curated chemical engineering content for use across universities and industry. Such sharing could reduce costs and advance broad access to high-quality content intended both for students and for professional engineers intending to further their education or change industries later in their careers.

Recommendation 9-6: Universities, industry, federal funding agencies, and professional societies should jointly develop and convene a summit to bring together existing practices across the ecosystem of stakeholders in chemical engineering professional development. Such a summit would explore the needs, barriers, and opportunities around creating a technology-enabled learning and innovation infrastructure for chemical engineering, extending from university education through to the workplace.

INTERNATIONAL LEADERSHIP

America's scholarly leadership in chemical engineering with respect to both the quantity of research, as measured by numbers of publications, and the quality of research, as measured by citation impact, has decreased significantly in the past 15 years, losing ground to international competitors, particularly China. The United States is in a leadership position in some areas of chemical engineering technology, but lags in many niches compared with various other countries.

The increase in research output from China is a result of large investments in a range of technology areas, many of which are either central or highly relevant to chemical engineering. Similar levels of investment in the U.S. research enterprise are imperative. This report outlines the numerous opportunities for chemical engineers to contribute in the areas of energy; water, food, and air; health and medicine; manufacturing; materials research; and tools development. Without a sustained investment by federal research agencies across these areas, as recommended above, it will be impossible for the United States to maintain its leadership position. At the same time, U.S. chemical engineering will be strengthened through increased coordination and collaboration across disciplines, sectors, and political boundaries. Almost all of the areas of research discussed in this report are multidisciplinary in nature and will require close collaboration between researchers in academic and government laboratories and industry practitioners to develop applications that are economically viable and scalable. Such cross-sector collaborations, as recommended throughout the report, will have additional benefits for graduate student education and faculty member development while also satisfying the need of industry to achieve rapid results.

Recommendation 10-1: Across all areas of chemical engineering, in addition to advancing fundamental understanding, research investments should be set aside for support of interdisciplinary, cross-sector, and international collaborations in the

areas of energy; water, food, and air; health and medicine; manufacturing; materials research; tools development; and beyond, with the goal of connecting U.S. research to points of strength in other countries.

1
Introduction

In 1988, the National Academies of Sciences, Engineering, and Medicine laid out an important vision for the field of chemical engineering in the report *Frontiers in Chemical Engineering: Research Needs and Opportunities*, also known as the "Amundson Report" (NRC, 1988). The report outlined a roadmap for making promising research opportunities a reality, and highlighted the remarkable potential of the profession to affect many aspects of American life and promote the scientific and industrial leadership of the United States. The report is widely recognized as having been a key driver for many advances in chemical engineering over the past 30 years. At the same time, tremendous changes have occurred in chemical engineering, in the scientific enterprise more generally, in technology, and in the relationship between science and the public. These changes have affected views on research priorities, education, and the practice of chemical engineering and will continue to do so.

PURPOSE OF THIS REPORT

At a 2016 American Institute of Chemical Engineers (AIChE) roundtable, leaders from the chemical engineering profession reached a major conclusion: the field of chemical engineering needs a new vision for the 21st century. Participants at that meeting, including both current and former AIChE presidents and multiple members of the National Academy of Engineering, underscored the transformative and lasting impact of the Amundson Report, unanimously supporting the need to update it. Perhaps more important than identifying this need, the community provided support for such a study: in addition to federal sponsors, more than 45 academic departments, private companies, and professional organizations offered financial contributions. This broad convergence of support culminated in the formation of the National Academies' Committee on Chemical Engineering in the 21st Century: Challenges and Opportunities. The committee's primary task was to outline an ambitious vision for chemical engineering research, innovation, and education that could guide the profession for the next 30 years (Box 1-1).

STUDY SCOPE AND APPROACH

The Committee was formed in late 2019 and completed its work over the course of 18 months. The original work plan included six in-person meetings, five of which would include information gathering sessions open to the public, to be held in locations across the United States to maximize participation of the chemical engineering community. Additional information-gathering webinars were to be held as needed to fulfill the committee's task. The committee's participation in several professional society meetings also was planned to increase community engagement and input opportunities.

BOX 1-1
Statement of Task

An ad hoc Committee will prepare a report that will articulate the status, challenges, and promising opportunities for chemical engineering in the United States. In particular, the report will:

- Describe major advances and changes in chemical engineering over the past three decades, including the importance and contributions of the field to society; technical progress and major achievements; principal changes in the practice of R&D; and economic and societal factors that have impacted the field.
- Address the future of chemical engineering over the next 10 to 30 years and offer guidance to the chemical engineering community:
    - Identify challenges and opportunities that chemical engineering faces now and may face in the next 10-30 years, including the broader impacts that chemical engineering can have on emerging technologies, national needs, and the wider science and engineering enterprise.
    - Identify a set of existing and new chemical engineering areas that offer promising intellectual and investment opportunities and new directions for the future, as well as areas that have major scientific gaps.
    - Identify aspects of undergraduate and graduate chemical engineering education that will require changes needed to prepare students and workers for the future landscape and diversity of the profession.
    - Consider recent trends in chemical engineering in the United States relative to similar research that is taking place internationally. Based on those trends, the report will recommend steps to enhance collaboration and coordination of research and education for identified subfields of chemical engineering.

The committee met in person in Washington, DC, in late February 2020 to identify information-gathering needs and begin its work. Shortly after that meeting, the COVID-19 pandemic necessitated a major adjustment to the committee's information-gathering plans. In its early deliberative sessions, the committee identified the societal and environmental areas in which chemical engineers have, or are likely to have in the future, the largest impact. The committee members divided into subgroups based on those areas (energy; water, food, and air; health and medicine; manufacturing and the circular economy; materials research; tools development; and education) to gather information and begin drafting what would become the main chapters of this report. Because of the limitations of meeting virtually, the committee shifted to shorter but more frequent virtual meetings. Information-gathering sessions were distributed across many meetings of both the full committee and subgroups from summer 2020 through spring 2021, during which time the committee met 42 times, with 27 of those meetings including a session that was open to the public. To receive additional input from the community in lieu of regional in-person meetings, the committee broadly distributed a questionnaire to chemical engineers from all sectors at any stage in their career. The committee chair and National Academies study director also led a town hall discussion at the 2019 AIChE fall meeting and participated in the April 2020 meeting of the AIChE Virtual Local Section to gather input from the broader chemical engineering community. Finally, the committee conducted an extensive literature review.

AUDIENCES FOR THIS REPORT

The primary audience for this report is the broad chemical engineering community, including researchers in academia and industry, educators, and students, as well as federal and state decision makers and program leaders. It is anticipated that the report will be used by

- students and faculty, to determine their research directions and design their programs;
- industrial scientists and engineers, in creating their research and development plans;
- universities and colleges, to improve undergraduate and graduate education and diversify their student populations; and
- government program leaders and other research sponsors, to design and justify their programs.

Researchers in both academia and industry, depending on their specific area of focus and expertise, will find the challenges and opportunities outlined in Chapters 3 through 8 of particular interest. Program managers at federal funding agencies will also find these chapters useful, as well as the final chapter on maintaining international leadership. Finally, faculty and leadership at colleges and universities, as well as professional organizations, will find strategies for improving undergraduate, graduate, and lifelong learning programs in Chapter 9. Chemical engineering departments can also look to Chapter 9 for ways to make their programs more accessible while maintaining the rigor that has made chemical engineering so successful as a discipline and a profession.

REPORT ORGANIZATION

To address the future of chemical engineering in the coming decades, the committee was tasked with identifying challenges and opportunities, as well as existing and new areas for intellectual and investment opportunities and scientific gaps. The committee chose to treat these elements collectively, as challenges and scientific gaps present exciting opportunities for both intellectual investment by academic and industrial researchers and funding investment by federal funding agencies and industrial research programs.

To provide a framework for the discussion in this report, the committee examined the role of chemical engineering in addressing the key challenges facing society. Several organizations have outlined grand challenges, but the committee chose to organize this report around the areas noted above: energy and the energy transition; water, food, and air; health and medicine; manufacturing and the circular economy; and materials.

Chapter 2 provides a brief history of chemical engineering as both a discipline and a profession, including key advances in training. This chapter focuses on some of the most important contributions of chemical engineering, as well as changes in the field in the practice of research and development (e.g., a shift from multiple–principal investigator [MPI] projects and reduced corporate investment in research and development) and the societal factors (e.g., the internet and global climate change) that have affected the field.

Major advances and changes in chemical engineering over the past three decades in specific societal and environmental challenge areas are detailed in Chapters 3 through 7. The committee envisions a future for the field of chemical engineering that is more collaborative with other disciplines and across sectors. Specific opportunities for collaboration are described throughout these chapters.

Chapter 3 presents opportunities for chemical engineers to contribute to decarbonization of current energy systems, describing the key research needs across the energy value chain, from sources to various end uses. Energy, water, and food are highly interconnected, and solutions in this complex system need to be both environmentally sustainable and economically viable. Chemical engineers have historically played a central role in the energy sector; their contributions in the space of water and food, as well as air quality, have been important but are growing, as described in Chapter 4. The development of disease treatments is a multidisciplinary enterprise, and Chapter 5 describes how chemical engineers can contribute to many aspects of medicine by applying systems biology to physiology, the discovery and development of molecules and materials, and process development and scale-up. Chapter 6 explores opportunities for chemical engineers to improve the sustainability of manufacturing by advancing the use of flexible feedstocks, process intensification and distributed manufacturing, and the transition from a linear to a circular economy. Specific materials applications are discussed throughout the report, but Chapter 7 describes the role of chemical engineers in discovery science and the development of new materials and materials processes, from the molecular to the macroscopic scale. Chapter 8 describes tools and techniques with the potential to enable future advances across all of the previously discussed challenge areas.

Chapter 9 focuses on education and the need to maintain those aspects of the curriculum that have made chemical engineering successful while allowing for innovation to ensure that the next generation of chemical engineers is more demographically diverse and receives training that ensures its ability to address the challenges facing society. The report concludes with Chapter 10, focused on international leadership. Appendix A lists acronyms used in this report; Appendix B provides journal titles used for the discussion of international leadership in Chapter 10; Appendix C summarizes the results of a questionnaire distributed to the chemical engineering community; and acknowledgments and biographical sketches of the committee and staff can be found in Appendixes D and E, respectively.

2
Chemical Engineering Today

Through discovery, design, creation, and transformation, chemical engineering is the engineering of systems at scales ranging from the molecular to the macroscopic that integrate chemical, physical, and biological elements in order to develop processes and produce materials and products for the benefit of society.

The work of chemical engineers has transformed societies and individual lives around the world, particularly in the United States. Chemical transformations are at the heart of the technologies that enable modern society (Box 2-1). Without synthetic fertilizers made with chemical engineering processes, Norman Borlaug could not have led a Green Revolution to feed the world. Without the invention of catalysts due to the leadership and vision of Karl Ziegler, a chemist, and Giulio Natta, the first chemical engineer to win a Nobel Prize, polyolefins would not exist, and the myriad benefits of plastics would not have been realized. Without the invention of tough, stable polymers by chemical engineers at the DuPont Company, including Roy Plunkett and Stephanie Kwolek, Teflon® and Kevlar® would not have been developed, nor would any of the commercial and medical devices made from those polymers. Without the contributions of chemical engineers such as Andy Grove at Intel and numerous others, the silicon chips, glass materials, and plastics that make up today's ubiquitous electronic devices would not have been created. Without the leadership and vision of Nobel Prize winner Frances Arnold, who built upon the discoveries of numerous biologists, biochemists, and chemical engineers, the tools of directed evolution would not have emerged. Without an army of chemical engineers, there would be no oil and gas industry to power the world. And without chemical engineers, a robust pharmaceutical industry would not have been able to discover and produce the therapies and vaccines needed for a long and healthy life. Figure 2-1 highlights some of the areas in which chemical engineers have made major contributions.

At the same time, however, chemical engineering is also responsible for unintended consequences, such as those resulting from the production of chemicals that will persist in the environment indefinitely, greenhouse gas emissions that contribute to climate change, plastic materials that accumulate in landfills and the oceans, and the chemicals of war that have inflicted long-term or permanent damage on humans and the environment. Thus the field of chemical engineering today faces challenges and opportunities not only to innovate for the future, but also to innovate in ways that repair the unintended consequences of the past.

BOX 2-1
The Chemical Industry

The U.S. chemical industry supports more than 25 percent of the country's gross domestic product. It is the country's number one exporter ($136 billion) with a $35 billion surplus, and directly touches more than 96 percent of manufactured products. The industry is divided roughly into basic chemicals, specialty chemicals, pharmaceuticals, agricultural chemicals, and consumer products (DHS, 2019).*

Basic chemicals. These chemicals are manufactured in large volumes to uniform composition specifications. The largest example is the transformation of hydrocarbon feedstocks or minerals into commodity products. The price margins for these chemicals are small, but the large volumes drive profits, and thus the sector is vulnerable to economic cycles. The sector is extremely energy intensive and has benefited from the shale gas boom in the United States, which provides cost advantages for U.S. feedstocks.

Specialty chemicals. These chemicals are manufactured in smaller volumes and are generally designed or engineered for a specific (and often demanding) application. Examples include flavors and fragrances, inks, catalysts, and electronic materials. The price per unit volume of some of these products is extremely high, and the manufacturing processes can be challenging because extremely high purities are required.

Pharmaceuticals. Therapeutic molecules for the prevention or treatment of disease are the most value-added chemicals. This sector requires a long time for drug development, manufacture, and testing, and products carry a premium price. As in the case of some specialty chemicals, the success of the products is completely dependent on the purity and function of the active molecules, and large profits can be realized from small amounts of material. Fewer than 10 percent of drugs that make it to clinical trials are approved by the U.S. Food and Drug Administration—another contributor to the high costs of these products (BIO, 2011).

Agricultural chemicals. Chemical agents, such as fungicides, insecticides, and chemicals that influence the growth cycle of plants are commonly used in agriculture along with various kinds of fertilizers. Other examples of agricultural chemicals are herbicides, rodenticides, and insect attractants or repellents. In all cases, the animal and human toxicity of these chemicals is a persistent concern. While these products are produced in smaller volumes than basic chemicals, annual shipments in 2019 were worth $41.1 billion in the United States (DHS, 2019).

Consumer products. Consumer products made from hydrocarbons or minerals are ubiquitous. The plastics economy has revolutionized the manufacturing of everything from cars to clothing, and concerns about its environmental impact are now driving the urgency of finding ways to reuse and recycle these materials. Similarly, concern about manufacturing of the largest-volume inorganic chemical—cement—is growing because the process accounts for about 8 percent of global greenhouse gas emissions (Andrew, 2018).

* The National Academies of Sciences, Engineering, and Medicine's Committee on Enhancing the U.S. Chemical Economy through Investments in Fundamental Research in the Chemical Sciences is currently examining the role of the chemical industry in the U.S economy.

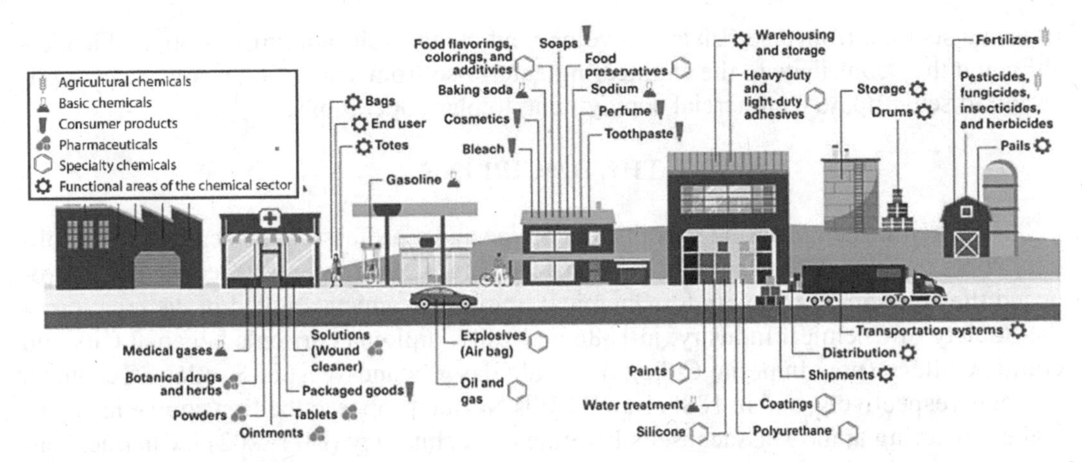

FIGURE 2-1 Schematic illustrating the broad impact of the chemical sector on several aspects of American life. Examples of basic chemicals, specialty chemicals, agricultural chemicals, pharmaceuticals, and consumer products are shown. SOURCE: DHS (2019).

Chemical engineering as a discipline brings together the three fundamental sciences of chemistry, physics, and biology, as well as mathematics. Chemical engineers are agnostic to the material used or the particular application; they work with all phases of matter—vapor, liquid, supercritical, solid, and plasma—and with multiphase mixtures and at interfaces. They work at and across all length scales, ranging from molecules to medicines to materials and even machines. Their work creatively transforms matter and products into higher-value materials and products using the principles and tools of thermodynamics, transport, kinetics, process control, and process design. Chemical engineering is the only field of engineering that takes advantage of chemical transformations, usually followed by separation and purification, to add value to products.

Chemical reactions and reacting systems are central to many transformations. The problems tackled by chemical engineers usually involve time dependence, transport phenomena, and nonequilibrium phenomena and feature many variables. Chemical engineers are trained to take a system-wide approach to solving problems, recognizing that a single step is rarely adequate to complete the desired transformation. They realize that a system of connected components needs to be optimized holistically to be most efficient and economic. Importantly, chemical engineers appreciate the balance among physical performance, fiscal profitability, and safety.

The education and training unique to chemical engineering produce intellectually versatile engineers who are comfortable with mathematics and computers, skilled in analytical thinking, and adept at solving open-ended problems. They draw on an education that has served them well in diverse areas of accomplishment. Accordingly, chemical engineers are found in interdisciplinary teams in the agricultural, biological, biomedical, chemical, environmental, food, health, energy, materials, petrochemical, nanotechnology, pharmaceutical, and semiconductor industries, as well as in the consulting, communications, software, financial, and insurance sectors. They engage at all levels in these industries and sectors, from research and development, to analysis, administration, and leadership, and they contribute to the engineering enterprise from scientific discovery, to application to scale-up, to commercial deployment, to plant operations.

THE DISCIPLINE

Chemical engineering is both a discipline and a profession. Its academic disciplinary roots trace back to the 1880s. In 1880, the English entrepreneur George Davis proposed the creation of a Society for Chemical Engineers, which resulted in the creation of the Society of Chemical Industry; in 1886 and 1887, diploma curricula began at City and Guilds College (now Imperial College) and at Glasgow and West of Scotland Technical College, respectively; and in 1888, Lewis Mills Norton proposed the first course in chemical engineering at the Massachusetts Institute of Technology (MIT, 2021). Chemical engineering evolved as a profession from the roots of industrial and applied chemistry, which in turn emerged from such ancient chemical processes as fermentation and leather tanning. Traditionally, industrial and economic needs and academic research have been tightly coupled, with many faculty members being engaged in consulting or joint research work with industry. To paraphrase L. E. Scriven, the academic tree takes its nourishment from the soil of practical challenges and periodically returns its many leaves of results and trained graduates to that soil (University of Minnesota, 2003).

This discipline is the only one in engineering with a central focus on molecules and their transformations—that is, chemistry. For much of its history, chemistry and physics were the core sciences of chemical engineering (along with mathematics). Because of its fundamental engagement with chemistry, it is natural for chemical engineering to be the principal home for the engineering of biological systems of all scales. The connection between chemical engineering and biology has a long history, and their interactions grew exponentially with the advent of methods for genetic engineering and the advancement of molecular biology from the late 20th century onward (Box 2-2). Indeed, numerous fields (e.g., biochemical engineering, metabolic engineering, tissue engineering, synthetic biology) have developed as outgrowths of chemical engineering or with key contributions from chemical engineers. The expansion of the influence of biology on the field has also been so significant that many academic departments have changed their names to incorporate some version of "bio" in addition to "chemical" engineering.

Because chemical engineers deal with both molecules and the enormous industrial plants that produce them, their work encompasses a large range of length and time scales, from the nanometer scales of chemical bonds and reactions to the kilometer scales of crude oil (petroleum) refineries, and from nanosecond chemical reactions to batch processes that take hours. Few fields of science or technology deal with changes of more than a dozen orders of magnitude in both length and time scales. The core curriculum of chemical engineering has lasted more than a century because it focuses on the analysis of linked processes regardless of scale.

The early history of chemical engineering has been well described (Colton, 1991; Furter, 1983). The growth of the core curriculum is marked by several events, the first of

BOX 2-2
The Growth of Biologically Related Content in Chemical Engineering

The birth of biochemical engineering occurred in the mid-1940s, following the successful application of antibiotics to treat infection during World War II and the realization that the United States could use such techniques more broadly if appropriately scaled. Merck Chemical Company was instrumental in designing improved aeration facilities, with consultation from Richard Wilhelm at Princeton and Elmer Gaden, a graduate student at Columbia, who, as part of his PhD thesis, examined the mass-transfer characteristics of fermentors. Merck, building on Wilhelm's work with fluidized beds, also improved the productivity of suspension cultures and chiral separation by crystallization. Margaret Hutchinson Rousseau designed and scaled up the first commercial penicillin production plant during World War II at Pfizer.

A major catalyst for growth followed the collaboration of Herbert Boyer and Robert Swanson to found Genentech in the early 1970s. Along with cocreators Stanley Cohen and Paul Berg, Boyer developed the technology used to "cut and paste" DNA from any organism into another in which it could be expressed. In 1977, Genentech succeeded in isolating the sequence for human insulin, and by 1982 it had marketed the first recombinant protein therapeutic. This continues to be a growing industrial area, and it is estimated that by 2024, the global market in protein therapeutics will be valued at $500–750 million annually (IndustryARC, 2021). Others have launched major changes in the field. Raymond F. Baddour worked at the interface of biotechnology and pharmaceuticals and cofounded Amgen in 1980. James E. Bailey was an American chemical engineer who played a major role in the development of metabolic engineering. Daniel Wang contributed to many aspects of biochemical engineering, notably enzyme technology and mammalian cell culture and bioreactors. And Frances Arnold pioneered the use of directed evolution to create enzymes with improved or novel functions, work recognized by the 2018 Nobel Prize in Chemistry.

which featured George E. Davis in England in the 1880s. Davis was an entrepreneur in chemical manufacturing who called for the creation of "chemical engineering" to address the rampant pollution generated by chemical processes (Cohen, 1996). The first course (Course X) developed by Lewis Mills Norton at MIT had its stops and starts, but by 1902, William Walker had accepted the leadership of Course X while remaining in consulting partnership with Arthur D. Little until 1905. Arthur D. Little, as chair of the visiting committee for chemical engineering at MIT, is credited with creating the concept of "unit operations" in 1915 to describe the basic physical (and later, chemical) operations used to transform raw materials into products (Flavell-While, 2011).

Throughout the 20th century, further academic advances in the curriculum took place across the United States. At the University of Wisconsin, Olaf A. Hougen and Kenneth M. Watson wrote the three-volume *Chemical Process Principles: Material and Energy Balances* (1943); *Thermodynamics* (1947b), and *Kinetics and Catalysis* (1947a). Also at the University of Wisconsin, R. Byron Bird, Warren E. Stewart, and Edwin N. Lightfoot published the paradigm-shifting textbook *Transport Phenomena* in 1960. That book transformed chemical engineering by bringing a strong mathematical approach to the unification of treatments of fluid mechanics and heat and mass transfer, both reinforcing and explaining the connections made by Allan P. Colburn and Thomas H. Chilton at the University of Delaware and the DuPont Company, respectively, in 1946 (Chilton and Colburn J-factor analogy). At the University of Minnesota, Neal Amundson and Rutherford Aris worked to develop the analytic and mathematical foundations for reaction engineering. At the University of California, Berkeley, John M. Prausnitz developed methods of molecular thermodynamics beginning in the 1950s and 1960s. Throughout and after this period, the role of computing in chemical engineering control and design grew steadily (Box 2-3). Thus by the mid-1960s, a canonical undergraduate program, rooted in a foundation of fundamentals from chemistry, physics, and mathematics, was formed in a way that is easily recognized today.

BOX 2-3
Control, Design, and Computer Applications in Chemical Engineering

The development of process control and process design coincided with the development of the use of computers in chemical engineering, as these topics were then, and remain today, heavily computational in nature. Much of the pioneering work in control was centered in industry, with Page S. Buckley at DuPont and Joel O. Hougen at Monsanto (and later at The University of Texas at Austin) at the forefront. F. Greg Shinsky and Edgar H. Bristol at Foxboro also contributed to the industrial practice of control from the perspective of hardware equipment manufacturing. The academic pioneers included Rutherford Aris at the University of Minnesota and Leon Lapidus at Princeton University, who contributed to the theory of optimal control in chemical processes.

The earliest textbooks on the topic were written by Daniel D. Perlmutter at the University of Pennsylvania (1975), Ernest F. Johnson at Princeton University (1967), and Donald R. Coughanowr and Lowell Koppel at Purdue University (1965). The first chemical engineering textbooks that highlighted computer applications were the numerical analysis books by Lapidus (1962) and Brice Carnahan and colleagues (1969) at the University of Michigan. The University of Michigan, motivated by the introduction of FORTRAN (Formula Translation), initiated a project under the direction of Donald L. Katz to study the use of computers in engineering, which resulted in a seminal 1966 report entitled *Computers in Engineering Design Education.* Chemical engineering education in general, but more specifically the computationally heavy process control and process design courses, owes much to the vision laid out by Katz in this report. Pioneering work in chemical process simulation and design was carried out by Rodolphe L. Motard and Ernest J. Henley at the University of Houston, who, along with J. D. "Bob" Seader at the University of Utah, wrote the dominant textbook on the topic (1998). The preeminent textbooks on computer applications in chemical engineering in general include those on process model building, process analysis, and process simulation written by David M. Himmelblau and Kenneth B. Bischoff (1968); on numerical methods by Brice Carnahan, H. A. Luther, and James O. Wilkes (1969); on process optimization, by Gordon S. G. Beveridge and Robert S. Schechter (1975); and on modeling and simulation, by Roger G. E. Franks (1972).

THE PROFESSION

Chemical engineering as an industrial profession has its roots in processes for materials extraction and transformation. Éleuthère Irénée du Pont de Nemours founded his eponymous company in 1802 to make gunpowder on the banks of the Brandywine

River in Wilmington, Delaware, at a site with a dam, a millrace, and saltpeter at hand. John D. Rockefeller became the world's first oil baron when he united several companies into the Standard Oil Company in 1870, exploiting wells in Ohio, Pennsylvania, and eventually elsewhere. Standard Oil quickly came to control about 90 percent of America's refining capacity (see Ohio History Central, 2021). Herbert Henry Dow founded his company in 1897 to extract bromine from underground brine in Midland, Michigan. Over time, each of these companies came to appreciate and need chemical engineering expertise and scientific advances to grow and diversify.

By the 1970s, the leading chemical engineering research efforts in U.S. industry were at Esso (later Exxon) and Mobil laboratories, DuPont Central Research and its Experimental Station, and Dow Central Research. Space does not allow enumeration of all of the advances in materials and processes that emerged from those laboratories and development efforts, or other important advances in research and product development from the laboratories at 3M, Shell, Universal Oil Products (later Honeywell UOP), Aramco, Standard Oil Company, Bell Labs, and elsewhere, but they all contributed in a significant and largely defining way to the quality of life enjoyed in the United States. Parallel developments also played out in chemical and other companies focused on materials. In 1953, for example, Lexan® polycarbonate was invented at the General Electric (GE) Company. This invention led to the creation of the GE Plastics Business, which grew to be a global business based on major contributions of chemical engineers (Plastics Hall of Fame, 2021).

Chemical engineering has also played an important role in electronic materials, a role amplified by the nearly ubiquitous role of electronic devices in today's society. The expansive proliferation and adoption of electronics globally has driven an explosion of uses for electronics well beyond the traditional uses of personal computers and cell phones. Today, the introduction of 5G, the internet of things (IoT), cloud computing, and autonomous driving is having a direct impact on how electronics are used. Looking forward, the further adoption of machine learning and artificial intelligence to expand computational capability and demands for electronic devices are at an inflection point. While these trends had been progressing for some time, they have been accelerated by the COVID-19 pandemic, which has made technical advancement more urgent than ever—similar to the way it was in early days of the internet and the introduction of personal computers.

TIMES OF CHANGE

The world has changed and continues to change at an increasingly rapid rate. Technology is transforming the way everything is done, disrupting established practices and organizations. The range of challenges facing engineers is evolving, and the challenges are becoming more difficult. Engineering is about solving problems, so it is natural to expect that as one problem is solved, another emerges or grows in importance. The goal of education in chemical engineering is to equip graduates with the intellectual tools needed to solve problems and the ability to adapt those tools as new problems emerge. Information used to be valuable in its own right, but the internet has made information essentially free. What is now more valuable than ever is the knowledge needed to curate data and synthesize new and novel solutions, and to do so more rapidly and at lower cost than competitors. In a real sense, an undergraduate chemical engineering degree is a platform upon which its holder can build, with a research degree, a professional degree, and/or a career filled with formal and informal learning.

There is of course nothing new about change. The last time the National Academies surveyed the challenges and opportunities for chemical engineering—in the 1980s (*Frontiers in Chemical Engineering: Research Needs and Opportunities*, better known as the "Amundson Report" [NRC, 1988])—the conclusions reached suggested an approach to developing new technologies and maintaining leadership in established ones. A Centennial Symposium of Chemical Engineering (Wei, 1991) laid out the challenges facing the field at that time, and included a rich discussion of the balance of revolution and evolution in chemical engineering. Comments at the time about the tension between teaching the established core of such topics as transport, thermodynamics, kinetics, and design and making room for education relevant to (then) new areas such as biotechnology and materials science resonate today in this report. Nonetheless, the rate of change driven by an interconnected world is without doubt faster today than it was yesterday, and will be even faster tomorrow.

As new advances combine to produce even newer advances, the pace accelerates. At the time of the Amundson Report in the 1980s, personal computing was just becoming common, under 1 million cell phones were in use in the United States (Tesar, 1996), the internet was used by a cognizant few, and Microsoft had just gone public. The top ten public companies in terms of revenue included Exxon (1, at $91 billion), Mobil (3), Texaco (4), DuPont (7), and Amoco (10) (*Fortune*, 1985). The undergraduate chemical engineering curriculum focused on the core fundamentals of thermodynamics, transport phenomena, reactors and kinetics, process control, and process design. In 2021, as this report was being written, the internets of information and things were affecting all aspects of life, 14 billion cell phones were in use worldwide (Statista, 2021), the human genome was known and could be edited, it was possible to "see" individual atoms, and the use of artificial intelligence and deep machine learning was growing rapidly. The top ten U.S. public companies in terms of revenue in 2020 were Walmart (1, $514 billion), ExxonMobil (2), Apple (3), Berkshire Hathaway (4), Amazon (5), United Health Group (6), McKesson (7), CVS Health (8), AT&T (9), and AmerisourceBergen (10). After ExxonMobil, the next largest oil and gas company was Chevron, at number 15. Dow Chemicals, the largest materials science and chemical company in the United States, was 78th with revenues of $43 billion, less than 10 percent of Walmart's (*Fortune*, 2020).

The undergraduate chemical engineering curriculum is still focused on the core fundamentals of thermodynamics, transport phenomena, reactors and kinetics, process control, and process design. The chemical industry, however, is a less important part of the overall economy than it once was, although the world can no more do without energy, materials, food, and water now than it could then. The top companies more than 100 years ago were all recognized industrial brands; today, technology companies have the largest market capitalizations. According to the Bureau of Labor Statistics (BLS), overall employment in chemical engineering dropped from 43,270 in 1997 to 30,120 in 2019, from almost 0.04 percent to shy of 0.02 percent of the national employment total (BLS, 2021). However, many people educated in chemical engineering work in areas not identified as such by BLS.

Over the last decade in the United States, the number of bachelor's and master's degrees awarded in chemical engineering more than doubled, a rate outpacing the 60–80 percent growth of engineering and STEM (science, technology, engineering, and mathematics) bachelor's and master's degrees. The number of doctorates awarded grew more modestly, with engineering growing most rapidly (Table 2-1).

Technology has also transformed the way people work. The linear industrial model of changing raw materials to products has shifted to a global interconnected platform model. That change, the increasing time-rate of change in society and business in general, and the substantial diminution of the economic significance of the chemical and oil and gas (but not the pharmaceutical) industries in the U.S. economy all have created substantial challenges and opportunities for chemical engineering.

The pace of change is now accelerating as a result of global connectivity, massive amounts of available data, artificial intelligence, sensors, robotics, and more. The background and training of chemical engineers are well suited to today's rapidly changing world, and many of them have found their way into companies leading change. For example, as the new era of artificial intelligence and machine learning advances, chemical engineers are well equipped to move into these fields because of their strong background in reaction (network) engineering and control, along with an understanding of how to deal with complex interconnected processes, in addition to their fluency in math and computing.

These longer-term and probably irreversible societal and business changes are augmented by a critical need to mitigate climate change resulting from greenhouse gas emissions. The ultimate measure of success in addressing climate change will be reducing greenhouse gas concentrations in the atmosphere while still delivering the energy that society needs. The necessary system solutions align with the core chemical engineering practices of designing, building, and extending major manufacturing assets. Chemical process engineers have a long history of building safe, resource-efficient processes at the lowest capital and operating cost.

Along with addressing climate change and the energy transition are parallel needs—to reduce raw material usage and increase recycling to create a more circular economy; deal with the need to generate and distribute food worldwide while conserving water and other resources; and create and scale the manufacturing and distribution of new medicines and therapeutics. Each of these needs is discussed more fully in later chapters. There are other needs as well. Complicating efforts to address all of these needs are inevitable shorter-term geopolitical and environmental issues related to the price and availability of energy and feedstocks, as well as existing and emerging security threats and the viability of the installed asset base (Box 2-4). In these times of change, chemical engineers also play a central role in advances in many new and important areas, including personalized medicine, the infrastructure needed to produce vaccines for a pandemic, the rapid conversion to an economy based on nonfossil fuels, the need for new materials, and many more.

TABLE 2-1 Bachelor's, Master's, and PhD Degrees Awarded in the United States between 2008 and 2019 in Chemical Engineering (ChE), All Engineering (Eng), and All STEM (Science, Technology, Engineering, and Mathematics) Fields

	Field	2008	2009	2010	2011	2012	2013	2014	2015	2016	2017	2018	2019
Bachelors	ChE	4919	5137	5838	6416	7176	7678	8202	9070	10032	11021	11653	11148
	Eng	70232	70991	74778	78502	83636	88201	94386	100316	109373	118379	124794	114818
	STEM	592,801	610,216	639,822	683,865	736,982	781,937	818,434	850,168	877,786	905,509	934,776	*
Masters	ChE	937	996	1051	1284	1395	1453	1521	1629	1701	1801	1921	2003
	Eng	33513	36909	38029	41751	43765	44037	46029	49855	55907	57754	56756	62682
	STEM	164,007	175,895	184,097	199,386	215,137	225,261	235,368	249,762	274,084	290,037	298,157	*
PhD	ChE	873	807	822	822	839	824	972	1002	920	930	981	1092
	Eng	7860	7637	7575	8024	8452	8998	9585	9875	9455	9771	10179	12372
	STEM	34,717	35,313	34,997	36,332	37,846	39,031	40,633	41,178	41,234	41,294	42,227	*

Data from ASEE (2020) and NCSES (2021).

BOX 2-4
Current and Future Threats: Cybersecurity

As chemical and biological process plants and other infrastructure have increasingly become connected to and become part of the internet, the threat of cyberattacks that can cripple a network, cause physical and/or life-threatening damage, or enable collection of a large ransom has increased dramatically. Most companies take cybersecurity across all business sectors very seriously. Many have large units devoted to cybersecurity across manufacturing facilities, research and development, and contractor monitoring. Most of these teams are made up of computer scientists, but they work closely with chemical engineers who are designing new processes or control systems. In this context, chemical engineers will not need to develop new cybersecurity tools, but they will need to be able to communicate effectively with computer scientists and other cybersecurity professionals. Universities need to make students aware of the importance of cybersecurity, as well as development of the skills needed to work with a diverse group of colleagues with deep expertise in various areas.

EDUCATIONAL CHALLENGES AND OPPORTUNITIES

While the core chemical engineering curriculum has apparently changed little over the preceding decades, the concepts are now taught in more modern ways. Initial unemployment for chemical engineering graduates is low and salaries remain high, although not at the highest levels relative to other fields, as was the case in the past. Thus, despite a course catalog that appears at first glance to be stuck in the past, the curriculum has been shown to yield graduates with the skills needed to adapt and succeed in the workplace. The hallmark of this continuing success appears to be the ability of chemical engineering graduates to think quantitatively, draw on data to guide the development of predictive models that can be expressed mathematically, and integrate pieces into a well-designed coherent system. In other words, "[the curriculum]…has endured not because it is frozen but because it has adapted dynamically to new ideas, emphases, challenges, and opportunities" (Luo et al., 2015). The undergraduate curriculum and chemical engineering education generally are discussed in more detail in Chapter 9.

New models of collaboration and integration are also emerging. Education has historically been designed as a linear flow from K-12 to college/university and then to a career. Today, global connectivity has led to new integrated learning and innovation practices. New education offerings are emerging from such online providers as EdX, Coursera, and Udacity. Online certificates and skills programs are finding wide application as the pace of change demands the constant refreshing of fundamental knowledge. The opportunity to build new shared content for use across universities is appealing, but would need careful consideration in light of the existing funding and reward systems within universities. Any such program needs to be available for lifelong acquisition and polishing of skills. Finally, assembling best-in-class content online for use as part of a university's offerings can make content available to smaller or underresourced institutions, which in turn could open up new paths for engaging different and more diverse talent pools with the concepts of chemical engineering.

For some, the undergraduate curriculum is a preamble to graduate studies. Graduate education is where the striking differences between the subjects undergraduates learn and the problems on which academics conduct research become clear. For example, undergraduates learn about small-molecule thermodynamics and some separation processes that have generally been highly specific to the oil and gas industry, although many bioseparation processes are now included in multiple courses. The oil and gas examples are obvious for historical reasons, but a vanishingly small amount of academic research is now occurring on those subjects. Thus, the real test for graduate students is their ability to adapt their critical thinking and analytic skills to problems in fields for which they probably lack both understanding of the vocabulary of the field and at least nontrivial basic science training.

This disconnect between the intellectual content of undergraduate and graduate work reflects the collective decisions of mainly federal but also other funding agencies that in large part support the work of graduate students. There is little federal support for work on basic thermodynamics or transport phenomena that an undergraduate would recognize. The disconnect is clear from even a casual look at the titles of grants funded most recently by the National Science Foundation's (NSF's) Division of Chemical, Bioengineering, Environmental, and Transport Systems, where many academic chemical engineers find at least some support for their research. Key phrases for funded proposals include "dynamic covalent junctions on block copolymer and network self-assembly," "sustainably derived high-performance nanofiltration membranes," "ultrahigh-resolution magnetic resonance spectroscopic imaging for label-free molecular imaging," and "photonic resonator hybrids." Funding of chemical engineers by the National Institutes of Health is of course even further disconnected from the traditional focus of the undergraduate curriculum.

Federal funding models are also driving a change in the way research is done. Through the 1990s, most research proposals in chemical engineering were written by individual principal investigators and generally were based on elements of intellectual curiosity balanced perhaps with a desire to solve a practical problem. Over the past 20 years, NSF in particular has increased the number of multi-investigator awards relative to single–principal investigator awards. Multiple–principal investigator proposals are often submitted in response to a call for proposals in a particular area (bioseparations or cybersecurity, for example). The data (Figure 2-2) show that this shift has been slow but steady. The impact on academic research has been positive in the sense that there are more interdisciplinary teams, and students are more likely to have multiple advisors or mentors, both of which improve educational outcomes. However, this shift also has led to a decline in funding for basic research by individuals on problems they find interesting. The new model supports innovation, frequently done by large teams, but not necessarily invention, which often comes from individuals (e.g., Wu et al., 2019). In addition to this shift toward large teams, the committee's sense is that funding opportunities are more prescriptive, or targeted, and less open-ended than in the past. In this context, funding agencies will need to consider the appropriate balance to meet their respective missions.

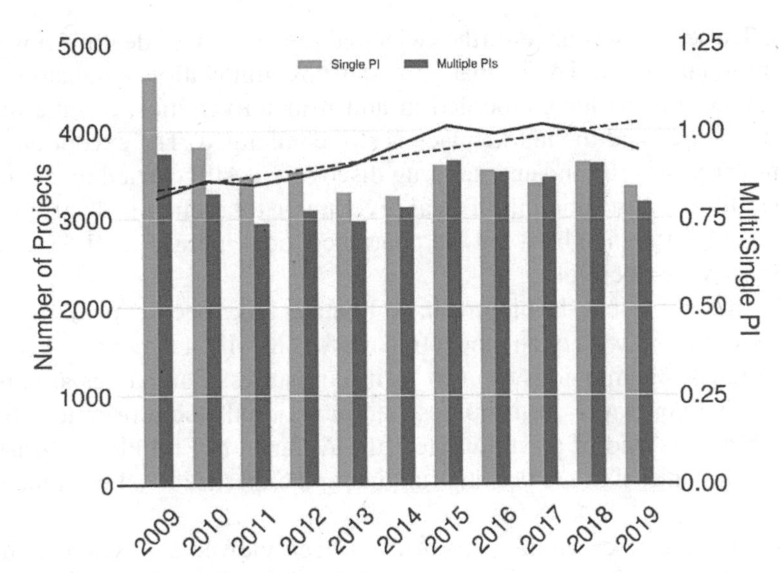

FIGURE 2-2 Number of new research projects funded by the National Science Foundation with a single principal investigator (SPI) (light blue; left) or multiple principal investigators (MPI) (dark blue; right) over the past 10 years. The ratio of MPI to SPI awards (solid black line with dashed black trend line) has increased. Data from NSF (2020).

This challenge is not unique to chemical engineering. Tension between basic and applied research characterizes most fields of science and engineering. Inventions cannot be planned and do not arrive on a schedule. Basic research is messy, nonlinear, and expensive. New knowledge relies on serendipity and the preparation of a fertile mind, and not all individuals have the ability to invent. The outcomes of basic research can fundamentally change the course of history, but for that to happen, a period of innovation after invention is usually required. Such innovation is often achieved through applied research or engineering.

On the industrial research front, the linear model of research, development, and commercialization has also been disrupted in the years since the Amundson Report was released. In the past, major companies deployed research and development (R&D) centers in support of their long-term growth objectives. Projects would then be transitioned to the appropriate business areas for commercialization, which often led to major new products and services. The R&D centers also served as a source of technical talent for the business areas because the research innovators would often transition to a business unit along with their technology developments. R&D centers maintained close connections to universities and academic programs as well. The overall system was a path from basic discovery at the university out to the market.

Priorities began to shift in the 1980s. Companies need to create new products or processes to maintain profitability. Large companies in the chemical and oil and gas industries have shifted their research activities toward supporting existing businesses. Companies have set shorter-term goals for corporate research, and their support for university research now focuses more on basic science and has grown only slightly since the 1980s

(NSB, 2020). These changes have further widened the "valley of death" between discovery and commercialization. To fill that gap, startups, innovation incubators, and many other intermediate models have emerged in and near universities, but the bidirectional connection of research and the marketplace is still challenged. The exception is the pharmaceutical industry, where fundamental drug discovery is still carried out. A comparison of R&D expenditures as a function of total revenue or total chemical sales (Figure 2-3) shows the marked difference between the pharmaceutical industry and the chemical and petroleum industry, respectively.

In a relatively recent development, companies have been expanding or replacing parts of their R&D effort with open innovation models in which they are seeking solutions from outside their R&D divisions or even their companies. Companies are also forming alliances with other companies, universities, and/or national laboratories to solve complex problems, such as the "end of plastic waste" (the Alliance to End Plastic Waste includes more than 40 companies that are pooling funding and expertise to address the problem of plastic waste).

Education and research have historically been viewed as separate activities. Today, there is an opportunity to design new models whereby learning and innovation are connected through new forms of public–private partnerships. Many opportunities exist to build a connected model in which students can take the latest ideas and technology to the marketplace, have experience with a business operation, and return with a perspective on the needs of the market to inspire new research initiatives. Many of the component steps for such a model exist, but new cross-sector collaborations to support and accelerate the growth of this model are needed.

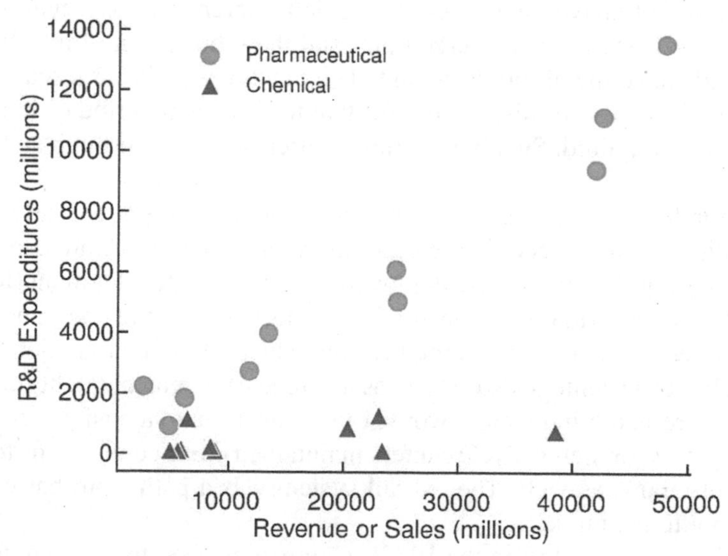

FIGURE 2-3 Research and development (R&D) expenditures in 2020 by the top 10 U.S.-based pharmaceutical (light blue circles) and chemical (dark blue triangles) companies (for which data were available), each compared with their revenue or chemical sales, respectively. Data from Buntz (2021); C&EN (2021); Macrotrends (2021a,b).

GROWTH OF INTERDISCIPLINARY WORK

The growth of interdisciplinary research reflects the increasing complexity of the problems to be solved and the growing sophistication of the tools needed to solve them. Nonetheless, such research also poses a danger. By definition, one cannot have interdisciplinary work without disciplines. The field of chemical engineering needs to recognize that its graduates are valued for the disciplinary skills they can bring to bear in working with others on a problem. Thus it is important to continue to educate students in the basic skills of chemical engineering, albeit with examples less reflective of an olefin-based business. At the same time, the field needs to be open to the influx of faculty members and practitioners who have not had a traditional chemical engineering education. Chemical engineering has benefited enormously from the influx of mathematicians, physical and other chemists, physicists and materials scientists, biologists, and others to its faculties over the years, a phenomenon expected to continue. One revealing example is that in the United States, chemical engineering is home to most programs in polymer science, a field largely founded in but not embraced by chemistry. This is no time for stasis, but instead a time for the field to expand and grow at its current frontiers while remaining true to the core that defines it. As societal challenges become increasingly complex, science and engineering solutions will necessarily come from connections across disciplines, and the boundaries between disciplines will continue to blur. To contribute to solutions for societal challenges in the coming decades, chemical engineers will need to become increasingly comfortable working across disciplines and as members of interdisciplinary teams. Indeed, this report highlights many areas in which chemical engineers will benefit from interdisciplinary collaborations.

3
Decarbonization of Energy Systems

- Addressing the existential threat of climate change will require decarbonization of current energy systems, a challenge rendered all the more difficult by the complexity and magnitude of the energy landscape and the resultant inability of any single energy carrier to meet the energy demands of all sectors in the foreseeable future.
- The field of chemical engineering continues to make important contributions to the scalability, delivery, systems integration, and optimization of the mix of energy carriers that will meet energy needs across different regions and sectors of society with lower carbon emissions and costs.
- Chemical engineers will enable technological advances at every point in the energy value chain, from sources to end uses, and bring to bear the systems-level thinking necessary to balance the economic and environmental trade-offs that will be necessary to transition to a low-carbon energy system.

Energy is a basic human need that is also essential for economic growth. The global energy demand today is about 155,000 TWh, 82 percent of which is currently supplied by fossil fuels (32 percent petroleum, 23 percent natural gas, and 27 percent coal) and the rest by nuclear (5 percent); wind and solar (3 percent); and other renewable energy sources, including hydroelectric, geothermal, and biomass (10 percent) (BP, 2019; IEA, 2021e). In 2018, the primary energy consumption by the various sectors was electricity generation (38 percent), transportation (28 percent), industrial processes (23 percent), and residential and commercial spaces (12 percent) (EIA, 2021h). The global energy supply mix will be altered substantially by the decarbonization efforts of various sectors, particularly in electricity generation, and by the greater adoption of electric vehicles (EVs) for light-duty transportation. Refineries are optimized for the production of gasoline or diesel, and substantial work and investment have gone into the use of biofuels for transportation. Widespread use of EVs will disrupt both the fossil and biofuels industries.

Chemical engineering has played an essential role in meeting society's demands for economical and energy-efficient conversion of natural resources into liquid and gaseous energy carriers while addressing environmental challenges associated with their production and use. Between now and 2050, the world population is expected to grow from 7.5 billion to well over 9 billion (OECD, 2012; UN, 2017), and increasingly prosperous populations will demand more energy; by 2050, the global demand for energy is forecast to increase by almost 50 percent (EIA, 2020b). Chemical engineering will continue to enable the equitable delivery of increasing amounts of reliable and affordable energy while supporting efforts to address the existential threat of global climate change (e.g., AIChE, 2020). Doing so will require the development and scale-up of renewable energy sources and carbon sequestration and utilization at an unprecedented rate and scale, as well as consideration of trade-offs in such areas as water consumption, cost, and environmental justice.

This chapter describes the impetus for prioritizing decarbonization of energy systems and the important role chemical engineers will have in advancing technologies that minimize the climate impact of the energy sector. The chapter is organized from sources to end uses. Opportunities for chemical engineers are explored in energy sources; energy carrier production; energy storage; energy conversion and efficiency; and carbon capture, use, and storage.

THE NEED FOR DECARBONIZATION

The international climate science community has established a link between global greenhouse gas (GHG) emissions and human actions and energy usage. The concentration of CO_2 in the atmosphere has tracked closely its rate of anthropogenic production since the start of the Industrial Revolution in 1750 (Figure 3-1). Emissions rose to about 5 billion metric tons per year in the mid-20th century before reaching more than 35 billion metric tons per year by 2000 (NOAA, 2020). In parallel, global surface air temperatures have increased by 1 °C. Oceans absorb a large amount of CO_2 released into the atmosphere. This absorbed CO_2 reacts with seawater to form carbonic acid. Thus, increased CO_2 levels in the atmosphere increase the acidity of the ocean, harming shellfish and other marine life.

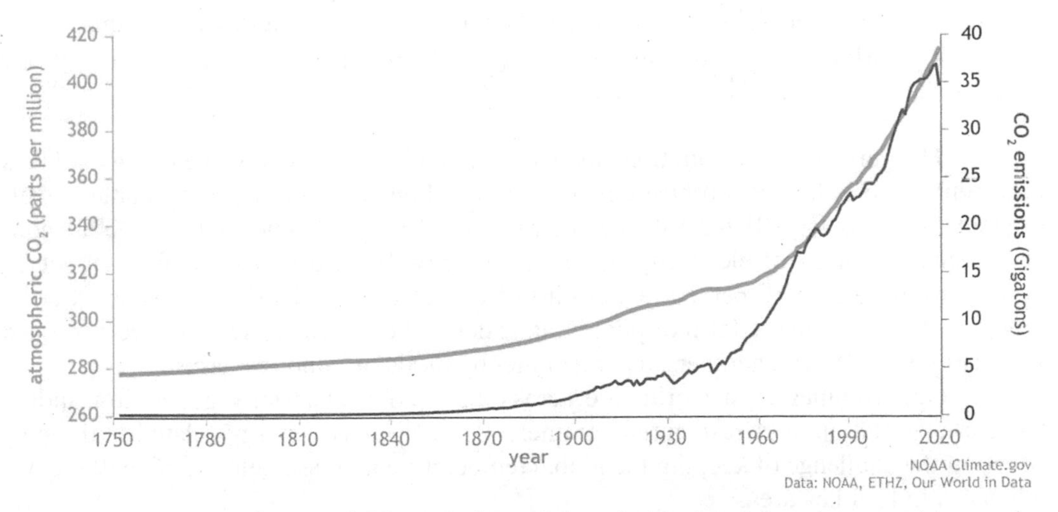

FIGURE 3-1 Increase in annual CO_2 emissions (right axis, blue) and the subsequent increase in the concentration of atmospheric CO_2 (left axis, magenta) since the Industrial Revolution. SOURCE: NOAA (2020).

International efforts to mitigate climate change began in 1992, culminating in 2015 with the Paris Agreement (UNFCC, 2015), whose primary goal is to keep the global temperature rise during this century well below 2 °C above preindustrial levels. This is a monumental challenge that will require decarbonization of the energy sector, net-zero emissions, and fast-paced removal of GHG from the atmosphere. Transitioning to net-zero emissions in the energy sector is likely to cost trillions of dollars and will require efforts at all levels of government and across all sectors (e.g., the coordinated, systems-level approach proposed as part of the Sustainable Energy Corps; Alger et al., 2021). It will take decades, and may never be complete. The time required to decarbonize energy systems will depend on technological advances, government policies, changing economics of energy carrier options, and essential modifications in consumer behavior. Given the magnitude and complexity of the global energy system, no single energy carrier will be able to satisfy requirements across all sectors in the foreseeable future. Thus, achieving the goals of the Paris Agreement will require a wide range of energy sources and carriers.

The question of viable energy mixes was addressed in a comprehensive multi-model study coordinated by the Energy Modeling Forum 27 (EMF27), which examined 13 scenarios for keeping the global temperature increase below 2 °C in this century (Kriegler et al., 2014). Because of the significant number of uncertainties and assumptions involved in these 13 scenarios, it is best to use a notional average of the scenario outcomes to approximate the various energy trends. A notional average view suggests that in 2040,

- petroleum and natural gas will continue to play a significant role in the energy mix;
- coal usage will decrease significantly;
- energy carriers from nonbiogenic renewables (e.g., solar, wind, hydroelectric power) and biogenic sources (bioenergy) will grow significantly; and
- carbon capture, use, and storage (CCUS) will become a key technology for decreasing CO_2 emissions.

The generation, distribution, and use of electrons from renewable energy sources represent some of the most robust enablers of decarbonization. In its Sustainable Development Scenario for 2019–2070, the International Energy Agency (IEA, 2020d) concluded that the share of electricity in end-use energy demand will grow from about 20 percent to more than 50 percent. One-third of that electricity demand is expected to be met by solar power in the form of photovoltaic devices deployed at scale in decentralized form (Figure 3-2), with another 20 percent met by modular wind resources.

Opportunities exist worldwide across all sectors (electricity generation, industrial, transportation, and residential/commercial) to decrease energy-related emissions. Meeting the challenge of keeping the global temperature increase below 2 °C will require advances in four key areas:

- energy efficiency;
- increased use of lower-carbon energy sources;
- development and deployment of novel energy and energy storage technologies; and
- government policies to promote cost-effective solutions.

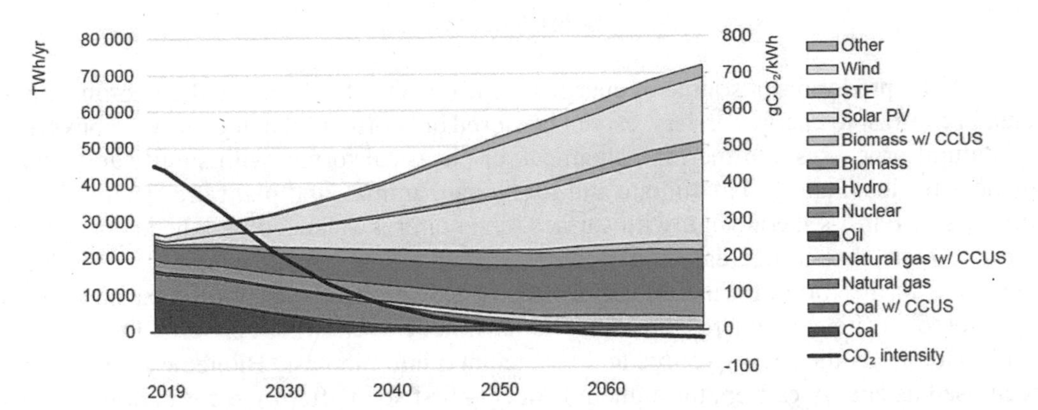

FIGURE 3-2 Projected global power generation by fuel technology type in the Sustainable Development Scenario for 2019–2070. The scenario assumes net-zero CO_2 emissions by 2060 due to increases in sustainable energy sources. NOTES: CCUS = carbon capture, utilization, and storage; PV = photovoltaic; STE = solar thermal electricity. "Other" includes geothermal power, ocean energy, and hydrogen. SOURCE: IEA (2020d).

This chapter describes the critical role of chemical engineering in the transition from fossil fuels to renewable energy, with contributions ranging from energy carrier generation, storage, and distribution to energy use and conversion across various sectors. Some sectors, such as electricity generation, light- and medium-duty transportation, and heating and cooling for residential and commercial buildings, are less challenging to decarbonize than others (e.g., heavy-duty long-haul ground, aviation, and marine transportation; energy-intensive cement and steel production). Any low-carbon energy transition "bridging" strategy will rely on the greater use of a mix of energy carriers: electrons, hydrogen, and lower-emission liquid fuels (e.g., advanced biofuels, synthetic liquid fuels; Santiesteban and Degnan, 2021). The energy transition will require hybrid systems that combine different energy carriers to address the challenges of different usage sectors, challenges that chemical engineers are uniquely positioned to address. At the same time, it should be noted that, while chemical engineers can develop technological solutions and improve the economic competitiveness of those solutions, many of today's barriers to addressing climate change are social and political, and chemical engineers work within that larger societal context.

ENERGY SOURCES

Chemical engineers play central roles in the recovery and development of energy sources and in energy distribution. This section focuses on the recovery and conversion of energy and opportunities for chemical engineering with respect to the two primary energy sources—solar and nuclear. The section on solar energy also includes opportunities for chemical engineers related to secondary sources (fossil fuels, biomass, and intermittent sources) derived from solar energy.

Solar Energy

The predominant source of energy for the planet is the flux of solar photons. Photons, in contrast to energy carriers, cannot be stored or "bottled." Their energy is converted via natural processes into thermal, chemical, or electrical forms, with significant consequences for local and global climate and for human, animal, and plant life. Thermal capture acts as Earth's thermostat, with surface temperatures balanced by albedo and greenhouse effects that also create the weather patterns from which energy is ultimately recovered in the form of wind and hydroelectric power. The energy of solar photons is also stored as chemical energy through photosynthetic cycles that convert CO_2 and H_2O (with photons as the energy source and coreactants) into biomass. Biogenic sources have been used as energy carriers throughout history—first soon after their formation as combustion fuels, but much more extensively in modern times as fossil fuels, well after geological chemical reactions have deoxygenated these photosynthetic residues and increased their energy density, forming natural gas, crude oil, and coal. The chemical reduction of primordial biomass that led to its deoxygenation and storage as fossil fuels is now being reversed, over much shorter periods of time, through their combustion and consequent conversion to CO_2 and H_2O, along with the release of heteroatoms sequestered within the biomass. These processes have had important local and global environmental consequences, but have also enabled the standard of living enjoyed today. This section describes solar energy as a primary energy source through the direct capture and conversion of photons to energy carriers (electrons; H_2, NH_3, or organic fuels; and heat), and as the source of fossil fuels (coal, natural gas, petroleum), biofuels (lignocellulosic and other sources), and intermittent sources (wind and marine).

Direct Capture and Conversion of Photons

Chemical engineering, the discipline most adept at transforming stored energy carriers into more convenient forms and into chemicals and materials, also brings an enabling skillset to the harvesting of photons and the storage of their energy until it is converted into its thermal, electrical, chemical, or mechanical forms. The discipline, closely collaborating with other disciplines, such as materials science, solid-state chemistry, and physics, will continue to contribute to the efficiency, durability, and reliability of photon capture materials and device components; the ability to manufacture and deploy them at scale; and the optimization, control, and systems-level strategies required for embedding modular devices within efficient electrical grids. Chemical engineers' expertise in chemistry and catalysis and ubiquitous transport processes has been applied throughout the discipline's history to improve the efficiency of chemical transformations. This expertise will be the enabling tool as photons are used, either directly or via electrons as intermediates, to affect the synthesis of chemicals or various energy carriers from CO_2 and water. The domains of surface catalysis, photocatalysis, and electrocatalysis are firmly planted within the chemical engineering discipline.

In the case of photons, their capture and conversion to energy carriers (electrons; heat; chemical energy as H_2, NH_3, or small hydrocarbons/alcohols) that can be stored and transported to consumers and markets are inseparable. The path to clean energy from photons, mediated by electrons and molecules that can be stored and transported, will require decentralized capture within "photon conversion factories," akin in concept to the integrated refineries used to transform fossil resources into fuels and chemicals today, but in much smaller and modular forms and deployed at diverse points of photon capture. These factories will produce heat, electrons, energy carriers, and chemicals as part of modular integrated systems designed to capture the largest possible fraction of the solar flux with high quantum efficiencies and convert it into useful forms of chemical energy.

Direct photon capture and conversion to electrons as energy carriers. There are several paths to photon capture and utilization. Direct solar conversion to electrons is carried out using semiconductors that first capture photons through electronic excitations across their band gap and then collect the excited electrons in the form of photovoltaic (PV) solar panels. Together, the decrease in the cost of PV panels and the expected increase in the electrification of energy systems are driving the deployment of global PV capacity at a rate that could not have been envisioned just a few years ago. This global capacity, only about 500 GW in 2018, will double by 2022 and is predicted to exceed 10 TW by 2030 and 30–70 TW by 2050 (the current global demand is 18 TW; Haegel et al., 2019).

Silicon (Si)-based PV cells represent about 80 percent of the currently installed solar capture capacity in the United States, the rest consisting of cadmium-telluride (CdTe) semiconducting thin-film PV cells (DOE, 2021a). High-purity amorphous and polycrystalline Si PV cells are manufactured using energy-intensive purification processes first developed for the processing of Si wafers for electronic devices. State-of-the-art Si PV cells operate at near-theoretical capture efficiencies (~30 percent), a limit set by the solar spectrum and the balance between the band gaps accessible by doping and the attainable current densities. The low-absorption cross-section of Si requires thick wafers, precluding the use of tandem devices designed to collect different components of the solar spectrum through systematic doping. Si PV cells will continue to evolve through incremental improvements in device architecture and design options at the cell/module scale, as well as through lower manufacturing costs; greater reliability/durability; and the development of infrastructure for their installation, maintenance, and seamless insertion into advanced electrical grids. Chemical engineers will continue to enable the evolution of Si PV cells, their deployment at a global scale, and the optimization and control strategies required to integrate them into the grid. Si PV cells represent the medium-term choice for deployment at scale in direct photon-to-electron conversion.

CdTe thin-film PV cells absorb photons at energies near the maximum flux in the solar spectrum. Recent improvements in efficiency and manufacturing costs have made them competitive with Si PV cells. These modules consist of micron-thick films of CdTe held within layers of conducting transparent oxides. Environmental concerns about the toxicity of the components in CdTe PV cells will need to be addressed through improvements in reliability and durability, thinner films, and higher efficiencies. These advances will enable greater market penetration as PV-based solar capture devices become more prevalent in practice. Such advances in manufacturing, cell/module architecture, and systems integration will be driven by chemical engineering as an enabling discipline, as illustrated by recent efforts to coordinate research and manufacturing capabilities through a National Renewable Energy Laboratory–led consortium.[1]

The parallel developments in dye-sensitized PV cells have recently been punctuated by the emergence of mesoscopic architectures, in which coatings of n-type semiconductor nanoparticles, such as titania, act as mesoporous anodes that provide 1,000-fold increases in dye-anode connectivity (Hardin et al., 2012; O'Regan and Grätzel, 1991). These molecular photovoltaics have emerged in parallel with perovskite solar cells (PSCs; Grätzel and Milić, 2019), leading to a significant disruption in the nature of research on PV cells and to very rapid advances in capture efficiencies. PSC devices also provide the benefits of roll-to-roll solution-based manufacturing processes, a tolerance for reagents lower in purity, and much smaller amounts of active materials relative to Si PV cells. PSC devices have evolved rapidly in photon capture efficiency, from 4 percent in 2010 to >25 percent in 2019 (Grätzel and Milić, 2019; Figure 3-3). These PSC devices are now approaching photocurrents near their theoretical maximum, but improvements in efficiency, open-circuit voltages, and long-term durability, as well as replacement of the toxic water-soluble components ubiquitous in the best-performing perovskite materials, need to be addressed before significant commercial deployment at scale can occur (Correa-Baena et al., 2017). These cells suffer from operational instabilities, short useful lives, and significant environmental and health concerns related to the long-term containment and ultimate disposal/recycling of their toxic constituents. These challenges are being addressed through significant funding from federal programs[2] and an influx of entrepreneurial capital.

Concerns about toxicity and long-term durability continue to prevent PSC devices from displacing Si PV cells in the marketplace. Recent developments have led to more stable perovskite compositions, to the identification and mitigation of extrinsic degradation mechanisms, and to device configurations that ensure more reliable containment to prevent the release of toxic components in the most efficient perovskites (e.g., methylammonium lead trihalides and formamidinium analogs). Durability and containment, however, remain formidable challenges (Correa-Baena et al., 2017; Rong et al., 2018). Significant ongoing research focuses on perovskite compositions that minimize intrinsic degradation processes and on modular device architectures that ensure reliable long-term operations. As in the case of Si PV cells, chemical engineering is well positioned to address and resolve these challenges, and to deploy the improvements in practice at scale. The development of advanced solution-based processes for perovskite cell manufacture, integration of PSC systems into existing grids, and life-cycle assessment (LCA) of the full environmental impact of these devices are also encompassed by fundamentals and practice of chemical engineering.

---

[1] See https://www.energy.gov/eere/solar/cadmium-telluride-research-and-development-consortium-coordination.

[2] See https://www.energy.gov/eere/solar/solar-energy-technologies-office-fiscal-year-2020-perovskite-funding-program.

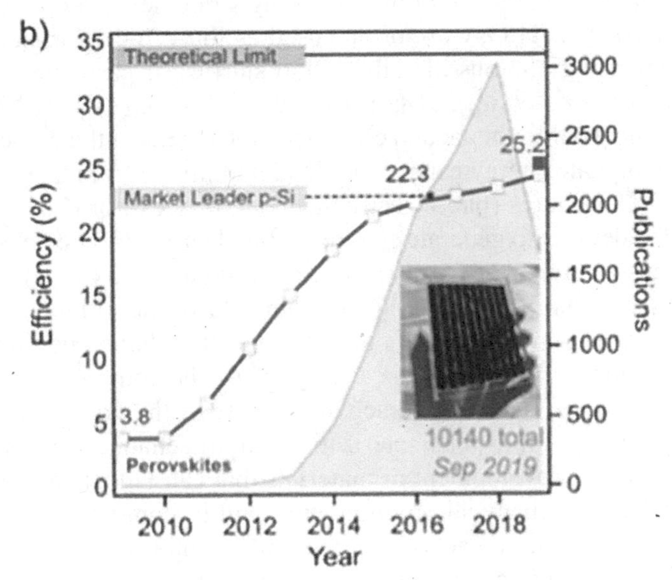

FIGURE 3-3 Improvements in capture efficiency of perovskite-based solar cells since 2010.
SOURCE: Grätzel and Milić (2019).

These PSC systems absorb light via direct electronic transitions, leading to high photon absorption cross-sections and to efficient capture using thin films, in contrast to the thick wafers required for Si PV cells, because of the low photon absorption cross-sections inherent in their indirect electronic transitions. Such thin films minimize the amounts of active components needed and provide significant opportunities for solution coating processes and for the scalable manufacturing of PSC devices (Li et al., 2018). Thin films also allow the synthesis of flexible devices suitable for curved surfaces; their transparent nature enables tandem cells with stacked layers of different perovskites designed to capture complementary wavelengths in the solar spectrum and a larger fraction of the impinging solar flux. A life-cycle analysis of Si-free tandem cells consisting of two perovskite layers recovered the energy required to manufacture them in 0.35 years, a much shorter period than the 1.44 years for perovskite-Si tandem cells (Tian et al., 2020a).

Conversion of photons to H_2, NH_3, or organic fuels as energy carriers. The previous section addresses the conversion of solar energy via electronic excitations and ejection and capture of the emitted electrons. In this context, electrons are transported to markets or stored as chemical energy within batteries to mitigate the intermittency of solar flux. The capture of the energy of photons as energetic molecules provides alternative paths for transporting the photon energies in a different form. At the point of capture, such strategies also deal with the intermittency issues inherent in solar capture. In all cases, these strategies require modular architectures and significant integration and intensification of the photochemical and electrochemical processes involved.

Implementation requires one of the following strategies: (1) direct reduction of a common molecule (such as H_2O, CO_2, or N_2) used as the vehicle for storing and transporting solar energy using photons directly within slurries of particulate photocatalysts (direct photocatalysis; Goto et al., 2018; Takata et al., 2020); (2) photoelectrochemical cells (PECs) that couple photovoltaic and electrochemical cells at the device scale to generate H_2 from H_2O, organic energy carriers or H_2–CO mixtures from CO_2–H_2O reactants, or NH_3 from N_2–H_2O mixtures (photoelectrochemical devices); or (3) spatially separate modules that use PV devices to generate electrons and electrocatalytic cells that use these electrons as reactants to reduce the carrier molecules (and their mixtures) to the end products listed in (2) (sequential processes). Such systems have the ultimate potential to deliver these energy carriers and chemicals at scale, but they have been demonstrated at practical scales only for H_2 production via strategy (3)—the combination of commercial PV cells and H_2O electrolysis modules, each at the state of the art. These strategies will require advances in the synthesis, characterization, and mechanistic assessment of catalytic solids, as well as the development of materials that can withstand severe chemical, photochemical, and electrochemical environments within complex hydrodynamics for systems that couple the required reactions through diffusional controls. Thus, the combination of, and advances in, various disciplines and such subdisciplines as catalysis, fluid mechanics, solid-state chemistry and physics, separations, advanced models and simulation at the microscopic and macroscale levels, process design, and process control will be important for future breakthroughs. These subdisciplines are all within the domains of chemical engineering research and practice.

H_2 generation via photocatalytic water splitting represents the most direct route to the capture of solar energy as chemicals for either transport to markets or a means of addressing intermittency at the point of capture. The state of the art and competitiveness of the three strategies described above are discussed in several reviews (Ardo et al., 2018; The Royal Society, 2018a).

Direct catalytic water photolysis uses particulate photocatalysts consisting of an absorber (e.g., $SrTiO_3$, Ta_2N_5) dispersed as aqueous suspensions. These photocatalysts generate electrons and holes that are collected separately at metal nanostructures present at their surfaces to form H_2 and O_2 at each location, in systems that are simple in design and applicable at larger scales than are possible with modular integrated photoelectrochemical devices. Collecting the H_2 and O_2 separately and preventing their recombination, extending the life of the metal-promoted semiconducting photocatalysts, and improving their capture efficiencies, however, pose significant safety, engineering, materials, and catalysis challenges (Ardo et al., 2018). This approach, and means of overcoming its challenges, are the current focus of the Japan Technological Research Association of Artificial Photosynthetic Chemical Process.

An extensive review of the challenges associated with reactor scale-up and synthesis, and of the efficiency of inorganic photoelectrodes provides the most detailed and up-to-date roadmap for the deployment of photoelectrochemical devices for water splitting. This review describes the trade-offs between efficiency and complexity as systems evolve from direct photocatalysis to integrated photoelectrochemical devices and ultimately to integrated systems with PV-electrolysis modules (Moss et al., 2021; Figure 3-4).

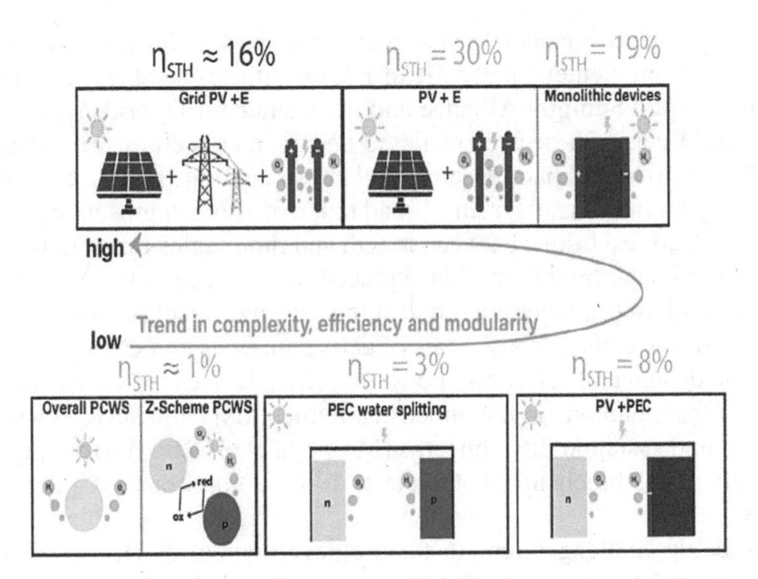

FIGURE 3-4 Illustration of several solar-driven water-splitting technologies, sorted from low to high complexity, efficiency, and modularity. NOTES: E = electricity; PCWS = public community water system; PEC = photoelectrochemical cell; PV – photovoltaic. SOURCE: Moss et al. (2021).

PEC devices, which combine photon capture and electron generation at the device scale, show higher capture efficiencies relative to direct photocatalytic water splitting. However, their complexity and modular architectures represent formidable hurdles for deployment at scale, as do the lifetime of the photoelectrodes and the delicate architectures required to integrate photon capture and electrolysis at the device scale. As in the case of direct photolysis, the efficiency of PEC devices decreases as more demands are placed on materials and interfaces to carry out the combined functions of photon capture, charge separation and collection, charge transport at a catalytic function, and the molecular-scale evolution of H_2 and O_2 via electron transfer at catalytic centers. Such PEC devices show very high photon capture efficiencies at the expense of greater cost and complexity; they represent solutions only for niche applications in the immediate future. Their ultimate use at scale remains uncertain, and any significant progress toward practical systems will require addressing engineering design, molecular and electronic transport, and materials discovery in concert.

These PEC devices for the synthesis of solar H_2-based fuels, as well as their architectural analogs for artificial photosynthesis strategies for converting CO_2–H_2O mixtures to CO and organic energy carriers, remain at the proof-of-concept stage (Lewis, 2016). They will require the development of materials and interfaces that can efficiently induce charge separation upon photon-induced excitations and transfer these charges to catalysts that can form H_2 and organic solar-derived fuels before recombination. These modules will need to be robust and relatively inexpensive for deployment at scale. Many of these challenges are being addressed as part of the work of large multidisciplinary centers, such as the Joint Center for Artificial Photosynthesis[3] and its recently announced successors, the Liquid Sunlight Alliance and the Center for Hybrid Approaches in Solar Energy to Liquid Fuels.[4] These centers aim to design in concert the different components required and develop hybrid photoelectrodes that can combine photon capture and molecular catalysis to generate carriers from a broad range of wavelengths in the solar spectrum. These advances require a bridge between length and time scales inherent in photon-driven excitation and molecular transformations induced by emitted photoelectrons at a catalytic function. Systems-based integration, control, and design; reaction-transport formalisms; and knowledge of the catalytic properties of active surfaces and centers will play an enabling role in the design and selection of cost-effective devices for the direct generation of energy carriers from photons in a manner that avoids toxic and scarce elements, as well as containment and sustainability concerns (Montoya et al., 2017). Bringing such considerations into chemical, biochemical, and electron-driven processes is a domain of chemical engineers.

Formidable challenges remain for the development of electrochemical systems for direct or sequential conversion of photon energies into chemical energy in the form of H_2 (from H_2O), CO (from CO_2), small alcohols and hydrocarbons or H_2–CO mixtures for subsequent thermochemical conversion to such molecular carriers (from CO_2–H_2O), and ammonia (from N_2–H_2O). The most enduring and significant of these challenges are

- the modular nature and complex interconnections among functions and the integration of electrical and chemical processes at scale;
- the ubiquitous requirement for scarce precious metals as electrodes, and toxic or rare elements as semiconductors, dye sensitizers, dopants, and connectors;
- the need for process intensification and high photon capture efficiencies limited by transport processes within electrolytes, electrodes, or semiconductors;
- the durability of modules during extended field use and their recyclability after their useful life;
- the costly extraction of dissolved product molecules from dilute aqueous media and the separators required to avoid the recombination of photocatalytic or electrocatalytic products; and
- the energy requirements in fabrication and recovery of the component elements after use.

These matters involve catalysis and kinetics in complex and nonideal liquid systems (thermodynamic and hydrodynamic); transport of molecules, ions, and electrons in fluids and solids; materials assembly with precise nanoscale and mesoscale architectures; process integration, control, and optimization; and LCA. The challenges, fundamentals, and coping/solution strategies lie firmly within the domain of chemical engineering,

[3] See https://solarfuelshub.org/.

[4] See https://www.energy.gov/articles/department-energy-announces-100-million-artificial-photosynthesis-research.

which has in the past adeptly tackled challenges of similar character for complex chemical, biochemical, and electrochemical conversion processes mediated by heterogeneous, molecular, or biological catalysts.

Electrolysis remains the proven technology for electrochemical generation of H_2 via modular systems based on acid polymer, liquid alkaline, or ionic-transport solid electrolytes (Ardo et al., 2018; Moss et al., 2021). Progress has recently been made in the scaling up of electrolysis, and chemical engineers have an opportunity to contribute to the development of applications at the scale required to disrupt the energy landscape. Electrolysis systems can be operated in acidic or alkaline regimes, although the rate-limiting nature of the O_2 evolution half-reaction (H_2O oxidation) has led to a preference for alkaline electrolyzers, which also avoid the platinum-based materials required to prevent electrode dissolution in acidic media, thus allowing the use of nonprecious metals (e.g., Ni, Fe, Cu) as electrodes. Acidic electrolyzers use polymer electrolyte membranes that minimize contact between H_2 and O_2 through fast proton transport and short anode–cathode distances. Alkaline electrolyzers have relied on microporous physical barriers that impose larger anode–cathode distances and ohmic losses. Recent developments in selective anion transport membranes prevent contact between H_2 and O_2 and have led to more compact membrane–electrode modules.

The challenges of deploying electrolyzers at scale include their efficient integration and control as multimodule stacks, the development of earth-abundant electrode materials, and thinner and more efficient separator membranes. The challenges are similar but even more formidable for electrochemical reduction of CO_2 via concurrent electrolysis with H_2O to form mixtures of H_2, CO, alcohols, carboxylic acids, and hydrocarbons (Hori, 2008; Nitopi et al., 2019). These products can be used directly as energy carriers or as precursors to such carriers on heterogeneous catalysts (e.g., H_2–CO conversion to liquid transportation fuels via Fischer-Tropsch or methanol synthesis). Electrochemical systems face similar challenges in meeting the scale required for impact:

- more efficient and robust electrodes based on earth-abundant elements;
- thinner and more selective membrane separators;
- higher-temperature electrolyzers;
- integration of electrocatalytic and thermocatalytic systems in sequence or within a single device to form more suitable energy carriers; and
- the development of supply chains to lower the costs of manufacturing and integrating the modular devices into molecular weight (MW)–scale distributed deployments for the synthesis of H_2 and other energy carriers.

Conversion of photons to heat as an energy carrier. The modular nature of PV and electrochemical cells, whether in separate or combined forms, poses significant challenges, including process integration and intensification and deployment at scale. These challenges can be addressed by using solar thermal strategies that capture the energy of photons as heat, which can be delivered to users as thermal energy, or converted to chemical energy for transport or for storage during diurnal or intermittent fluctuations in solar flux. The end use of the captured thermal energy depends on the temperatures accessible through solar collectors and heat transfer media and on the location of markets relative to the point of capture. The specific end-use option of generating H_2 at scale has been highlighted in recent reports because of its relevance to a hydrogen economy and its inherent advantages over electrolyzers in large-scale deployment (Gonzalez-Portillo et al., 2021; Moss et al., 2021; NREL, 2017).

This capture and storage of photon energies as heat can be used directly in ambient temperature control in commercial or residential spaces; as process heat in existing chemical processing plants or manufacturing operations; or for conversion into electrons or hydrogen, typically through the generation of high-pressure steam or through conversion cycles commonly termed "chemical looping." The latter approach involves the design of reactors based on the chemical engineering principles of kinetics, transport, chemical absorbents, and construction materials that can withstand the extreme temperatures required for thermal and chemical efficiencies.

The achievement of very high temperatures (~1400 °C) has been enabled by advances in parabolic solar thermal concentrators (Sargent & Lundy LLC Consulting Group, 2003); these systems have in turn allowed the generation of very high–pressure steam for power generation in high-temperature turbines. At such high temperatures, chemical looping using redox-active oxides enables the cycling of such solids between a reduced state that reacts with water to form H_2 and an oxidized state that evolves O_2 in a spatially separate stage. Such processes use oxides of earth-abundant elements and form separate streams of H_2 and O_2 from H_2O, thus avoiding the need for gas separators and the use of costly metals as electrodes, as well as the transport limitations inherent in the use of liquids as electrolytes.

These cyclic processes can be deployed via large-scale devices that allow process integration and intensification strategies unavailable for modular systems but ubiquitous in conventional refining, chemical manufacturing, and power generation processes, albeit at somewhat lower temperatures. The high temperatures required for efficient solar thermal capture pose significant challenges with respect to the design, use, and handling of the heat transfer media and the durability of the required redox-active oxides. Molten salts are typically used as heat transfer fluids, but solid particles in fluidized systems have recently emerged as attractive alternatives (Gonzalez-Portillo et al., 2021).

Fossil Fuels

Fossil fuels (coal, petroleum, and natural gas) have been essential for society's development and progress, having powered the Industrial Revolution and shaped the modern world. Technological advances based largely on the ingenuity of chemical engineers have enabled the efficient extraction, processing, and conversion of fossil fuel raw materials into useful products. The expertise of chemical engineers remains essential in enabling a transition from the current energy landscape to one based on renewable and sustainable energy sources. However, most studies have concluded that fossil fuels will continue to play a key role in the energy mix at least until 2050, and in the meantime, the imperative is to reduce the carbon footprint of fossil fuels—a key opportunity for chemical engineers.

Coal. Coal is the most abundant and least expensive of all fossil fuel resources. It can be converted into gas or liquids to produce chemicals and fuels, but is used primarily for direct combustion. As a solid fuel, coal generates more CO_2 per unit energy than other fossil fuels—from approximately 30 percent more than diesel to nearly twice as much as natural gas (EIA, 2021i).[5] Coal resources are used mainly to generate electricity, and while reliance on coal has increased in low- and middle-income countries (e.g., China and India), the opposite has been true in higher-income countries (e.g., United States, European Union, Japan). In the United States, electricity generation from coal has been declining since 2008, with the biggest drop (~16 percent) taking place in 2019; in contrast, the use of natural gas has increased dramatically since 1995, and the use of renewables has increased since 2005, albeit at a slower pace (Figure 3-5).[6]

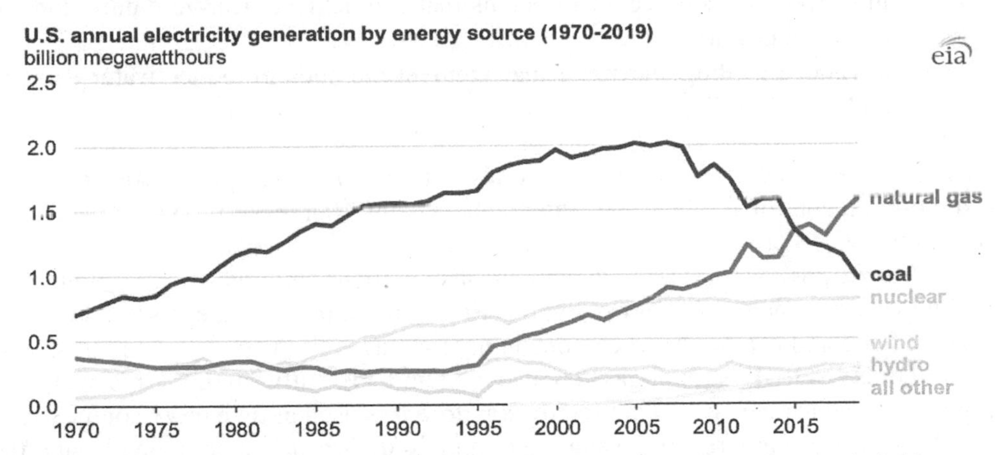

FIGURE 3-5 Annual electricity generation by different sources (natural gas—dark blue; coal—dark red; nuclear—yellow; wind—green; hydro—light blue; all other sources—gray) in the United States, 1970–2019. SOURCE: EIA (2020d).

Most investments in coal-fired plants in 2019 (almost 90 percent) were for higher-efficiency (supercritical and ultrasupercritical) plants; the remaining small portion of investments were in inefficient subcritical plants, mainly in Indonesia (IEA, 2020c). High-efficiency coal-fired power plants use water at high, above-critical temperatures and pressures (373 °C and 220 bar, respectively). The efficiency gains thus achieved reduce by about 20 percent both the amount of coal needed and CO_2 emissions. These plants also emit substantially lower amounts of nitrogen oxides and sulfur oxides (IEA, 2012).

---

[5] Pounds of CO_2 emitted per million Btu (gJ): coal, 215 (227); diesel, 161 (170); gasoline, 157 (166); propane, 139 (147); natural gas, 117 (123) (EIA, 2021).

[6] Preliminary IEA analysis indicates a sharp drop in power-sector demand in 2020 as a result of the COVID-19 pandemic, with demand for coal having the greatest uncertainty of all fuels used for power.

To capitalize on coal's advantages and help mitigate its disadvantages, research and development (R&D) is needed to increase thermal efficiency, demonstrate cost-effective and secure carbon capture and storage, further improve emission controls, and reduce water consumption. Meeting these challenges will require research to improve existing and develop new breakthrough technologies. The Electric Power Research Institute has recommended the following key goals for such efforts, all areas in which chemical engineers can play a role (Maxson and Phillips, 2011):

- improved plant efficiency via high-temperature materials and higher turbine inlet temperatures;
- cost-effective and scalable CO_2 capture in new or retrofitted applications;
- environmentally safe and permanent storage of CO_2;
- improved emission control systems that can achieve near-zero emissions of all pollutants; and
- advanced cooling and water management methods to reduce water demand and pollutant discharges.

Progress has been made in several of these areas (e.g., improved plant efficiency, improved emission control systems, and water management), but less so in the implementation of viable CCUS processes.

Natural gas. Natural gas contains mostly methane, but also small amounts of ethane and varying amounts of heavier hydrocarbons, including propane, butane, and pentane. The ethane and heavier hydrocarbons in natural gas are typically referred to as natural gas liquids (NGLs). Natural gas can also contain CO_2, sulfur, helium, nitrogen, hydrogen sulfide, and water, which are removed before it is used as an energy source.

Processing plants remove water vapor and nonhydrocarbon compounds, and the NGLs are separated from the wet gas and sold separately. The separated NGLs are called natural gas plant liquids, while the processed natural gas is called dry, consumer-grade, or pipeline-quality natural gas. Some natural gas is dry enough to satisfy pipeline transportation standards without processing. Odorants (light mercaptan compounds) are added to aid in the detection of pipeline leaks. Pipelines transport dry natural gas to underground storage fields or to distribution companies and eventually to consumers (EIA, 2021g; Figure 3-6).

In the mid-2000s, a step change in natural gas and shale oil production occurred in the United States. Often referred to as the shale revolution, this innovation was made possible by a combination of hydraulic fracturing and horizontal drilling techniques that enabled economical oil and gas production from shale formations. The result has been a near doubling of U.S. natural gas production, from 18 trillion cubic feet in 2005 to ~34 trillion cubic feet in 2019. The United States is now a world leader in natural gas and oil production, and a global supplier. In the United States, natural gas is used primarily for electricity generation (power sector) and for heating (industry, residential, and commercial), with a small fraction used for transportation. Figure 3-7 shows natural gas consumption by the various sectors in 2019 and its evolution over the period 1950–2019.

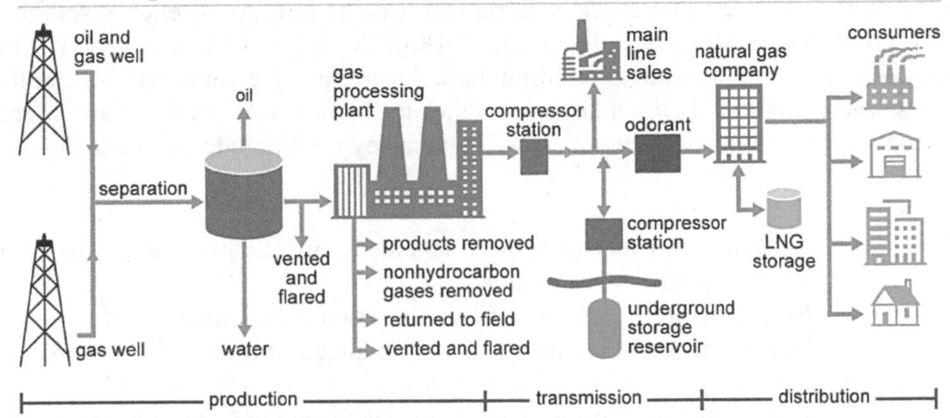

FIGURE 3-6 Schematic diagram of natural gas production, processing, and delivery to end users. NOTE: LNG = liquified natural gas. SOURCE: EIA (2021f).

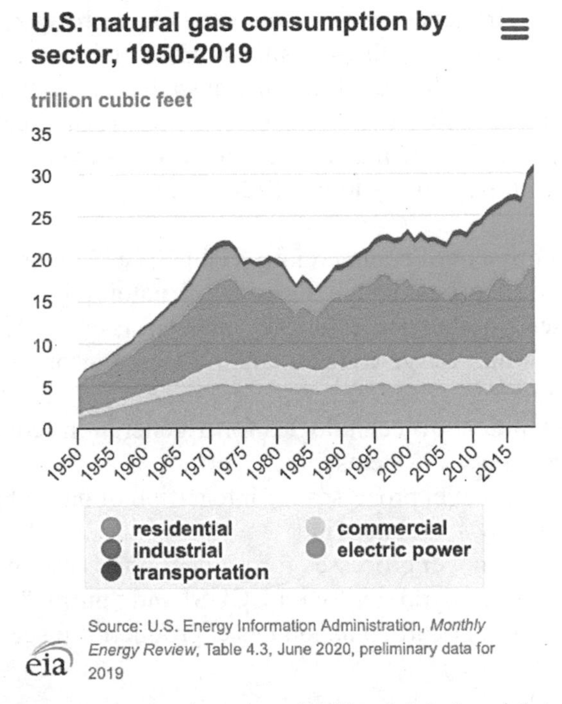

FIGURE 3-7 Natural gas consumption by sector (residential, industrial, transportation, commercial, and electric power), 1950–2019. A net increase in natural gas consumption occurred over this time period across all sectors. SOURCE: EIA (2021e).

It is generally accepted that, relative to other fossil fuels, natural gas provides a cleaner bridge to a renewable energy future, and it is the only fossil energy source projected to grow in the coming decades (DOE, 2018b). However, the longer-term future for natural gas is less certain. Innovation throughout the entire value chain will be required if natural gas is to continue being a key contributor to the future of the low-carbon energy mix. Areas in which chemical engineers will play a key role include the following:

- Production
  - Advances in water-quality management; water recycling for shale or unconventional gas production
  - Further reduction of methane venting to the atmosphere
  - Accelerated development of CO_2 to replace water as a fracturing agent
- Processing
  - Development of low-energy processes for natural gas separation and purification
- Storage and transportation
  - Methane leakage control
  - Higher-efficiency compressors and heat exchangers
  - Smart sensors for pipeline operational efficiency
  - Materials for intercontinental transport via pipeline versus the current practice, which involves liquefaction and regasification
  - Low-cost pipeline materials to enable cotransport of natural gas and high concentrations of hydrogen (>20 percent)
- Use
  - Development of commercially viable CCUS technologies
  - Improved efficiency of the overall natural gas system, including increased combustion efficiency and waste-heat recovery, and development of innovative controls and low-cost sensors that enable data-driven operations
  - Development of technologies for trigeneration (combined cooling, heating, and power systems
  - Design of novel processes for integration of natural gas with renewables, particularly solar and wind
  - Design of novel processes for production of low- or zero-CO_2 hydrogen (e.g., "blue" hydrogen [with CCUS] and "purple" hydrogen [with black carbon and/or carbon nanotubes coproduction]; see the discussion of hydrogen below)

Petroleum. Chemical engineers have played a central role in the oil industry from its beginning, initially converting crude oil into useful products in small and simple refineries, and subsequently optimizing large and integrated refineries to address energy efficiency and environmental concerns in the manufacture and use of transportation fuels and chemicals. Chemical engineers have also worked closely with geologists to maximize the recovery of conventional and unconventional fossil resources.

The oil industry and chemical engineering evolved together. Many advances in chemical engineering science and technology were driven by the needs of the oil industry, and these advances in turn have spurred growth in the oil industry. With increasing global focus on the need to accelerate the transition to low-carbon energy to mitigate climate change, the expertise of chemical engineers is required now more than ever to help the oil industry minimize its carbon footprint. As discussed previously, the energy system is enormous and complex, and the transition to a low-carbon energy mix will take decades; in the near term, the need for petroleum and its derivative products will continue.

Petroleum is the largest energy source in the United States, used both as a fuel for transportation (road, aviation, marine, and rail) and as a feedstock for the manufacturing of such products as plastics, fibers, lubricants, paints, and solvents. In 2020, U.S. crude oil consumption averaged about 18 million barrels per day, with the transportation sector accounting for 66 percent and the industrial sector for 28 percent of this total (EIA, 2021k).

U.S. oil production totaled 9.6 million barrels per day in 1970 and declined over the subsequent 35 years. Production in 2005 was 5.2 million barrels per day, and imports reached more than 10 million barrels per day—just under 50 percent of total U.S. crude oil consumption. Since 2010, however, the combination of hydraulic fracturing and horizontal drilling has enabled access to oil trapped in shale, making the United States a top world producer of crude oil (Figure 3-8).

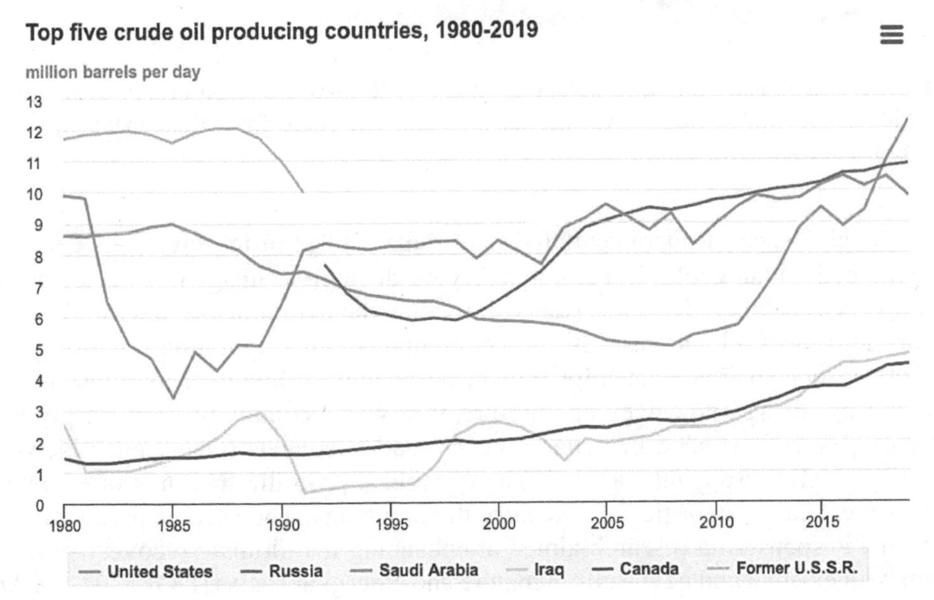

FIGURE 3-8 Amount of crude oil produced in millions of barrels per day for the top five oil-producing countries (United States, former U.S.S.R./Russia, Saudi Arabia, Iraq, Canada), 1980–2019. SOURCE: EIA (2021j).

In 2020, the United States produced more than 11 million barrels of crude oil per day, with tight/shale oil accounting for about 65 percent of this total (EIA, 2021c). The U.S. Energy Information Administration, projects that tight/shale oil will remain the main source of crude oil produced in the United States (EIA, 2021b; Figure 3-9). Relative to conventional crude oil, tight/shale oils contain lighter hydrocarbons, have higher H/C ratios, and are generally very light crude (API [American Petroleum Institute] gravity 45–50) and sweet (<0.1 percent sulfur). They also require less energy to process into desired products. Thus, they are positioned to play an important role in the oil industry's efforts to minimize its carbon footprint.

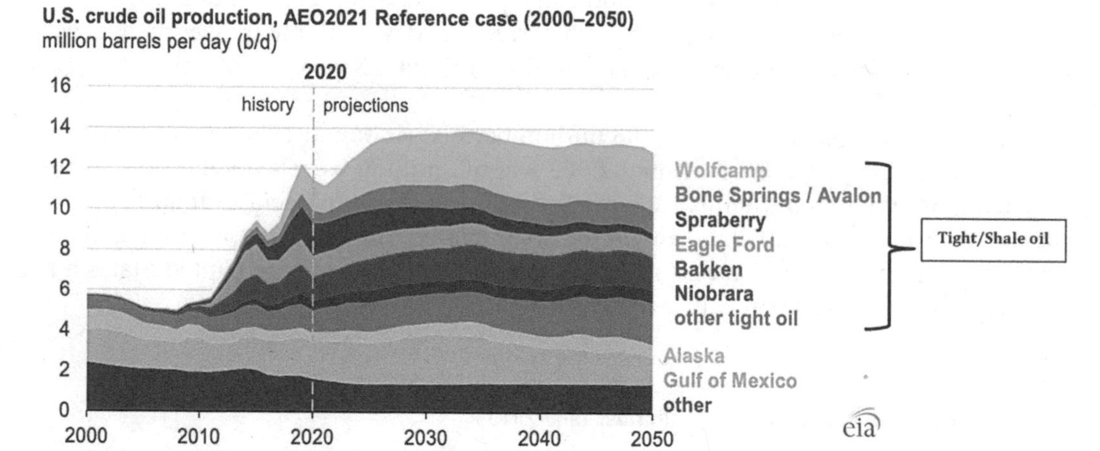

FIGURE 3-9 Historical crude oil production, 2000–2020, and projected crude oil production, 2020–2050, in the United States. An increasing share will come from tight/shale oil sources. SOURCE: EIA (2021b).

The challenge of unconventional tight/shale oil lies in improving its extraction, as it is stranded within geological features that are difficult to image and access because of their low permeability. Thus, the reservoirs need to be hydraulically fractured to create paths for the flow of oil and gas. This process requires either hydraulic fracturing of the geological systems to create paths for flow or horizontal drilling over long distances before fracturing, using explosives or high-pressure water containing various proppants (small particles such as sand or ceramic beads) and chemicals (Geoscience News and Information, 2021). Proppants, as their name implies, prop the fractures open, and the chemicals create a viscoelastic fluid to carry the proppants. The U.S. Department of Energy (DOE) is sponsoring research aimed at enhancing the ultimate recovery of oil and gas from both existing and new wells in mature and emerging basins (DOE, 2021b). Areas in which chemical engineers can contribute to innovation in tight/shale oil production are described below.

Improved water management. As discussed in the above section on natural gas, extraction of tight/shale oil and gas requires a large amount of water and produces a large amount of water that requires treatment or disposal. Low-cost technologies for produced-water treatment are required to maximize water usage recycling, thus minimizing fresh-water usage. Produced water may contain injected chemicals plus naturally occurring materials such as brines, metals, radionuclides, and hydrocarbons. The flowback and produced water are usually stored in tanks or pits before treatment or disposal, often through underground injection (DOE, 2021b; EPA, 2021b). Potential options for either replacing water as a working fluid or minimizing its use include the use of liquefied propane gas (LPG; e.g., API, 2021), supercritical CO_2 (Song et al., 2019), or microwave fracking (e.g., Aresco, 2021).

Increased recovery to extend well life. The amount of oil produced in primary recovery from an unconventional reservoir is much smaller than that produced from a conventional reservoir. In addition, production rates from unconventional wells often decline by more than 50 percent in the first year. Improved fracturing technology to create more efficient and durable oil and gas flow pathways is therefore needed. Chemical engineers can contribute to meeting this challenge by applying their understanding of mass transport in porous materials.

Data-driven approaches. The DOE national laboratories, in collaboration with universities and industry, are leading an effort to integrate physics-informed statistical models; inverse models, such as neural networks; natural language processing; big data analytics; and other emerging artificial intelligence/machine learning (AI/ML) technologies to draw meaningful insights from reservoir data for real-time rapid visualization and prediction to enable effective decision making (DOE, 2021b).

Methane management. The atmospheric concentration of methane, a more potent GHG than CO_2, has risen steadily for more than a decade (Nisbet et al., 2019). This trend reflects the increased production of shale oil and gas, as well as the natural (e.g., from wetlands and other flood zones) and biogenic (e.g., from agriculture or waste) emissions that also play a role. As a result of regulations, the oil industry has made good progress in reducing methane emissions; however, more progress is needed. The main source of fugitive methane emissions is well venting, followed by pneumatic devices that use natural gas as the operating fluid, as well as storage and transport venting and leaks.

Chemical engineers can enable significant reductions in methane emissions by

- developing low-cost modular technologies for conversion of methane to liquid products to replace venting;
- developing methods for using air instead of natural gas as the operating fluid for pneumatic controllers;
- improving techniques and developing smart sensors for methane leakage control and detection; and
- improving methods for deploying higher-efficiency compressors.

Biofuels

The production of biofuels from biogenic carbon, such as waste plant matter, algae, and organic waste, has long been heralded as a means of offsetting GHG emissions from the combustion of fossil fuels (e.g., Lynd et al., 1991; Pacala and Socolow, 2004). The combustion of waste plant matter—lignocellulose—and other biogenic carbon for cooking and home heating has been practiced since before recorded history. The well-known conversion of carbohydrates into fuel ethanol has also been pursued for more than a century. The use of ethanol from starch-based sugars as a fuel was advocated by Henry Ford in the early years of the automobile industry, an approach superseded by the development of crude oil production and refining. Fuel ethanol has been produced successfully at scale since the 1970s in Brazil and later in other countries, predominantly from sugars derived from sugarcane and cornstarch. Annual production in 2019 reached nearly 18 billion gallons (430 million bbl) in the United States (EIA, 2020a) and 29 billion gallons (690 million bbl) worldwide (EIA, 2021a).

By 2019, the annual production and use of biodiesel, mainly from waste oils and fats, had risen to about 2.5 billion gallons (60 million bbl) per year in the United States, with a global production of 10.9 billion gallons (260 million bbl; EIA, 2021a). The COVID-19 pandemic notwithstanding, the contribution of biofuels to the global transportation sector has increased each year over the last two decades. Though reasonably mature today, starch- and sugar-based ethanol and biodiesel still pose challenges for chemical engineers in the areas of extracting value from by-products (e.g., glycerol in biodiesel production), capturing and sequestering CO_2 from ethanol production processes, and expanding the range of fuel products beyond ethanol at a scale of impact. These challenges remain significant barriers to improving the economic feasibility of at-scale biofuel production.

Interest in biofuel production as a potential strategy for offseting GHG emissions and decreasing dependence on fossil resources has seen a resurgence, prompting substantial debate about the long-term viability and utility of biofuels. From a technical perspective, most biogenic feedstocks have lower energy content than their fossil-based counterparts because of the high oxygen content of lignin and plant-derived polysaccharides (e.g., sugars contain about 54 percent oxygen by mass; Figure 3-10). Thus, for nearly all processes for converting biogenic, oxygenated compounds to energy-dense liquid fuels, the production of high-density fuels at a reasonable cost relative to the amortized, technologically mature petroleum industry represents a significant challenge.

The above technological challenges combine with the potential environmental consequences of harvesting crops for energy use to generate considerable controversy (Searchinger et al., 2008). The environmental cost of clearing agricultural land and its loss for growing food crops, complications related to water use, and the potential for an uncertain landscape and diverse mix of political and tax boundaries around the world make the future of biofuels uncertain. Furthermore, the potential rapid growth in the adoption of EVs may reduce demand for transportation fuels more broadly. The impact of biofuels in the energy sector will depend critically on bringing judicious, rigorous, and transparent economic, environmental, and technical analyses—hallmarks of the chemical engineering profession—to bear on the selection of viable options for biofuel production.

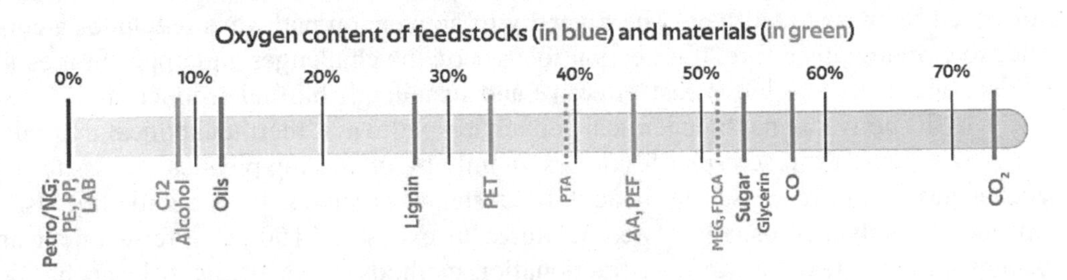

FIGURE 3-10 Representative oxygen content of various feedstocks and materials ranges from 0 percent to more than 70 percent (weight percent). The oxygen content of energy sources varies widely. The higher oxygen content of biofuels results in a lower energy content relative to fossil-based energy sources. NOTES: AA = acrylic acid; FDCA = furan-2,5-dicarboxylic acid; LAB = linear alkyl benzene; MEG = monoethylene glycol; NG = natural gas; PE = polyethylene; PEF = polyethylene furanoate; PET = polyethylene terephthalate; PP = polypropylene; PTA = purified terephthalic acid.

Lignocellulosic feedstocks. An estimated 1 billion tons of lignocellulose could be sustainably harvested for biofuel production annually in North America (DOE, 2016). The conversion of lignocellulosic biomass into biofuels and other organic coproducts—which represents the foundation of the biorefinery concept—poses considerable challenges for chemical engineers. Lignocellulose is a complex and heterogeneous composite material whose carbon content is predominantly in the form of two polysaccharides—cellulose and hemicellulose—as well as the aromatic polymer lignin. Numerous processing options have been considered over many decades; they entail either fractionating biomass into its constituents (thus enabling selective processing options, akin conceptually to the methods used in petroleum processing) or processing biomass directly into liquid or gaseous intermediates for subsequent conversion to transportation fuels and chemicals. Unlike petroleum, biomass is a solid, polymeric material that can be substantially heterogeneous, a feature that presents feedstock-associated challenges beyond the processing of liquids and gases that form the bedrock of the chemical engineering discipline.

The combined challenges of converting lignocellulosic biomass and offsetting fossil resources as a feedstock for transportation fuels create considerable opportunities for chemical engineering to continue making enabling contributions to the at-scale conversion of lignocellulose into biofuels and other products (flexible feedstocks are discussed in Chapter 6). These opportunities begin with the crops themselves. Now that biology is a core component of the chemical engineering discipline, modern molecular biology techniques are now ubiquitous in the toolkit of many chemical engineers. These techniques can be used to modify plants so they can be more efficient photon collectors (Kromdijk et al., 2016), to produce target chemicals *in planta* (Yang et al., 2020), and to reduce the recalcitrance of plant components in chemical conversion in a biorefinery (Chen and Dixon, 2007).

Once plants have been harvested and transported to a biorefinery or centralized depot, there are many opportunities for overcoming recalcitrance to enable conversion of plant-based biomass to biofuels. The diversity of conversion pathways precludes a comprehensive review; therefore, this section focuses on the challenges and opportunities for chemical engineers to achieve cost-effective and sustainable biofuel production.

In the conventional biochemical conversion pathways, thermochemical treatment processes increase the reactivity of biomass, mainly by improving physical access to polysaccharides. These processes apply acid, base, steam, organic solvents, ionic liquids, or deep eutectic solvents, usually at temperatures in excess of 100 °C. Pretreatment approaches can also take the form of fractionation methods that separate polysaccharides from lignin for more direct and selective processing in parallel process trains. Polysaccharides are subsequently converted into monomeric sugars via carbohydrate-active enzymes or sugars and dehydration products, such as furanics or levulinic acid, through further use of acid catalysts. Soluble carbohydrates and derivatives are then converted into fuel molecules or precursors through biological and/or chemical catalysis. In the pioneer cellulosic ethanol plants built in the 2000s, lignin is commonly used to provide heat and power via on-site combustion.

Many attempts have been made to bring biochemical conversion–focused biorefineries to scale (1,000 to 2,000 metric tons/day), especially with the aim of converting nonfood crops or agricultural residues (e.g., wheat straw, corn stover), supported by substantial government investments. Yet the formidable challenges of economical feedstock collection, feedstock handling, biomass pretreatment, aseptic solid–liquid separation, and sterile bioconversion continue to prevent these facilities from achieving the required capacity factors for economic viability at scale of impact.

The lessons learned from these early facilities inform the many opportunities for chemical engineers to advance this field in moving from the process paradigm described above. For example, because on-site lignin combustion is estimated to be the most expensive single unit operation in a biorefinery, as well as a major source of non-GHG emissions, the development of alternative uses for lignin, 40 percent of which can be made of biomass carbon, is a major opportunity for research (Davis et al., 2013; Eberle et al., 2017). Successful conversion of lignin into value-added biofuels or biorefinery coproducts represents a major frontier for the chemical engineering community. Process intensification (PI) through the consolidation of biomass deconstruction into fewer unit operations and a focus on the elimination of costly steps is critical. An important component of PI involves separations, which are often key cost drivers in biorefineries. The challenges in lignocellulosic separations differ from those in petroleum processes because they involve the handling of solids and because biomass-derived compounds consist of high–boiling point oxygenates. Thus, separation technologies that operate wholly in the condensed phase will likely be necessary for the biorefinery.

Beyond biochemical conversion strategies, other lignocellulosic conversion routes employ fast, thermal deconstruction of the whole biomass or fractions thereof. Hydrothermal liquefaction uses liquid water at temperatures above 250 °C to produce a liquid biocrude stream that can be catalytically converted into biofuel molecules. Alternative biomass pyrolysis routes use oxygen-free environments at or above 500 °C to produce bio-oil, light gases, and char from lignocellulosic biomass; some of these products are deoxygenated catalytically, either in the pyrolysis reactor or in subsequent process steps. At an even higher temperature (>700 °C), synthesis gas (CO and H_2) can be produced from biomass via gasification in mildly oxidizing hydrothermal environments. Research opportunities common to these high-temperature biomass conversion processes include the need to understand the complex reaction networks, the design of catalysts and catalytic processes for substantial deoxygenation and operation in the presence of common catalyst poisons entrained in and originating from biomass, and the challenges of operating continuous high-pressure processes.

In most biofuel production processes, chemical coproducts are often invoked as a requirement for economic viability, with the associated challenges of the very large–scale disparity between these two value streams. Even ethylene, the chemical produced in largest amounts from fossil sources, is produced in amounts approximately an order of magnitude smaller than diesel and gasoline, while other commodity-scale chemicals are dwarfed by the scale of ethylene. Ultimately, expensive, small-volume coproducts cannot serve as adequate justification for expensive biofuel production processes, and chemical engineers will play an important role in finding realistic, scalable solutions. It is important to note the annual scale of global petroleum production: 5.0–5.5 billion metric tons (100.69 million bbl per day) of crude oil and 4.1 trillion cubic meters (3.6 billion tons of oil equivalent) of natural gas (IEA, 2021c,d; EIA, 2021d). Petroleum refineries have scales ranging from 10 million to 130 million metric tons per year globally. For biofuels to compete economically with fossil fuels, they need to be produced at similar scales, and even that may not be sufficient because of their unfavorable (oxygen) stoichiometry. Nonetheless, government regulations and/or the implementation of a carbon tax may bring the production of biofuels to a scale of impact, as in the recent case of ethanol in the United States and Brazil.

Feedstocks beyond lignocellulose. Waste plant biomass is not the only source of renewable or waste carbon for producing biofuels, as is evident from global efforts to use algae, which can grow on marginal lands and in ocean or brackish water. The economical conversion of algae to lipids and other biofuel precursors has not been achieved, however, despite extensive research over decades. Many engineering challenges remain, including cost-effective cultivation in open ponds or controlled photobioreactors, greater CO_2 and photon capture efficiency, separation of target products from cells, and catalysts and processes for the downstream conversion of algae-based intermediates (e.g., fatty acids and carbohydrates) to biofuels.

Some organic waste feedstocks are also of potential use in biofuel production (see the discussion of feedstock flexibility in Chapter 6). Given that municipal solid waste (MSW) contains substantial organic matter (e.g., food waste, paper, and cardboard), chemical engineering has many opportunities to increase the use of such feedstocks through research in fractionation and separations, as well as combined conversion approaches. Similarly, industrial and consumer-based food production yields substantial, often highly reduced (deoxygenated) waste feedstocks, such as oils and fats, that can be catalytically converted to biofuels, although that process poses substantial challenges. In moving toward a zero-waste society, these organic waste feedstocks, among others, offer substantial opportunities for chemical process development.

Lastly, it is noteworthy that the conversion of CO_2 or gaseous mixtures, such as flue gas from coal-fired power plants or gases from steel processing, is of interest to the chemical engineering community. Considerable effort is currently devoted to realizing the potential of CO_2 and other gas conversions via biological, electro-, and thermal catalysis routes (e.g., Ye et al., 2019). It will be critical to consider substrate concentrations (e.g., direct CO_2 air capture and conversion is a major challenge), the source of reducing equivalents, and the cost and sustainability of any type of process in this vein, as discussed earlier in this chapter in the context of photon capture.

Overall, biofuels will play a role in reducing GHG emissions associated with transportation fuels. However, judicious analyses of process feasibility, economics, and environmental impact will be critical for deciding among the many options; LCA will provide the rigor needed for such analyses. The challenge for chemical engineers is to identify those options that will ultimately be economically successful and sustainable at the scale required to meet society's fuel needs.

Intermittent Energy Sources

Wind. Wind has provided a source of power for centuries. Three main types of wind turbines are used today:

- distributed or "small" wind turbines (<100 kW), which are used to power a home, farm, or small business directly and are not integrated into the electrical grid;
- utility-scale wind turbines (100 kW to several MW), which deliver electricity to the grid for distribution to end users; and
- larger offshore wind turbines (up to 15 MW).

Wind energy, a niche option a few decades ago, is now the largest source of renewable electricity in the United States. In 2019, wind energy output represented about 7 percent of the U.S. electricity supply, with 100 GW of capacity—equivalent to powering about 32 million homes (AWEA, 2020). On a global scale, wind energy accounts for about 5 percent of electricity demand (IEA, 2020d).

Chemical engineers are involved in several areas of wind energy production (Veers et al., 2019). Specific challenges in materials research, development, and implementation include

- carbon composites and/or recycled materials for turbine blades;
- cement and steel manufacturing with lower CO_2 emissions for wind turbine structural units, such as motors and gear boxes; and
- metallurgy and lubricants for state-of-the-art wind turbines.

The large diameter of modern turbines poses significant manufacturing and transportation challenges and the need for modular manufacturing and on-site assembly for both onshore and offshore installations. The decentralized deployment of installations for capturing wind energy also requires local energy storage and robust sensor and control systems, and often creates environmental concerns regarding the impact on coastal ecosystems and land and ocean animal life (NREL, 2020). Finally, as with all renewable energy sources, challenges exist with respect to the integration of wind energy into chemical production (Centi et al., 2019) and end-of-life considerations for turbine components.

Marine. Marine energy includes energy derived from ocean waves, tidal movements, ocean and river currents, salinity gradients (i.e., where a river empties into the sea), and thermal conversion (i.e., based on the temperature difference between surface seawater and deep [~1 km] seawater). Economical production of tidal energy requires tidal waves larger than 3 m. The United States has several demonstration projects in tidal-energy power production, but none are producing power at commercial scale. Overall, marine energy's development level is similar to that of wind energy roughly 30 years ago, which is to say that wind and solar energy are at commercial scale, while wave energy is at precommercial scale (see IRENA, 2020, for marine energy status and prospects). Chemical engineers can potentially make contributions to marine energy through the development of

- materials capable of withstanding seawater corrosion;
- flexible materials capable of handling the fatigue loads imparted by waves with their fast cycles of 8–10 s;
- antifouling coatings for submerged equipment; and
- electroactive polymers—polymers that generate electricity from mechanical stimuli (e.g., dielectric elastomers, piezoelectric materials, ionic polymer metal composites, and triboelectric materials)

Nuclear Energy

Nuclear power plants use heat produced during nuclear fission to produce steam, which is used to spin large turbines that generate electricity. The current technology is based on nuclear fission in pressurized water-moderated reactors (light water reactors). Fast breeder reactors have been in development for several decades, and some are now in commercial operation in Russia. While there is renewed interest in nuclear fusion, its potential commercial deployment is still decades away, and the role of chemical engineering in this area is likely to be marginal and is therefore not discussed here.

The United States is the world's largest producer of nuclear power, accounting for more than 30 percent of worldwide nuclear electricity generation. Nuclear power has contributed almost 20 percent of electricity generation in the United States reliably and economically over the past two decades. It has been the single-largest contributor (more than 70 percent) of U.S. non-GHG-emitting electric power generation (DOE, 2021c). However, the actual and/or perceived hazards of nuclear power plants and the public's negative perception of nuclear power have contributed to its slow growth. As of January

2021, the United States had 94 operable reactors (96,550 MW); 39 inactive reactors (18,140 MW); and two new reactors under construction in Georgia, with a planned electricity generation capacity of about 1,100 MW each (WNA, 2021).

Advanced nuclear industrial cogeneration offers a potential pathway with sufficient heat and energy intensity to address the problems of industrial emissions at scale. Traditional nuclear reactors rely on large light water reactors operating at maximum temperatures below 300 °C—a temperature high enough to make steam for power generation but too low to drive industrial processes. Consequently, the nuclear power industry is currently focused solely on power generation. Advanced reactors have higher output temperatures relative to light water reactors—up to 600 °C for molten salt reactors and 900 °C for high-temperature gas reactors. These higher temperatures are sufficient to drive most petrochemical processes. During the past decade, DOE has explored using this heat for industrial processes with the Next Generation Nuclear Plant.

Advanced nuclear reactors have the potential to provide the heat required by various industrial processes. Significant cost reduction is required for this advanced technology to be affordable for industrial heat generation, but with some new concepts based on low-cost natural gas, competitive, cost-effective solutions are not out of reach. Efforts to drive down cost are focused in three areas:

- New qualified fuels—TRISO (TRi-structural ISOtropic) particle fuel and molten salts—offer fundamentally better process safety profiles. Each TRISO particle is made up of a uranium, carbon, and oxygen fuel kernel, which is encapsulated by three layers of carbon- and ceramic-based materials that prevent the release of radioactive fission products. These fuels are designed so that processing shuts down automatically if they overheat, thus allowing for inherently safer reactors that are much simpler to operate relative to traditional reactors.
- Safer fuel allows for extensive or complete automation, which significantly lowers operating costs.
- Well-supervised factory production of standard reactors attempts to drive down unit costs by applying the fixed manufacturing facility costs over many units and driving annual improvements in efficiency. This factory-built approach has been used to achieve dramatic cost improvements in the wind and solar industries.

Increased demand for nuclear reactors and efficiency gains associated with the corresponding manufacturing learning curve could lower the cost of nuclear reactors. The petrochemical industry sector is capable of driving demand for these units for decades. High-temperature reactors provide high-quality heat directly to industrial facilities, and the integration of this heat will require chemical engineers working within an interdisciplinary team that understands process safety, integration, and intensification.

While existing as a separate discipline, nuclear engineering borrows heavily from physics, as well as from mechanical and chemical engineering. Setting aside the particle physics associated with fission and fusion reactions, a nuclear reactor is effectively a chemical reactor that takes in fuel as a feed, produces fission or fusion products as waste, and produces heat as a product. Process integration and process design are necessary to extract energy most efficiently from the steam that is generated by a nuclear reactor. Viewed through the lens of the fundamental pillars of chemical engineering (transport phenomena, reaction engineering, thermodynamics, and applied mathematics), the design and safe operation of nuclear power plants are a good match for the skillset of a well-trained chemical engineer. Optimal thermodynamics and heat transfer are key to an efficient process design, as is process control to operate a power plant effectively and safely. Further development of advanced reactor designs, as well as storage solutions for nuclear waste, will also benefit from the same chemical engineering fundamentals. Advances in nuclear energy present a clear opportunity for chemical engineers to collaborate with nuclear and other engineering disciplines.

ENERGY CARRIER PRODUCTION

Energy carriers are intermediates in the energy-supply chain, located between primary and/or secondary sources (Thollander et al., 2020) and end-use applications. For convenience and economy, energy carriers have shifted continually from solids to liquids, recently from liquids to gases, and more recently to electricity, a trend that is expected to continue and even accelerate to address climate change concerns. Currently, about one-third of final energy carriers reach consumers in solid form (as coal and biomass), one-third in liquid form (consisting primarily of oil products used in transportation), and one-third through distribution grids in the form of electricity and gas. It is projected that the share of all grid-oriented energy carriers could increase to about 50 percent of all consumer energy by 2100 (Sims et al., 2007).

The following sections describe the opportunities for chemical engineers to contribute to electricity generation and the production of low-carbon fossil fuels, hydrogen, and synthetic fuels; production of advanced liquid biofuels was discussed previously in this chapter.

Electricity

Reduction of GHG emissions during electricity generation, as well as electrification of light- and medium-duty vehicles and residential/commercial heating, is crucial for decarbonization of the energy sector in the near and medium terms.

Electricity generation from coal-fired plants, the largest CO_2 emitters, has decreased in the United States since 2009, while electricity generation from both natural gas and renewable energy sources has increased. In 2020, about 4,000 TWh of electricity was generated at utility-scale electricity generation facilities in the United States, with about 60 percent of that total being generated from fossil fuels—coal, natural gas, petroleum, and other gases; about 20 percent from nuclear energy; and about 20 percent from renewable energy sources (EIA, 2021b).

The shift from coal to natural gas and renewables was made possible by a steep decline in the cost of key technologies associated with shale gas, wind power, solar power, and grid-connected electricity storage (DOE, 2015). New wind and solar technologies offer the lowest levelized cost[7] of electricity over most of the Earth's surface (IRENA, 2020). Since 2009, the levelized cost of wind has declined by 70 percent and that of solar photovoltaics by almost 90 percent, providing an important means of supplying electricity with no direct CO_2 emissions (Lazard, 2019).

Recent decarbonization studies (e.g., IEA, 2021b; NASEM, 2021a) indicate that deep decarbonization of electricity generation can be accelerated, but further innovation is required. Chemical engineers are contributing to the scale-up, cost reduction, and reliability of improved and novel technologies, particularly in the solar and wind energy sectors. (Specific opportunities in these sectors were discussed earlier in this chapter.)

Low-Carbon Liquid Fossil Fuels

Liquid hydrocarbons from crude oil have been the preferred energy carrier for the transportation sector because of their high energy density, easy distribution and storage, low cost, and well-established and extensive infrastructure along the value chain. If they are to play a role in the low-carbon energy mix of the future, however their carbon footprint will need to be significantly reduced (Figure 3-11).

In 2019, the global demand for liquid hydrocarbons was about 100 million barrels per day, approximately 58 percent of which was for the transportation sector, 14 percent for feedstock for chemicals, 12 percent for power generation/residential/buildings, and 16 percent for other industrial use (ExxonMobil, 2019). Demand is expected to increase until at least 2040, although not uniformly across all sectors. Demand for hydrocarbon liquid fuels for industrial use and for power generation, residential uses, and buildings is projected to decrease, being replaced by energy carriers from renewable sources. In the chemical sector, demand for liquid hydrocarbons used as feedstock to manufacture consumer products is expected to increase. In the transportation sector, overall demand is projected to grow, but not uniformly across transportation types. Gasoline demand for light-duty vehicles is projected to decrease as a result of greater market penetration of EVs, while demand for liquid fuels in commercial transportation (heavy-duty, aviation, and marine) is expected to increase, particularly in the heavy-duty long-haul sector. Changes in global demand for oil, based on new policy scenarios, are projected to follow similar trends (IEA, 2021a).

Petroleum Refining

Most of the CO_2 emissions associated with liquid fuels come from the use/combustion of the fuel itself, but there are also opportunities to reduce emissions during oil production, extraction, and refining. The majority of current refineries were designed and optimized to manufacture primarily gasoline; diesel, aviation, and other heavy fuels and

---

[7] Levelized cost is the sum of total lifetime costs divided by the amount of energy produced, and thus represents the present value of the total cost of building and operating a power plant over an assumed lifetime.

chemicals are normally secondary products. To maintain a low cost of production, the amount of lower-cost heavy (high C/H ratio) and "dirty" crude oils in the feedstock mix is maximized. Refineries will require significant reconfiguration not only to satisfy product demand and meet challenges associated with GHG emissions, but also to maintain low production costs for liquid fuels. Chemical engineering is already playing a central role in addressing these challenges.

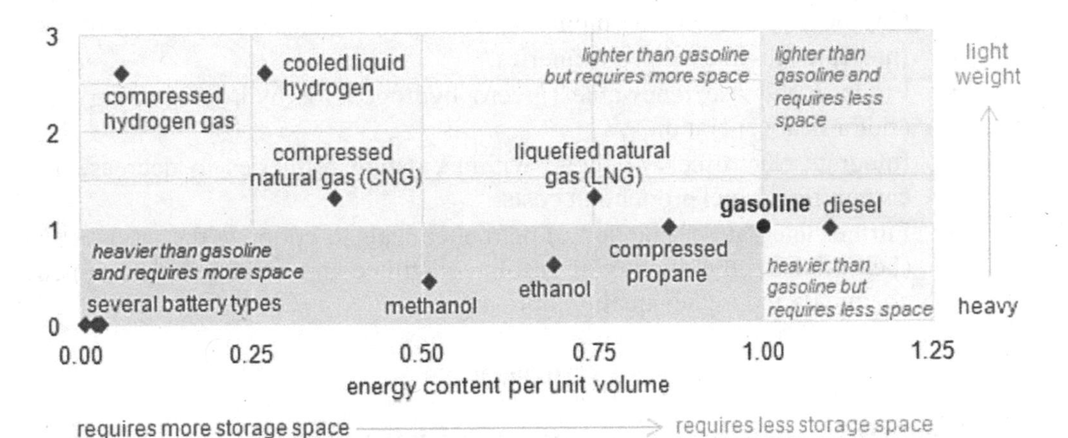

FIGURE 3-11 Energy density of transportation fuel types, indexed to gasoline = 1. The data points represent the energy content per unit volume or weight of the fuels themselves, not including the storage tanks or other equipment they require. For instance, compressed fuels require heavy storage tanks, while cooled fuels require equipment to maintain low temperatures. SOURCE: EIA (2013).

Opportunities to reduce CO_2 emissions during refinery operations include the following:

- Increase energy efficiency through further improvement of energy management systems to ensure that refineries are run according to the most energy-efficient standards.
- Replace combustion of liquid fuel for heat generation with renewable sources (e.g., green electricity).
- Produce renewable (green) hydrogen with electrolyzers using imported or self-generated renewable electricity.
- Use low-grade heat generated during operations to produce electricity for internal and external use.
- Eliminate flaring in refineries.
- Replace steam-driven rotating machines and fired heaters with electric counterparts.
- Deploy CCUS from refinery flue gases.

Opportunities for chemical engineers to contribute to reductions in the carbon footprint of refinery feedstocks include the following:

- Use lighter and sweeter crude oil (higher H/C ratio and fewer heteroatom contaminants, such as sulfur and nitrogen compounds) instead of carbon-intensive and harder-to-process heavy oils (lower H/C ratio and more heteroatom contaminants).
- Coprocess crude oil with biomass.
- Integrate bio- and oil and refineries.
- Produce and use renewable (green) hydrogen for hydroprocessing needs and/or heat generation.
- Integrate electrofuels (e-fuels) within existing refineries to decrease low-carbon liquid fuel production costs.
- Further integrate production of petrochemicals to optimize the product slate (i.e., minimize production of gasoline/distillate and maximize that of petrochemicals and higher-quality lubricants).

Hydrogen

Hydrogen is a versatile energy carrier with significant potential to contribute to a clean, low-carbon energy system. To realize this potential, chemical engineering and related fields can contribute in the following ways:

- Increase production from nonfossil sources.
- Significantly increase clean hydrogen production via water hydrolysis using clean electricity.
- Develop a hydrogen infrastructure for distribution and storage at scale to satisfy demand.

Globally, 96 percent of hydrogen is produced from fossil sources (48 percent natural gas, 30 percent liquid hydrocarbons, and 18 percent coal), with only about 4 percent produced from electrolysis of water (IRENA, 2018). In 2018, the global demand for pure hydrogen was above 70 million metric tons, and the demand for hydrogen as part of a mixture of gases, such as synthesis gas, was about 45 million metric tons. The vast majority of this hydrogen (~95 percent) was used for production of chemicals (mainly ammonia for fertilizers and methanol) and for oil refining (IEA, 2019). Hydrogen is currently produced in large quantities, but a vast expansion of the world's production capacity would be needed for hydrogen to replace a significant fraction of oil in transportation. Current hydrogen demand corresponds to ~8.4 EJ (2,333 TWh), or about 6 percent of the annual global transportation energy demand (IRENA, 2021).

A growing number of countries have policies that directly support investment in low-carbon hydrogen technologies, and global demand for hydrogen as an energy carrier is projected to increase significantly after 2030 (IEA, 2020e). This increase is projected mainly in sectors that are relatively more challenging to decarbonize and that have needs that cannot be met by electrification, including fuel for heavy-duty long-haul transportation; synthetic fuels for aviation and shipping, for which available low-carbon fuel options are limited; ammonia as fuel for shipping; and a source for heat generation in the industrial and buildings sector. Hydrogen is a promising option for storing renewable energy.

Today, hydrogen is produced primarily by steam reforming of natural gas; partial oxidation (catalytic and not) and autothermal reforming of natural gas technologies are also used to a lesser extent. In some countries, particularly in Asia, gasification is used commercially for hydrogen production from coal. Production by water electrolysis is also a commercial technology, but at a much smaller scale and higher cost. Methane pyrolysis, or methane splitting of natural gas using renewable electricity to produce hydrogen and black carbon, is currently in the commercial demonstration scale.[8] Many other routes to low-carbon hydrogen, such as thermochemical water splitting, direct photocatalysis, and biological production from microorganisms, are in various stages of R&D.

It is generally accepted that hydrogen from fossil fuels without CCUS will remain the main source of hydrogen production. After 2030, it is estimated that almost all of the growth in hydrogen production will come from low-carbon hydrogen (IEA, 2020d) made from renewables-based electricity or from fossil fuels, particularly natural gas, in combination with CCUS. By 2070, electrolytic hydrogen is projected to account for nearly 60 percent of global hydrogen production (IEA, 2020d).

Hydrogen produced from different feedstocks is identified by colors. "Black," "gray," and "brown" refer to hydrogen produced from coal, natural gas, and biomass, respectively. "Blue" is commonly used for hydrogen produced from fossil fuels, with CO_2 emissions reduced by the use of CCUS. "Green" hydrogen is produced from water electrolysis using renewable electricity.

Blue and green hydrogen have a path to competitiveness with gray hydrogen (Hydrogen Council, 2019). The competitiveness of blue hydrogen depends primarily on scale-up of CCUS facilities and the value attributed to sequestered CO_2. Carbon prices or taxes of USD 50/ton—a figure consistent with near-term milestones of major economies with net-zero commitments (e.g., in the European Union by 2030; Argus, 2020)—would make blue hydrogen competitive. The competitiveness of green hydrogen will require steep cost reductions for electrolyzers, as well as reductions in renewable energy costs. The production of green hydrogen is projected to break even with that of gray hydrogen before 2030 in regions with low renewable-energy costs, and before 2035 in regions with average renewable-energy costs (Hydrogen Council, 2019). A combination of green and blue hydrogen production pathways will be required to satisfy the potential demand for low-carbon hydrogen. Their supply mix will depend on a range of technical and societal factors, production costs, existing infrastructure (such as power supply and transmission networks), and emerging hydrogen trade routes.

Many challenges and opportunities related to the production of low-carbon hydrogen can be addressed by chemical engineers. For blue hydrogen, the challenges and opportunities center on improved CCUS (i.e., at lower cost) and assessment of geological

[8] See www.monolithmaterials.com.

sites for long-term CO_2 storage; for green hydrogen, they center on the need for a source of sustainable, low-cost, renewable electricity and access to freshwater.

Larger-scale electrolysis plants, whose development will depend on achieving lower costs and improved electrical efficiency for the electrolyzers, are also needed. Three main electrolyzer technologies exist today: alkaline electrolysis, proton exchange membrane (PEM) electrolysis, and solid oxide electrolysis cells (SOECs). Alkaline electrolysis is a mature, commercial-scale technology with relatively low capital costs. PEM electrolyzers use pure water as an electrolyte solution and avoid the recovery and recycling of electrolyte solution necessary with alkaline electrolysis; however, they use expensive electrode catalysts and membrane materials. Lifetimes of PEM electrolyzers are currently shorter than those of alkaline electrolyzers, and overall costs are higher. SOECs are the least-developed electrolysis technology and not yet commercialized. Ceramics serve as the electrolyte, resulting in lower material costs. Because steam is used for electrolysis, SOECs require a heat source. If the hydrogen produced were to be used for the production of synthetic hydrocarbons (in power-to-liquid or power-to-gas schemes), the waste heat from these synthesis processes (e.g., Fischer-Tropsch synthesis and methanation) could be recovered to produce steam for further SOEC electrolysis (IEA, 2019).

Synthetic Fuels

Synthetic fuels are produced by converting hydrogen and a carbon source into compounds that can be used as energy carriers, such as methane; methanol; ethanol; and such higher-carbon-number products as gasoline, diesel, and aviation fuel. Ammonia is also increasingly seen as a non–CO_2-generating synthetic fuel, although ammonia production itself is currently a major source of CO_2 emissions. Electrification and advances in electrochemical routes to ammonia synthesis may increase the potential for ammonia as a promising fuel. Low-carbon or carbon-neutral synthetic fuels require green hydrogen and carbon from a bioenergy source or from CO_2 captured from flue gases or the atmosphere using direct air capture (DAC) technologies. In the case of DAC, however, significant energy and technology advances will be necessary to make this route viable. The low-carbon or carbon-neutral synthetic fuels are referred as power-to-fuels (PtF) or electrofuels (e-fuels).

The production of e-fuels requires significant amounts of electricity, and from a thermodynamic point of view, the electricity should be used directly. For example, 25 kWh of energy is required to produce 1 L of synthetic kerosene from electrolytic hydrogen together with CO_2 captured through DAC. More than 80 percent of the energy is used to produce hydrogen; around 15 percent is used for the capture of CO_2 through DAC; and the rest is used in the Fischer-Tropsch synthesis. Currently, only about 40 percent of the energy input is stored in the final liquid product, although with process optimization, the overall conversion efficiency could potentially increase beyond 45 percent (IEA, 2020d). However, e-fuels have the potential to help some sectors—particularly the aviation transportation sector, which is difficult to decarbonize and for which electrification is not an option because of high energy-density requirements for the energy carrier. These considerations would need to be important enough to justify the thermodynamic inefficiencies, however.

A few aviation e-fuel demonstration plants have been announced. For example, the Norsk e-fuel project is planning a first plant in Herøya, Norway. That plant is expected to become operational in 2023, with a production capacity of 10 million L annually, scaling up to 100 million L annually by 2026.[9] Cost reduction for e-fuel production is expected to benefit from the economy of scale and experienced manufacturing learning curve. It should be emphasized that DAC technology is in its infancy, with substantial room for technology and process improvement and cost reduction. For e-fuels to be competitive with conventional fossil fuels and even bio–jet fuel, a combination of low electricity costs and a high CO_2 tax or cost will be needed.

ENERGY STORAGE

Energy storage occurs when an energy carrier is held in a fixed location until it can be deployed. When the energy carrier uses chemical bonds or potential energy, as is the case for liquid fuels and pumped hydropower, respectively, energy storage is conceptually simple. When the energy carrier is electrons, storage requires batteries or capacitors or conversion of the electrical energy into another energy carrier. Because many of these are mature technologies, chemical engineers have limited potential to impact some modes of energy storage, especially those involving mechanical energy, such as pumped hydropower and compressed gas storage. For energy carriers that involve electrons or chemical bonds, however, chemical engineers can play a critical role in the development and deployment of scalable and economical energy storage.

The changing nature of the world's energy system has continued to create significant demand for energy storage, and this trend is likely to accelerate in the coming years. One obvious example involves intermittent renewable electricity sources such as solar and wind, whose full contribution to decarbonization of electricity generation can be realized only if they are coupled with grid-scale storage. A second example is the trend toward electrification of light-duty vehicles, which will require massive deployment of on-board batteries.

The properties of an energy storage system can vary with both scale and application. In vehicle applications, for example, the weight and size of batteries are critical; in grid-scale storage, by contrast, the battery weight is unimportant. The typical storage time and desired charging and discharging rates are also very different for these two applications. This example illustrates why future energy storage needs will be met by a wide array of technologies—there is no one-size-fits-all approach to energy storage. The rapid deployment of electrical storage systems is underpinned by the commercial sector's large investments in technology development. If academic research is to have an impact in this environment, researchers will have to seek solutions in frontier areas in which well-

[9] See https://www.norsk-e-fuel.com/en.

funded development efforts in industry are less likely to overwhelm the scale of any academic effort.

An important distinction between energy storage in electrons and chemical bonds lies in the number of times a particular piece of storage medium can be cycled. For fuels based on chemical bonds, storage simply involves a tank, and little to no change in the storage medium is expected even after thousands of cycles. Here, chemical engineers have a role to play in understanding evaporation and erosion, as well as improving leak detection methods. Batteries, however, tend to degrade during cycling. Many of these degradation processes stem from chemical reactions or nucleation and growth of phases that are driven by fundamental chemical engineering principles. This observation indicates that chemical engineers can play a key role in developing concepts that increase battery lifetimes. At the same time, for batteries to be a core part of a truly sustainable energy system, their end-of-life disposal or recycling needs to be considered. Similarly, the global availability of battery components (e.g., lithium) is an important consideration in LCA comparisons of competing technologies. The intentional design of batteries for end-of-life disposal and the use of earth-abundant elements are areas in which chemical engineering researchers are poised to play an important role.

As discussed in the previous section, hydrogen is a versatile energy carrier with significant potential to contribute to a clean, low-carbon energy system. Given its low energy density, however, its long-distance distribution and storage poses challenges, particularly if it needs to be shipped overseas. It has been estimated (IEA, 2019) that for distances above ~1,000 miles, shipping hydrogen as ammonia or liquid organic hydrogen carriers (LOHCs) is likely to be more cost-effective than shipping liquified hydrogen. Chemical engineers have opportunities to significantly lower the cost of conversion before export and reconversion back to hydrogen in the case of LOHCs, and before consumption or the direct use of ammonia as fuel.

Vast sectors of the global energy system rely on liquid or solid energy carriers, such as gasoline and coal. The energy density and ease of transporting these carriers relative to gases or electrons give them strong intrinsic advantages. Chemical engineering played a central role in what might be termed the hydrocarbon economy of the 20th century, and the discipline will also be central to the deployment of more sustainable and environmentally benign liquid energy carriers in the 21st century. Unlike high-value-added products such as pharmaceuticals, new or improved liquid energy carriers need to be competitive with low-cost alternatives and deployable at massive scale in order to be viable. In many cases, achieving these goals will require integrated process development rather than a singular focus on improved catalysts or similar chemical steps. The ability of chemical engineers to use LCAs and technoeconomic assessments (TEAs) to focus research on approaches with a plausible path to economic viability ensures that their contributions will have impact in the domain of energy storage.

ENERGY CONVERSION AND EFFICIENCY

Ultimately, whatever the source and carrier, energy is converted to heat or work, commonly referred to as energy use or consumption. This section is organized by application within the transportation, industry, residential, and commercial sectors, and focuses on opportunities for chemical engineers to contribute to energy efficiency and decarbonization in those sectors.

Transportation Sector

Transportation is a large and diverse sector that encompasses road (passenger and freight vehicles), aviation, marine, and rail transportation. In 2018, the transportation sector accounted for nearly a quarter of global CO_2 emissions (IEA, 2021a), and efforts to decarbonize the transportation sector are therefore critical to achieving the goals of the Paris Agreement. As noted previously, the transition to a net-zero emissions transportation sector will take decades and cost hundreds of billions of dollars, and may never be complete (Ogden et al., 2016). Net-zero emissions aside, just reducing transportation CO_2 emissions significantly in the coming decades is a formidable challenge. Tackling this challenge will require structural shifts in the transportation of people and freight, much larger gains in energy efficiency, major advances in technology, effective government policies, significant levels of investment in infrastructure for low-carbon energy carriers, and the manufacture of low-carbon and zero-emission vehicles.

No single energy carrier can, in the foreseeable future, satisfy the requirements across all segments of the transportation sector (EPA, 2021a). Some segments of the sector are easier to decarbonize (light- and medium-duty vehicles) than others that require high-energy-density fuels (heavy-duty long-haul, aviation, marine). A "bridging" low-carbon energy transition strategy will rely on the combined increased use of such energy carriers as electrons, hydrogen, and low-carbon liquid fuels, particularly advanced biofuels and synthetic liquid fuels (Santiesteban and Degnan, 2021). This section focuses on opportunities for chemical engineers to contribute to technology improvements in EVs, hydrogen-powered fuel cell engines, and internal combustion engines.

Electric Vehicles

The global EV market has grown over the past decade as a result of both technological advances and supportive government policies. About 7.2 million EVs were in use in 2019, and this number is predicted to increase to nearly 140 million within the next decade (IEA, 2020a). Many automobile manufacturers have announced plans to stop production of internal combustion engines by 2030. The key technological enabler of the growth in EVs is the progress made in lithium-ion batteries, in terms of both improved performance and cost reduction (BloombergNEF, 2020; Figure 3-12). To accelerate market penetration of EVs, several remaining challenges for lithium-ion batteries need to be overcome, many of which chemical engineers are poised to help address, as described below.

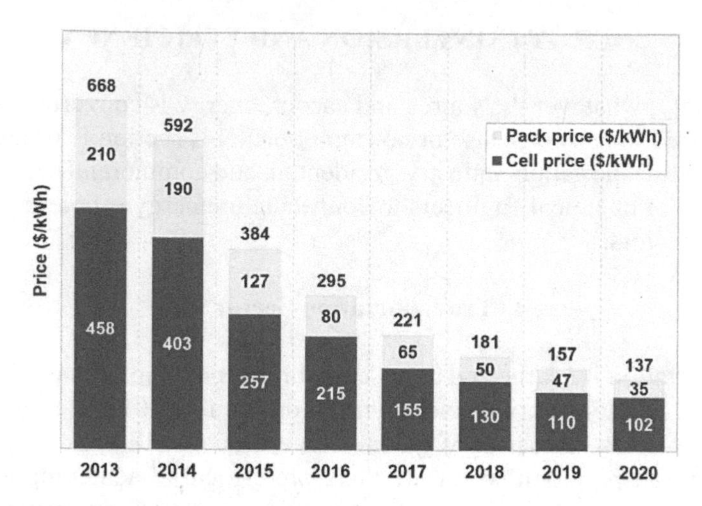

FIGURE 3-12 Survey of lithium-ion cell and pack integration prices for automotive years 2013–2020, adjusted to real 2020 $/kWh. SOURCE: BloombergNEF (2020).

The chemical industry has played a critical role in the successful development and manufacturing of lithium-ion–based batteries and continues to drive innovation to meet booming demand. Improvements are being pursued in mining, metals processing and purification, battery design and manufacturing, battery chemistries, and performance.

The advances needed for lithium-ion batteries in automotive applications are described by Masias and colleagues (2021):

- further increases in energy density per unit weight and volume;
- continued cost reduction on a $/kWh basis, which is both challenging and necessary to increase the opportunity for wide-scale adoption of electric transportation;
- the ability to validate and predict long battery life quickly and accurately;
- fast recharging while preserving long battery life and overall safety; and
- direct recycling of cathodes and anodes to reduce both the costly geological extraction and initial material processing steps of rare elements.

Lithium-ion–based batteries are approaching the theoretical energy density limit imposed by their inherent chemistry; therefore, they may be unable to satisfy future EV demand, and several next-generation lithium batteries are being investigated (Wu et al., 2020), the three main technologies being lithium-air, lithium-sulfur, and lithium-metal. Despite some research progress, however, development and scale-up have been slow. The lithium-air battery has faced battery life and energy efficiency challenges. The lithium-sulfur battery has shown more promise than the lithium-air, but it, too, has been limited by life and volume concerns because of its low energy density. Lithium-metal batteries have advanced the furthest in the past decade but are still in the development stage (Masias et al., 2021).

All of the lithium battery types described above use a liquid electrolyte. One strategy for improving overall performance is the use of a solid electrolyte material. Such batteries are referred to as solid-state or all-solid-state batteries (ASSBs). The discovery of highly conductive solid-state electrolytes has led to tremendous progress in the development of ASSBs (Tan et al., 2020a), making it possible to overcome such technical challenges as poor interfacial stability, scalability, preservation of high energy density, production safety, and cost reduction.

Solid-state batteries with lithium-metal anodes have the potential to achieve high energy densities. Several start-ups, many in close collaboration with large automakers, are focusing on the development of technologies required to optimize solid-state electrolytes. Substantial increases in the energy density of solid-state batteries, together with improvements in other performance metrics, such as cost and durability, could make electrification a viable and more attractive commercial option for medium- and heavy-duty regional and long-haul vehicles. These advances could also have a significant impact on short-distance and small-freight marine and aviation transportation.

Hydrogen-Powered Fuel Cell Vehicles

The two primary options for zero-emissions transportation are electric drivetrains powered by batteries (battery electric vehicles [BEVs]) and by hydrogen fuel cells (fuel cell electric vehicles [FCEVs]). Both are used for light-, medium-, and heavy-duty vehicles. These two technologies have complementary strengths and meet different application and customer needs. BEVs are emerging as the technology of choice for light- and medium-duty vehicles, while FCEVs are better suited for heavy-duty commercial vehicles that are driven long distances, carry heavy loads, and require relatively quick refueling.

PEM fuel cells are used in FCEVs because they offer higher power density, lower overall weight, and lower total volume compared with other fuel cell types (NRC, 2015). They are also easily scaled for different vehicle classes. Because these cells operate at fairly low temperatures (<100 °C), they have a short warm-up time and better durability relative to other fuel cell types. These features also contribute to much greater overall efficiency compared with high-temperature solid oxide or molten carbonate fuel cells in a frequent start-up/shutdown vehicle application.

In a PEM fuel cell, the anode–separator–cathode structure is known as the membrane electrode assembly (MEA). It consists of a solid polymer electrolyte membrane (e.g., Nafion®) with a catalyst layer and a gas diffusion layer on each side. The catalyst layers are platinum-group metal nanoparticles (usually platinum or platinum alloy) on a porous carbon/ionomer support. The gas diffusion layers are multilayer carbon fiber/polytetrafluorethylene paper sheets that facilitate mass transfer of reactants and products. The MEA is sandwiched between bipolar plates that have channels for gas flow and conduct the electric current. Like batteries in an EV, the fuel cell unit in the vehicle is actually a stack of many individual fuel cells arranged in series to provide sufficient voltage. The stack is rated by maximum power output.

The key challenges and opportunities to which chemical engineers are contributing in this domain are as follows:

- Reducing costs—The cost of platinum-based catalysts is the primary driver of fuel cell costs, a fact that has motivated research on technologies for reducing the platinum content of MEA catalysts and even developing platinum-free MEA architectures.
- Increasing durability—The on-road durability of PEM fuel cells is currently less than desirable for large-scale commercial deployment. The primary cause of degradation is loss of catalyst activity (e.g., due to sintering) and deterioration of the PEM. One important lever for increasing durability is to increase catalyst loading (i.e., add more platinum), but doing so increases cost. A better alternative is to design sintering-resistant catalysts or replace platinum with lower-cost metals.
- Increasing efficiency—PEM fuel cells for on-road FCEVs have a peak power efficiency of up to 60 percent in terms of converting the available fuel energy of hydrogen into electrical energy, well below the theoretical maximum efficiency of approximately 80 percent.

Hydrogen is stored on the vehicle as a compressed gas at a pressure of 700 bar in one or two carbon-fiber composite tanks. Since even highly compressed hydrogen has a much lower energy density than gasoline and diesel fuel, the volume of the tank(s) needs to be large enough to enable a driving range comparable to that of conventional internal combustion engine (ICE) vehicles. Research, development, and demonstration (RD&D) in this area targets primarily the cost of the composite tanks (e.g., cheaper carbon fibers or other materials and cheaper construction methods) and auxiliary components. Early-stage research also is being conducted on other technologies for storing hydrogen on FCEVs, such as

- adsorbents (e.g., metal organic framework materials [MOFs] and metal hydrides) that would allow lower hydrogen pressures;
- cold/cryocompressed conditions in insulated tanks (e.g., −75 °C/500 bar or −235 °C/70 bar); and
- chemical storage, in which a hydrogen-dense chemical (e.g., NH_3BH_3, methylcyclohexane) is loaded into the tank, H_2 is disassociated through heat or chemical reaction, and then spent chemical is collected and regenerated at a central facility.

Without a breakthrough, vehicle manufacturers would be unlikely to opt for these alternatives in the foreseeable future because of the need to develop additional technologies and/or fueling infrastructures to accommodate them.

Internal Combustion Engine Powertrain Vehicles

For both heavy- and light-duty ICE powertrain vehicles, chemical engineers can play important roles in advancing technology for increasing fuel efficiency, thereby reducing CO_2 emissions. Advances in ICE technology have traditionally been led by mechanical engineers, making this a key opportunity for chemical engineers to collaborate with another engineering discipline. Areas for such collaboration include

- advanced combustion schemes (e.g., low-temperature combustion);
- electrified accessories and waste-heat recovery;
- hybridization;
- lightweighting, with new, aerodynamic cabs and trailer components;
- advanced communication and logistics;
- energy-efficient tires and monitoring systems; and
- advanced GPS-based predictive control technology.

GHG emissions could be reduced in ICE-powered heavy-duty vehicles by the use of low-carbon liquid fuels. Codevelopment of more efficient fuels and vehicle engines is a major collaborative opportunity for chemical and mechanical engineers, with chemical engineers bringing expertise in reaction mechanisms and transport and mechanical engineers bringing expertise in transport, computational fluid dynamics, and engine design. Opportunities in this area include

- blends of high-quality, low-carbon fossil-diesel fuel with advanced biofuels and/or synthetic fuels (e.g., e-fuels) to achieve specific tailpipe GHG emission levels; and
- increased control of the variability of diesel-fuel properties.

Industrial Sector

The industrial sector produces the goods and raw materials people use every day. Industry is energy intensive—nearly half of the world's energy is dedicated to industrial activity; accordingly, it is also responsible for a large portion of global CO_2 emissions. Production of cement, steel, and chemicals accounts for the largest portion of industrial CO_2 emissions—about 70 percent (IEA, 2020d). Reducing manufacturing-related emissions in the United States and China will be critical to reducing CO_2 global emissions from manufacturing (Figure 3-13).

The lack of commercially available and scalable low-carbon alternatives to fossil fuels makes deep reductions in CO_2 emissions from industry highly challenging in the short and medium terms. This fact is reflected in the Sustainable Development Scenario projections (IEA, 2020d), in which the industrial sector emerges as the second-largest CO_2 emitter in 2070, after the transportation sector, accounting for around 40 percent of residual emissions, even though its emissions are projected to be 90 percent lower overall than in 2019 (Figure 3-13). Currently, energy inputs to the industrial sector are approximately 70 percent from fossil fuels (IEA, 2020d). In the Sustainable Development Scenario projections, the use of fossil fuels in industry would be reduced by more than 60 percent by

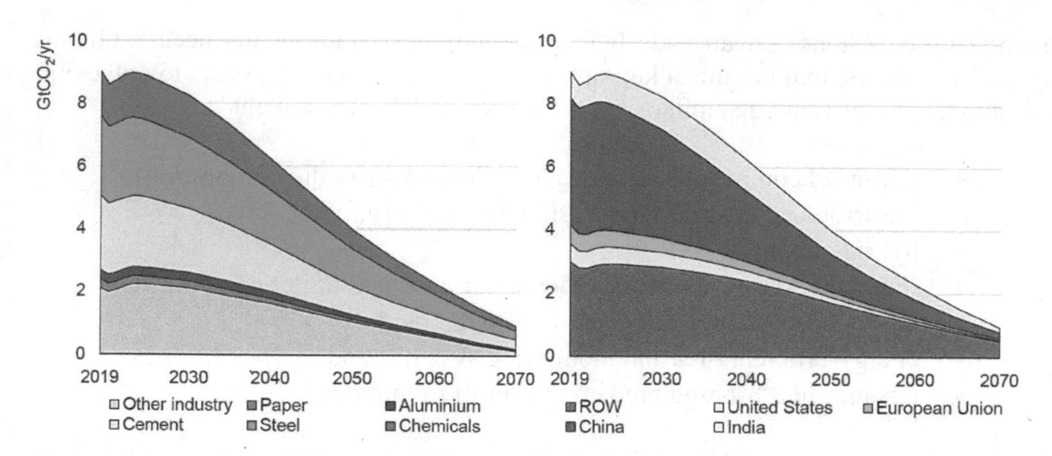

FIGURE 3-13 Projected decrease in global direct CO_2 emissions of industry by subsector (paper, aluminum, cement, steel, chemicals, other) and region (United States, European Union, China, India, rest of world [ROW]) in the International Energy Agency's Sustainable Development Scenario, 2019–2070. SOURCE: IEA (2020d).

2070, being replaced primarily by electricity and bioenergy, while more than 75 percent of the remaining CO_2 emissions would be captured and stored permanently. The following sections focus on opportunities for chemical engineers in the cement, steel, and chemical production subsectors, as well as cross-cutting approaches for decarbonization of the industrial sector.

Cement Production

More than 70 percent of the energy used in the U.S. cement industry comes from coal and petroleum coke. More than 50 percent of the total CO_2 emissions attributable to cement production are process related (from calcination of limestone in the kiln), not energy related. To advance decarbonization of the cement industry, in addition to energy-efficiency improvements, these process-related CO_2 emissions will need to be reduced. Key strategies for deep decarbonization of the cement industry are clinker substitution (supplementary cementitious materials [SCMs]), a switch to lower-carbon fuels, CCUS, low-carbon cement and concrete chemistries, and kiln electrification. While several of these strategies can be combined to approach net-zero emissions, decarbonization efforts related to demand reduction, use of SCMs, and waste-carbon utilization will also be needed. For example, Hasanbeigi and Springer (2019) recently showed that, compared with 2015, total CO_2 emissions from California's cement industry could decrease by 68 percent by 2040 even though the state's cement production is projected to increase by 42 percent (Figure 3-14).

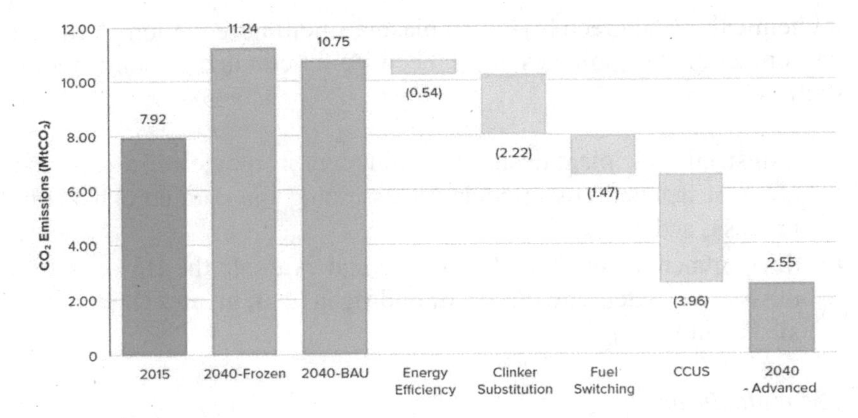

FIGURE 3-14 Impact of various options for CO_2 emissions reduction in California's cement industry. NOTES: BAU = business as usual; CCUS = carbon capture, use, and storage. SOURCE: Hasanbeigi and Springer (2019).

Further RD&D is necessary to decarbonize cement processes, including technologies to

- improve concrete performance and durability using alternative raw materials while meeting construction codes and standards, especially for natural SCMs (e.g., pozzolans, calcined clay);
- develop innovative ways of using large-scale nonpurified CO_2 for different applications and product streams with low energy penalty and cost;
- develop electrified kilns capable of operating at very high temperatures suitable for producing cement;
- advance the use of waste biomass and green hydrogen in cement kilns and improve understanding of resulting effects on the kiln and final product;
- achieve high carbon-capture efficiency (more than 90 percent) on retrofitted and new cement plants;
- develop better catalysts and process designs to deliver higher efficiency levels, reduce costs, and lower material consumption or waste production for CCUS in the cement plant; and
- improve CO_2 transportation and storage infrastructure for CCUS.

Steel Production

Around 70 percent of steel in the United States is produced by electric arc furnaces; the remainder is produced by blast furnaces (BFs) or basic oxygen furnaces. Key technologies needed to decarbonize the steel industry include improvements in energy efficiency; a switch to low-carbon fuels; use of green hydrogen instead of natural gas in direct reduction of iron (DRI); production of iron by electrolysis of iron ore using only renewable electrical energy; postcombustion CCUS (such as top-gas recycling in BFs); DRI with CCUS; HIsarna (smelting reduction) with CCUS; carbon utilization (carbon to ethanol or chemicals); and green hydrogen plasma smelting reduction. Although some of these decarbonization technologies have been commercialized, some require further RD&D, such as

- industrial-scale plant design for producing iron by electrolysis, and development of detailed cost models for assessing the commercial viability of the process; and
- development of methods to analyze and evaluate the BF operation continuously when extending the use of hydrogen to all tuyeres (injection nozzles for air/H_2) in BFs.

Chemical Manufacturing

The chemical industry is highly diverse, producing more than 70,000 products globally. Yet 18 large-volume chemicals—including light olefins, ammonia, BTX (benzene, toluene, xylene) aromatics, and methanol—account for 80 percent of the energy demand and 75 percent of the total GHG emissions attributable to global chemical manufacturing (IEA, ICCA, and DECHEMA, 2013). For perspective, the global production volumes in 2012 were 220 million metric tons for ethylene and propylene, 198 million metric tons for ammonia, 58 million metric tons for methanol, and 43 million metric tons for benzene. Together, production of these four products used about 7.1 EJ of energy per year (or a specific energy consumption of about 13.7 MJ/kg of product), and the top 18 large-volume products used a total of about 9.4 EJ of energy per year (Figure 3-15).

Because about 90 percent of chemical manufacturing processes use catalysts, catalyst and catalyst-related process improvements could reduce the energy intensity of these 18 products by 20–40 percent by 2050, amounting to energy savings of about 13 EJ and a CO_2 emissions reduction of 1 metric gigaton (IEA, ICCA, and DECHEMA, 2013). Incremental improvements will suffice in the short to medium term; in the longer term, however, the deployment of new technologies, such as biomass feedstocks and green hydrogen, will be necessary. The challenges for chemical engineers are clear: to identify top catalyst and catalyst-related process opportunities for these 18 and other chemicals.

Key technologies for decarbonization of the chemical industry include energy systems engineering; fuel switching; novel catalytic processes (e.g., olefin production via catalytic cracking of naphtha or renewable methanol); advanced separation processes; electrification; transitions to low-carbon feedstocks and processes (e.g., biomass, hydrogen-based production of ammonia and methanol, artificial photosynthesis, renewable-energy electrochemistry, biobased plastics production, gas-to-liquids gas); and CCUS. Chemical engineers can contribute to the RD&D necessary to achieve these technological advances, including the following examples:

- Achieve viability for natural gas crackers and improve catalyst production; improve the efficiency of olefin production via catalytic cracking of naphtha or via methanol; and address environmental issues for used catalysts, including regeneration and catalyst selectivity and lifetime.

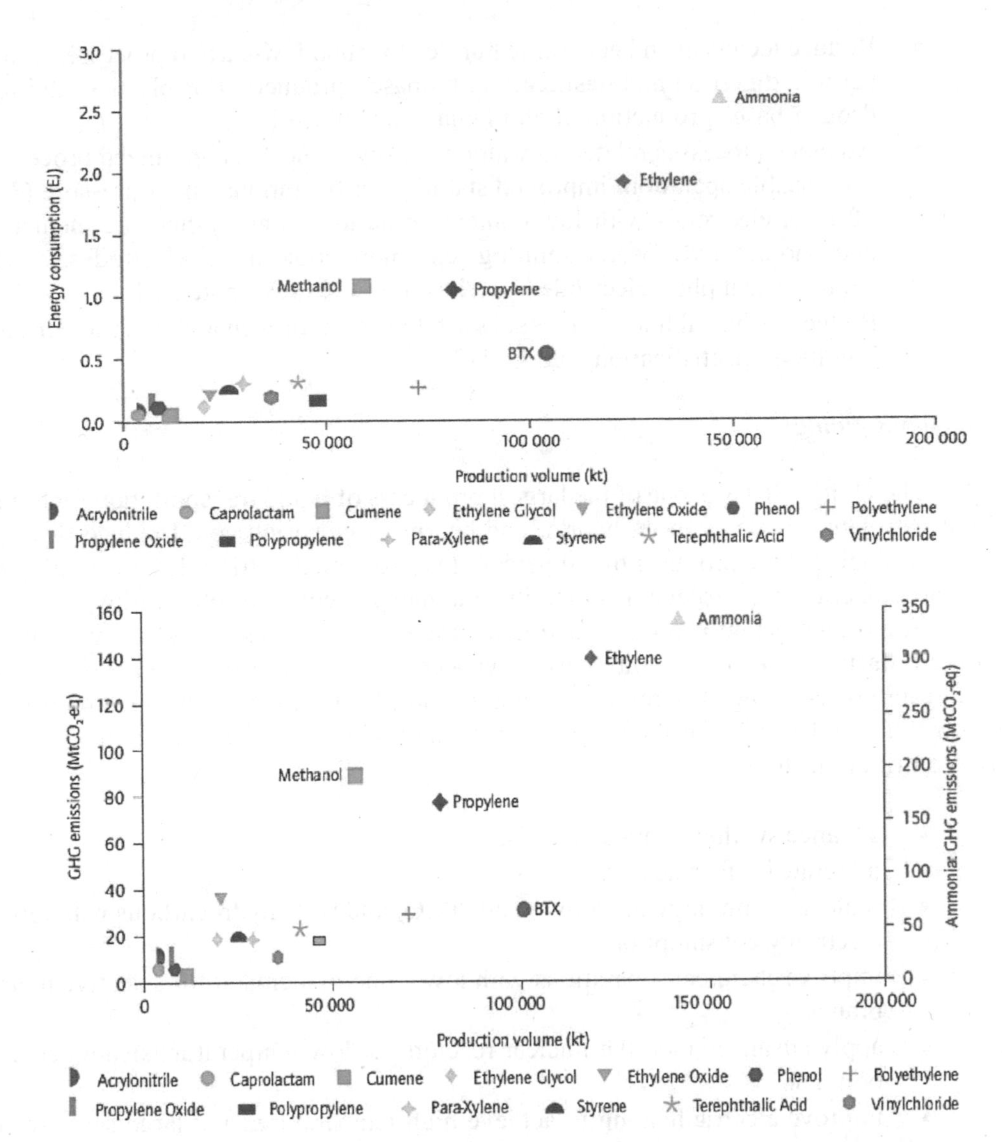

FIGURE 3-15 (a) Global energy consumption versus production volumes, and (b) global greenhouse gas (GHG) emissions versus production volumes for the top 18 large-volume chemicals in 2010. NOTE: BTX = benzene, toluene, xylene. SOURCE: IEA, ICCA, and DECHEMA (2013).

- Improve the hydrogen peroxide propylene oxide process—oxidation of propylene with hydrogen peroxide yields propylene oxide and water as a by-product; its energy consumption could be about 35 percent lower than that of the traditional process.
- Resolve waste management issues for selective membranes, including membrane washing/cleaning, and drive step-change advances in separations, including the use of ceramic membranes.

- Reduce technical and economic hurdles for such low-carbon processes as anaerobic digestion and gasification, biobased production of plastics, and hydrogen-based production of ammonia and methanol.
- Advance processes related to water electrolysis, such as optimized processes for variable operation, improved stability for operations under pressure (30–40 bar), electrodes with low-content noble metals and other rare elements, and photocatalytic water splitting (e.g., non–noble metal electrodes, corrosion-resistant photoelectrode materials improved over potential).
- Reduce technical hurdles for fuel switching to hydrogen and other lower-carbon fuels, electrification, and CCUS.

Petroleum Refining

The United States is one of the largest producers of liquid transportation fuels and refined petroleum products in the world, and energy engineering could reduce the fuel used in producing these products by 50 percent (Morrow et al., 2015). Key technologies for decarbonization of petroleum refining include fuel switching, electrification, biomass hydrothermal liquefaction and biocrude oil, synthetic fuel synthesis, reduction or elimination of flaring, selective membranes, advanced control and improved monitoring, CCUS, catalytic cracking, progressive distillation, self-heat recuperation, and biodesulfurization. Further RD&D is critical for these technologies, and chemical engineers can make contributions to

- advance synthetic fuel synthesis;
- integrate biofeedstocks;
- scale up technology for conversion of CO_2 and H_2 to hydrocarbons with lower electricity consumption;
- improve chemical separations with lower energy demand for selective membranes;
- apply advanced modular nuclear reactors for low-temperature steam generation; and
- improve electric heating to achieve high temperatures and large scales efficiently and economically.

Food and Beverage Industry

The food and beverage industry is the third-largest consumer of energy in the United States; its energy consumption is dominated by mechanical systems, compressed air, refrigeration, and process heat in the moderate-to-low temperature range. Key technologies for decarbonization of the food and beverage industry include fuel switching to lower-GHG sources, such as electricity and renewables, and plant efficiency measures, particularly because many of these plants tend to be operated by small- and medium-sized manufacturers with limited energy-management capacity. Electrification of dewatering, drying, and process-heating applications using heat pumps, hybrid boilers, induction heating, dielectric heating, and advanced cooling/refrigeration represents an important opportunity for this subsector. In addition, advanced processing and preservation to reduce degradation in processing, along with improvements at the supply chain and consumer levels, are important because on a global scale, about a quarter of the food supply is wasted (see Chapter 4; Buzby et al., 2014; D'Odorico et al., 2018; Finley and Seiber, 2014). A key challenge for this subsector is that food and beverage processors must comply with multiple regulations that complicate the implementation and slow the adoption of new technologies. Although some decarbonization technologies are commercially available, further RD&D is needed in such areas as

- shifting from steam and fossil fuels to electric and solar heating technologies;
- demonstrating and certifying alternative processing technologies to reduce GHG emissions while maintaining product safety and quality, and reducing degradation in the supply chain;
- using waste to produce bioenergy; and
- adopting fuel switching and expanded implementation of energy efficiency.

Cross-Cutting Approaches for Decarbonization of the Industrial Sector

To achieve the net-zero emissions goal for industry, five broadly applicable decarbonization pillars require vigorous pursuit in parallel over the next decades: demand reduction, energy efficiency, fuel switching and electrification, transformative technologies in sectors, and abatement. The applicability and selection of these pillars will vary across sectors, with weighing of trade-offs in costs and accessible resources (de Pee et al., 2018). Except for sector-specific transformative technologies, these approaches can be considered cross-cutting and can facilitate reduction of GHG emissions across multiple sectors.

Barriers abound across the landscape of deployment, development, scale-up, and whole-system integration of current, emerging, and transformative technologies that can advance these cross-cutting approaches. Nonetheless, chemical engineers have opportunities to help overcome these barriers (e.g., in the areas of competitiveness, carbon capture and use, and advanced materials). Many low-carbon technologies are in the early stages of development and will require extensive RD&D for effective deployment. RD&D needs range from advancing cross-cutting technologies (e.g., improved electrolysis of water to lower-cost H_2) to making radical changes (e.g., applying high-temperature heat for ethane crackers). Figure 3-16 summarizes a more comprehensive set of recommendations from a recent study by the National Academies of Sciences, Engineering, and Medicine on low-carbon technologies, approaches, and infrastructure needing RD&D investment in the 2020–2050 period (NASEM, 2021a). A portfolio of collaborative RD&D initiatives, multigeneration plans, agile management, and durable support will be needed to face these challenges successfully and drive progress going forward (NASEM, 2019f, 2021b).

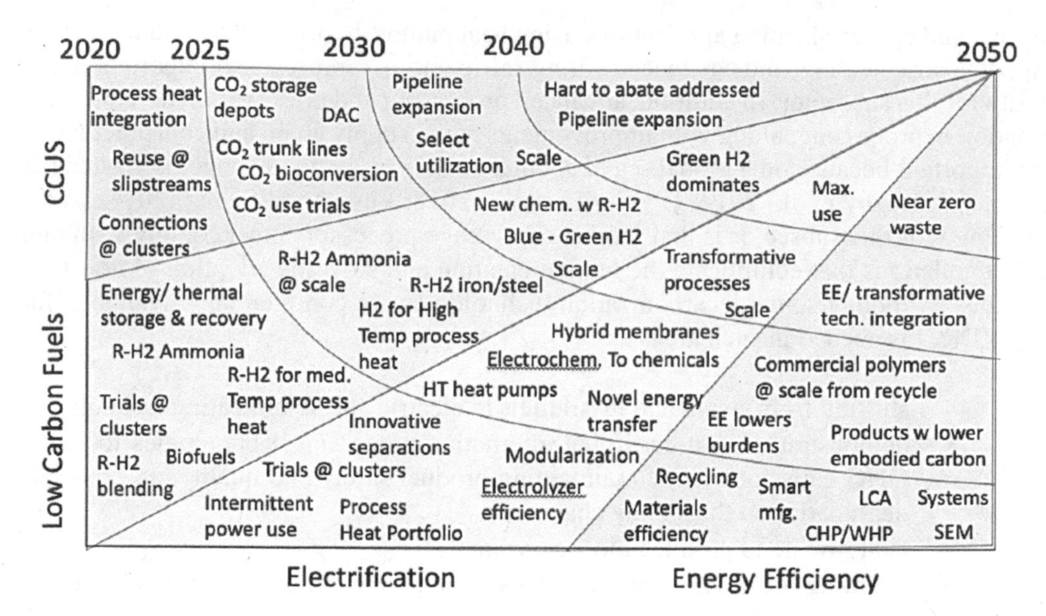

FIGURE 3-16 Research, development, and demonstration investment needs to advance low-carbon technologies (carbon capture, use, and storage [CCUS]; low-carbon fuels; electrification; and energy efficiency [EE]) and achieve decarbonization over the period 2020–2050. NOTES: CHP/WHP = combined heat and power/waste heat to power; DAC = direct air capture; HT = high-temperature; LCA = life-cycle assessment; SEM = strategic energy management. SOURCE: NASEM (2021b).

Commercial and Residential Sectors

The impact of the commercial and residential sectors on net energy use and GHG emissions in the United States is similar in scale to that of the transportation sector. Although significant opportunities exist for improving energy efficiency in commercial and residential settings, the highly dispersed nature of this sector creates challenges not shared by the industrial subsectors discussed in the previous section, which have large, fixed facilities. A recent National Academies report on decarbonization identifies key strategies that employ existing technologies, including electrification of energy use and dramatically increased use of electric heat pumps for heating and hot water (NASEM, 2021a). The future energy needs of residential buildings in the United States and elsewhere in the world are quite different. In the United States, the critical need is for more efficient use of energy for heating and air conditioning as existing systems for these purposes are replaced or updated. Low- and middle-income nations, on the other hand, are likely to see enormous growth in the use of air conditioning, driven by rising prosperity and, to a lesser extent, by the impact of climate change.

To contribute to significantly reducing energy use in the commercial and residential sectors, chemical engineers will need to work closely with other engineers, including mechanical and civil engineers, who are more closely affiliated with these sectors, in the development of technologies that can drive progress. As is the case with other topics discussed in this chapter, putative technologies will need to be deployable at low cost with high reliability to have any chance of success.

Although systems for refrigeration and air conditioning are ubiquitous, improvements in their efficiency and sustainability will require overcoming key technological challenges. Chemical engineers can play a pivotal role in the development of new heat-transfer fluids that are nontoxic and nonflammable and lack the very high GHG intensity of many chlorofluorocarbons (CFCs) and other halocarbons. Nontraditional heat-transfer cycles, such as adsorption cooling, and materials with caloric properties (e.g., Moya and Mathur, 2020) also have considerable potential that intersects strongly with chemical engineering applications in other domains.

Improving the properties of building materials is a key path toward improving commercial and residential energy efficiency. Opportunities include the development of passive materials, such as paints for roofs that reflect the solar spectrum more effectively and, as a result, reduce the cooling load required for buildings (Li et al., 2021a); and active materials, such as "smart glass," which adapts to external sunlight to reduce the net energy demands in buildings (Alias et al., 2019). Chemical engineers have many opportunities to contribute creatively in these areas, provided that researchers remain focused on meeting the cultural and economic needs of end users rather than on fashioning "pure technology" solutions.

CARBON CAPTURE, USE, AND STORAGE

CCUS will be centrally important to controlling the concentration of carbon in the atmosphere. Achieving net-zero emissions to halt growth in the concentration of atmospheric CO_2 will not require achieving zero anthropogenic CO_2 emissions, but rather balancing anthropogenic CO_2 emissions with natural and anthropogenic CO_2 sinks. An extensive portfolio of mitigation strategies for GHG emissions could contribute to the achievement of net-zero emissions (Figure 3-17). Six negative emissions technologies (NETs) remove carbon from the atmosphere and sequester it (Fuss et al., 2018; NASEM, 2019b): coastal blue carbon, terrestrial carbon removal and sequestration, bioenergy with carbon capture and sequestration, DAC, carbon mineralization, and geological sequestration. Other approaches have been proposed, as well, such as cloud alkalinity, biomass burial, enhanced ocean upwelling and downwelling, DAC by freezing, marine bioenergy with carbon capture and sequestration, and electrochemical lining (The Royal Society, 2018b). The removal of other gases, such as methane, N_2O, and CFCs with significant global warming potential will also be important in the overall effort; however, the concentration of these gases in the atmosphere is much smaller than that of CO_2, making them very difficult to remove once they have been released.

The broad area of CCUS is a very rich one for chemical engineering; it presents numerous opportunities for major contributions, as well as challenges to address over the next years and decades, in the technical, economic, LCA, and integrated assessment management (IAM) areas. IAM is a quantitative tool for combining information from diverse fields (i.e., science, economics, and policy) to assess the impact of emissions or their reduction. To address these challenges effectively, however, chemical engineers will need to partner with engineers from other disciplines, such as environmental engineering (e.g., NASEM, 2019c).

DOE has identified priority research directions for CCUS (DOE, 2018c). Of these priorities, chemical engineers could contribute in the following areas:

- Capture—designing high-performance solvents and developing environmentally friendly solvent processes, designing sorbent materials and integrated processes, developing membranes and related processes, and producing hydrogen from fossil fuels with CO_2 capture.
- Utilization—designing interfaces for enhanced hydrocarbon recovery with carbon storage; valorizing CO_2 from catalytic, electrochemical, and photochemical transformations to fuels, chemicals, and new materials; and tailoring microbial and bioinspired approaches to CO_2 conversion.
- Storage—advancing multiphysics and multiscale fluid flow to achieve Gt-per-year capacity; locating, evaluating, and remediating existing and abandoned wells; and optimizing injection of CO_2.
- Cross-cutting—integrating experiments, simulations, and machine learning across multiple length scales to guide the development of materials and processes; intensifying CCUS processes; incorporating social aspects into decision making; and integrating LCA and technoeconomic assessment (TEA), along with environmental and social considerations, to guide technology portfolio optimization.

Near-zero- and positive-emissions technologies emit almost zero GHGs or emit GHGs to a lesser extent compared with alternative technologies, respectively. They include enhanced energy efficiency, clean or renewable electrification, bioenergy, hydrogen and hydrogen-based fuels, and CCUS technologies. LCA and TEA (including scalability of the specific technology) are two primary tools used to evaluate and rank these technology options. Chemical engineers can play a role in the development of many, technologies that have been reviewed extensively elsewhere (e.g., Bui et al., 2018; Fuss et al., 2018; Hepburn et al., 2019; IEA, 2020d; NASEM, 2019b). Opportunities for chemical engineers in the areas of DAC and CO_2 and CH_4 utilization are briefly summarized in this section.

Direct Air Capture

DAC appears to be a relatively easy fix for climate change and has the additional advantage that it can be located close to the sequestration reservoir, thus avoiding the need to use a pipeline for CO_2 transportation. However, dilute systems, such as those with CO_2 in the air at a concentration of about 400 ppm, require more energy than concentrated systems, such as those with CO_2 in flue gas from ammonia manufacture at a concentration exceeding 98 percent; from coal-fired power plants at a concentration of 12–15 percent; from cement, iron/steel, and glass production at a concentration of 20–35 percent; and from natural gas–fired power plants at a concentration of 3–4 percent on a volume basis (NASEM, 2019a). Thus, the concentration of CO_2 in flue gases is between almost 100 and 300 times that of CO_2 in the air, and as a result, the energy required to capture CO_2 from the air is 2 to 3 times greater than that required to capture CO_2 from the flue gases (Bui et al., 2018).

Areas on which chemical engineers will focus in the future include the development of low-cost solid sorbents, highly CO_2-selective materials that require reduced regeneration energy, materials that are highly active in ambient conditions, and processes with increased mass-transfer coefficient and high throughput and low pressure drop (NASEM, 2019b). Other areas of focus will include packing designs, process intensification, catalytic additives, and long-term stability of sorbent matrices.

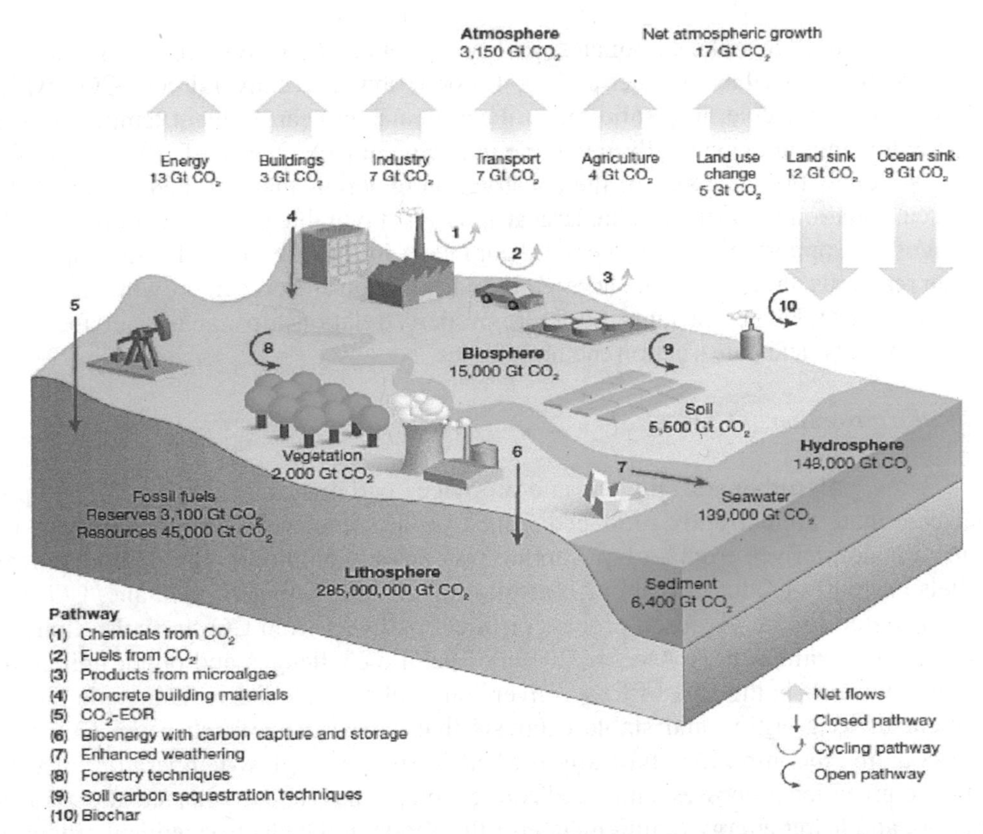

FIGURE 3-17 Stocks and net flows of CO_2, including potential uses and removal pathways. The numbers 1 through 10 represent various pathways for CO_2 use and removal: 1 = chemicals from CO_2; 2 = fuels from CO_2; 3 = products from microalgae; 4 = concrete building materials; 5 = CO_2 in enhanced oil recovery (EOR); 6 = bioenergy with carbon capture and storage; 7 = enhanced weathering; 8 = forestry techniques; 9 = soil carbon sequestration techniques; and 10 = biochar. SOURCE: Hepburn et al. (2019).

Carbon Utilization

Use of CO_2 and CH_4 as feedstocks is an important mechanism for CCUS. However, only a small amount of CO_2 and CH_4 emitted each year is currently being captured and used. The main pathways for CO_2 utilization include mineral carbonation, chemical utilization, and biological utilization, while the main pathways for CH_4 utilization are chemical utilization, biological utilization, and direct uses as fuel. Most carbon utilization technologies are early in their development phase. A number of research areas are described in the above-referenced National Academies report (NASEM, 2019a); specific opportunities for chemical engineers are highlighted in the following sections.

Mineral Carbonation

CO_2 is used to make carbonates, such as cement and concrete, for use in the construction sector, as well as in paper and food production. The conversion of CO_2, which is a low-energy molecule, into solid mineral carbonates in near-ambient temperatures is one of the few thermodynamically favorable reactions of CO_2 (NASEM, 2019a). For this reason, as well as the sheer size of the construction materials market, the use of CO_2 for mineral carbonation is considered the largest and most favorable CO_2 utilization pathway. Challenges and opportunities for chemical engineers in this area include control of carbonation reactions, process design, accelerated carbonation and crystal growth, green synthesis routes for alkaline reactants, structure property relationships, analytical and characterization tools, and construction methodologies.

Chemical Conversion of CO_2

Urea, polycarbonate, ethylene and propylene carbonates, salicylic acid, and polyether carbonate are currently produced from CO_2 at commercial scales; however, the CO_2 used is not derived from carbon capture processes. Commodity and fine chemicals and fuels currently produced from CO_2 at pilot plants are methanol, methane, CO, fuel via a CO_2-based Fischer-Tropsch process or direct pathway from CO_2 to fuels, diphenyl carbonate, and oxalic acid (NASEM., 2019a). Specific challenges and opportunities for chemical engineers in the area of CO_2 conversion to chemicals and fuels include the development of long-lasting and stable catalysts that can also work when the CO_2 feed stream contains the impurities typically present in flue gases, low-temperature electrochemical conversion processes, enhanced conversion per pass and avoidance of carbonate formation, and lower energy requirements for the anode in the electrochemical reduction of CO_2.

Chemical Conversion of CH_4

Sources of methane waste gas include emissions from oil and gas plants, landfills, sewage, manure, and other waste operations. The methane waste gas from oil and gas plants is primarily methane, whereas that from waste management operations, called biogas, is a mixture of CO_2 and methane. In contrast with CO_2, methane is a high-value and high-energy chemical and has no equivalent pathways to mineral carbonation. Because of its high energy, it is used primarily as fuel, and thus any conversion to chemicals needs to compete with the fuel value of methane. The cost trade-offs are likely to change with increased use of CCUS technologies. Challenges and opportunities for chemical engineers in this area include the development of catalysts, integration of catalyst and reactor technology, and identification of new chemical targets. Tools such as LCA and TEA will also be critical in identifying situations in which methane conversion is cost-competitive with the use of methane as fuel.

Biological Conversion of CO_2

Biological conversion of CO_2 holds great promise because some microorganisms have a natural ability to capture and covert CO_2. Photosynthetic pathways to CO_2 utilization include approaches using algae (products include biofuels, dietary protein and food additives, and commodity and specialized chemicals), green algae (products include biodiesel, dietary protein, polyunsaturated fatty acids, pigments, lipids, and terpenoids), and cyanobacteria (products include ethanol, butanol, fatty acids, heptadecane, limonene, bisabolene, ethylene, isoprene, squalene, and farnesene). By using CO_2 as feedstock, these photosynthetic pathways mitigate the problem of the high-cost sugar feedstocks needed for microbial pathways; however, the slow growth rates of algae and cyanobacteria prevent them from achieving industrially relevant productivity and scale-up. Challenges and opportunities for chemical engineers in this area include bioreactor and cultivation optimization, analytical and monitoring tools, genome-scale modeling and improvement of metabolic efficiency, bioprospecting, valorization of coproducts, genetic tools, and pathways to new products.

Biological Conversion of CH_4

Methanotrophs can use methane as their carbon and energy source. However, significant challenges arise, such as the risk of contamination during fermentation, buildup of toxic intermediates, and the high cost of additives. Some commercial activity has taken place in this space. Calysta™ has commercialized FeedKind® protein as an alternative feed for fish, livestock, and pets, using no agricultural land and less water than is required for similar agricultural products. Intrexon™ has used methanotrophs to produce high-value chemicals, such as isobutanol and farnesene. And Mango Materials™ plans to convert biogas into polyhydroxyalkanoate (PHA), which is a biodegradable plastic. Challenges and opportunities for chemical engineers in this area are the same as those for biological conversion of CO_2.

CHALLENGES AND OPPORTUNITIES

In the energy sector, the overarching challenge for chemical engineers is to address the environmental and climate impacts of current energy systems, particularly the use of fossil-based energy sources. The transition to a low-carbon energy system will require a bridging strategy that relies on a hybrid system with a mix of energy carriers. Chemical engineering is rooted in the transformation of stored energy carriers into more convenient forms and into chemicals and materials. Chemical engineers have an important opportunity to continue to apply their skillsets to non–fossil-fuel-based energy sources and carriers and thus contribute to the decarbonization of the energy sector.

In the long term, achieving net-zero carbon emissions will require significant advances in photochemistry, electrochemistry, and engineering to enable efficient use of the predominant source of energy for Earth—the solar flux. To this end, novel systems will be required to improve the efficiency of photon capture and conversion to electrons; improve the storage of electrons; and advance the direct and/or sequential conversion of photons to energy carriers via reactions with H_2O, N_2, and CO_2 to produce H_2, NH_3, and liquid fuels, respectively.

Specifically, greater market penetration of PV solar panels will require a continued decrease in the cost of these panels, as well as increased electrification of energy systems. Chemical engineers have an opportunity to play an enabling role in addressing this challenge by advancing incremental improvements in device architecture and design, lowering manufacturing costs, and improving reliability and durability. Beyond PV technologies, critical challenges hindering the greater use of PSC systems include operational instabilities, short useful lives, and the containment and ultimate disposal of some toxic components. Research conducted by chemical engineers and others will be critical to developing perovskite compositions that minimize degradation and ensure reliable long-term operation. Conversion of photons to H_2, NH_3, or organic fuels will require advances in the synthesis, characterization, and mechanistic assessment of catalytic solids, as well as the development of materials that can withstand severe chemical, photochemical, and electrochemical environments within complex hydrodynamics for systems that couple the required reactions through diffusional controls, all of which are research opportunities for chemical engineers.

Successfully mitigating climate change will require a long-term transition to renewable and sustainable sources of energy. In the short term, however, chemical engineers have many opportunities to reduce the carbon footprint of fossil fuels. For coal, these opportunities include research and technological advances to increase thermal efficiency, further improve emission controls, and reduce water consumption. While natural gas is a cleaner bridge fuel compared with other fossil fuels, innovations are still needed throughout the value chain. Chemical engineers can enable advances that will minimize or replace water use as a fracturing agent, improve storage and transportation, and better integrate natural gas with renewable energy sources. For petroleum, challenges and related opportunities for chemical engineers include improved water management, increased recovery to extend well life, data-driven approaches to reservoir management, and improved methane management. Decreasing the GHG emissions associated with all types of fossil fuels will require the demonstration of cost-effective and secure carbon capture and storage methods, a critical opportunity for chemical engineers.

Because most biogenic feedstocks have lower energy content than their fossil-based counterparts, the greatest challenge for increased use of biofuels is the production of high-density fuels at a reasonable cost that is competitive with that of existing, fossil-based fuels. This challenge, combined with the need to account for the environmental consequences of harvesting crops for energy use, creates opportunities for chemical engineers to use systems-level economic, environmental, and technical analyses to select the most viable biofuel options. Furthermore, increasing the market penetration of EVs for personal transportation will require reimagining petroleum refineries that were designed to produce gasoline or diesel fuel as their main products. Refineries will require reconfiguration to shift their product slate toward petrochemicals and low-carbon liquid fuels needed for the difficult-to-decarbonize commercial transportation sector (e.g., heavy-duty long-haul ground, aviation, and marine transportation). Full integration of existing petroleum refinery assets with biorefineries and the greater use of renewable energy will enable significant reductions in carbon footprints and lower-cost low-carbon liquid fuels, another area that presents considerable opportunities for chemical engineers.

For intermittent energy sources, chemical engineers can contribute to the development of advanced materials that can increase the viability of wind and marine energy. Chemical engineering research will also be critical to advancing low-carbon fuels; improving petroleum refining; advancing clean hydrogen production; and developing improved synthetic fuels for sectors, such as aviation, that are difficult to decarbonize. A successful transition to low-carbon energy systems presents the challenge of energy storage. Chemical engineers can enable the development of new battery materials, as well as contribute to LCAs of competing battery technologies and the design of batteries for safe end-of-life disposal.

For end uses, the production of cement, steel, and chemicals presents the clearest opportunities for chemical engineers to contribute to decarbonization of the industrial sector. To achieve the net-zero emissions goal for industry, five broadly applicable decarbonization pillars require vigorous pursuit in parallel over the next decades: demand reduction, energy efficiency, fuel switching and electrification, transformative technologies in sectors, and abatement. All will benefit from the contributions of chemical engineers.

Finally, CCUS will be centrally important to controlling the concentration of carbon in the atmosphere. This broad area is rich with opportunities for chemical engineers, including LCA, integrated assessment management, and the science and technology advances necessary to advance direct air capture and carbon utilization.

Recommendation 3-1: Across the energy value chain, federal research funding should be directed to advancing technologies that shift the energy mix to lower-carbon-intensity sources; developing novel low- or zero-carbon energy technologies; advancing the field of photochemistry; minimizing water use associated with energy

systems; and developing cost-effective and secure carbon capture, use, and storage methods.

Recommendation 3-2: Researchers in academic and government laboratories and industry practitioners should form interdisciplinary, cross-sector collaborations focused on pilot- and demonstration-scale projects and modeling and analysis for low-carbon energy technologies.

4
Sustainable Engineering Solutions for Environmental Systems

- While water, food, and air have historically been the focus of other disciplines, chemical engineers bring both molecular- and systems-level thinking to pioneering efforts in this highly interconnected space.
- The positive impact of chemical engineers will be magnified as they adapt to thinking beyond the traditional unit operation scale to focus at a global scale on the water–energy–food nexus.
- The Earth's atmosphere, with its large range of spatial and temporal scales, presents intriguing challenges for chemical engineers, who will continue to aid in improving air quality.

By 2050, the Earth's population is projected to grow to more than 9 billion, leading to a 60 percent increase in food demand, an 80 percent increase in energy demand, and a 55 percent increase in water demand (OECD, 2012; UN, 2017). Food, energy, and water are highly interconnected, with production or consumption of one usually directly linked to production or consumption of another. Agricultural crops produce biofuels and provide food for animals and humans. Energy is used to purify, transport, heat, or cool water; to produce fertilizers; and to power farm equipment, food processing, and cooking. Land and water diverted for energy production are no longer available for food production and vice versa. Water is used for the production of fuels and electricity, and for agriculture, food processing, livestock, and cooking. In addition to these complex relationships, air quality affects or is affected by all three sectors and has a direct impact on human health and well-being. Although water is a renewable resource that is conserved in the Earth, freshwater can be depleted locally, and the policies for its local allocation are set in a highly political context.

The concept of a water–energy–food (WEF) nexus was first introduced in 2011 by the World Economic Forum in *Water Security: The Water-Energy-Food-Climate Nexus*. To better contextualize this WEF nexus, it is important to intertwine an additional nexus, one that considers sustainability and environmental conditions (including climate), as well as economic and social (including human health) components, sometimes referred to as the "triple bottom line" (Das and Cabezas, 2018; Figure 4-1).

This chapter explores the role of chemical engineers in ensuring adequate food supplies and clean water and air. After a discussion of the WEF nexus, scientific gaps in understanding of the fundamental properties of water, as well as the need for engineering solutions for water quality and supply, are reviewed. Opportunities for chemical engineers to both pioneer and contribute to multidisciplinary efforts to advance agricultural and food processing technologies are then described, followed by a discussion of the research needs for understanding and improving air quality.

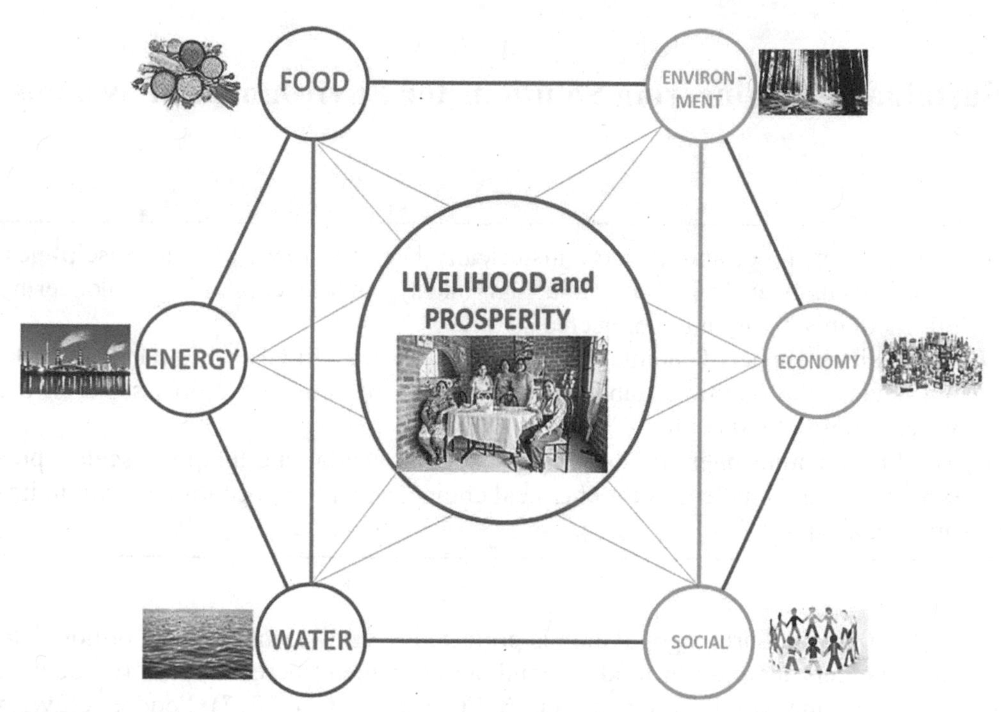

FIGURE 4-1 The concept of the water–energy–food nexus is further contextualized to understand solutions to global problems and support human livelihoods and prosperity by showing its interconnectedness with the environment–economy–social nexus.

THE WATER–ENERGY–FOOD NEXUS

Solutions in the WEF nexus need to be sustainable; thus, environmental, economic, and social factors need to be carefully considered. Increased resource demand presents a monumental challenge for chemical engineers. They have historically played a central role in the energy sector, and their contributions in the interconnected space of water, food, and air quality have been important and are now growing. Fundamental insights from chemical engineering disciplines form the foundational knowledge necessary to understand and create solutions in the WEF nexus, which inherently spans multiple disciplines. Examples include the structure and dynamics of water, the nature and physics of aerosol particles, and the scaling of synthetic protein production. Additionally, chemical engineers' knowledge of biochemical engineering and its applications to agriculture, and of separations with applications to water and air pollution, as well as their systems-level understanding, is critical to solving global problems.

Fossil fuels (petroleum, natural gas, and coal) make up most of the landscape of primary energy sources, contributing globally about 82 percent of the total energy produced in 2019 (IEA, 2021e). The continuing reliance on fossil fuels contributes to greenhouse gas (GHG) emissions and air pollution, as well as associated human health, environmental, and climate problems (D'Odorico et al., 2018). Today, one in five people lack access to modern electricity in their homes, and 3 billion people use wood, coal, charcoal, or animal waste for heating and cooking, all of which have deleterious impacts on air quality (D'Odorico et al., 2018). Note that energy generation and consumption are covered extensively in Chapter 3 and are discussed in the remainder of this chapter only in the context of interdependencies with food, water, and air resources.

Over the past 50 years, global crop production has increased by more than 300 percent, and animal production by more than 250 percent. Dairy and meat production is expected to increase by more than 60 percent by 2050 (D'Odorico et al., 2018). All of these increases are attributable to the wider use of fertilizers and higher-yielding crop varieties, which enable greater food production. Of the global land surface, 38 percent is used for agriculture, with roughly one-third of that total used for crops and the other two-thirds used for animal grazing (FAO, 2020).

Food waste is a major problem around the world. In the United States, the percentage of food wasted ranges from 16 percent for meat to ~25–30 percent for grains, dairy, and eggs (Finley and Seiber, 2014). On a global scale, about one-quarter of the food produced for human consumption is lost or wasted in the supply chain; therefore, the percentage of arable land and freshwater resources used to produce that food is also wasted (D'Odorico et al., 2018).

The interdependency within the WEF nexus is massive (Figure 4-2). Approximately 15 percent of global water withdrawals are used for energy, 70 percent for agriculture, and the remaining 15 percent for other applications. About 8 percent of global energy is used to transport, purify, and pump water, and about 30 percent to produce food. About 1 percent of all food is used to produce energy (Garcia and You, 2016). The interconnection between water and energy in various sectors of the U.S. economy in 2011 is depicted, from withdrawal and extraction through use, in the Sankey diagram in Figure 4-3 (DOE, 2014). The competition of water use for energy and food is at the core of the WEF nexus landscape: the growing population and the larger number of people now in the middle class drive demands for food, energy, and water, while at the same time freshwater resources are fixed and limited. This situation generates challenges for the environmental, economic, and social aspects of life on Earth (Albrecht et al., 2018; DOE, 2014; Simpson and Jewitt, 2019).

In 2016, the United Nations estimated that 800 million people suffer from food insecurity, approximately 1.2 billion people lack access to electricity, and 800 million people lack access to safe drinking water (Scanlon et al., 2017). The world's population is projected to increase to well over 9 billion in 2050, and the global middle class will continue to expand. The resulting projected drain on WEF resources is so great that it will be necessary to address the nexus holistically rather than as separate sectors. Future resource security (defined as the uninterrupted availability of resources at an affordable price) will require not only the integrated management of water, energy, and food resources but also a transition to a circular economy, with special attention to challenges of sustainability and climate (Biggs et al., 2015; D'Odorico et al., 2018; see Chapter 6 for further discussion of the circular economy). This integrated management approach is at the core of the capabilities and focus of chemical engineers, from the chemical to the system scale, with respect to reducing demand (conservation), increasing supplies, and managing storage and transport (i.e., managing the spatial and temporal imbalances of production and consumption) while working toward a circular economy in the WEF nexus. The U.S. Department of Energy (DOE, 2014) frames integrated solutions for the WEF nexus around six pillars:

- optimizing the freshwater efficiency of energy production, electricity generation, and end-use systems;
- optimizing the energy efficiency of water management, treatment, distribution, and end-use systems;
- enhancing the reliability and resilience of energy and water systems;
- increasing safe and productive use of nontraditional water sources;
- promoting responsible energy operations with respect to water quality, ecosystem, and seismic impacts; and
- exploiting productive synergies among water and energy systems.

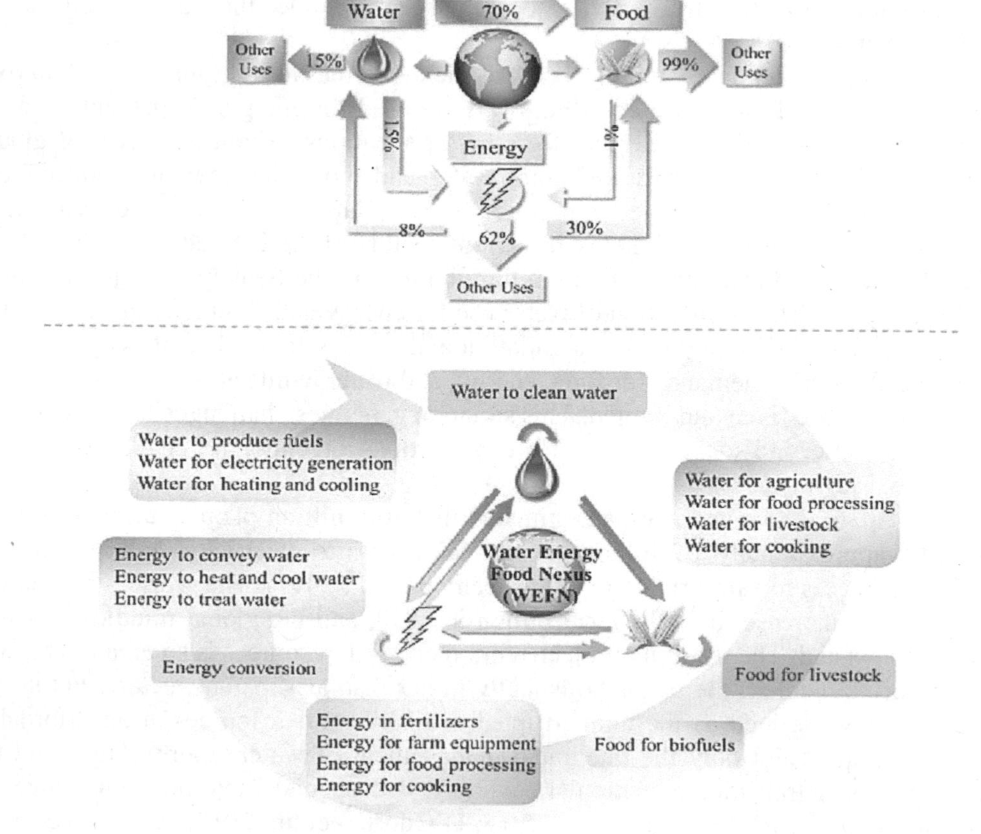

FIGURE 4-2 The water–energy–food nexus emphasizes the interconnectivity of these three major resources. The top illustration quantifies the reliance of each on the other two, while the bottom illustration describes their connectedness and how they contribute to one another. SOURCE: Garcia and You (2016).

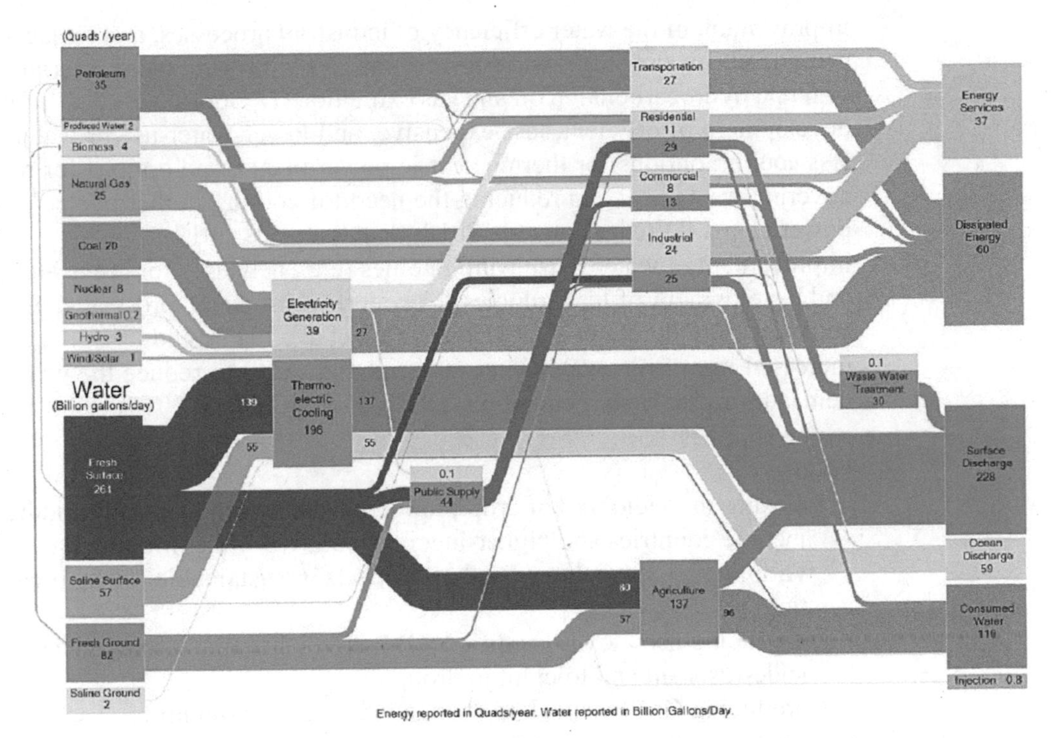

FIGURE 4-3 This Sankey diagram shows energy (top) and water (bottom) consumption associated with various end uses. Numerical values shown are based on 2011 data. SOURCE: DOE (2014).

Chemical engineers can lead and contribute to advances in the many technology vectors required within each of these pillars. Some key examples include the following:

- Reducing water demand (conservation)
  - reduction of food waste and development of technologies that reduce spoilage, and use of food waste to produce chemicals (e.g., bio-oils) and to feed livestock (Balicka, 2020);
  - implementation of a "more-crop-per-drop" (i.e., water productivity in agriculture) approach, use of engineered crops with higher water efficiency and/or drought tolerance, and development of better pesticides (Scanlon et al., 2017);
  - use of brackish groundwater instead of freshwater for energy (D'Odorico et al., 2018; Scanlon et al., 2017);
  - development of advanced sensors to avoid waste and improve process reliability, as well as use of relevant data collection, analysis, and reporting (DOE, 2014);
  - development of alternative fluids to replace freshwater in various processes (D'Odorico et al., 2018; Scanlon et al., 2017);

- improvement of the water efficiency of industrial processes, and replacement of freshwater (e.g., with supercritical CO_2, nitrogen, nanomaterials, or liquid hydrocarbons) in oil and gas extraction (D'Odorico et al., 2018);
- development of efficient, less expensive, and lower-water-use or waterless cooling options for thermoelectric power plants, and options for recovering waste heat and reducing the need for cooling in thermoelectric power plants (D'Odorico et al., 2018; Scanlon et al., 2017);
- improvement of water reuse within homes (e.g., new uses for greywater) and development of technologies for waterless products, processes, and activities (D'Odorico et al., 2018; Scanlon et al., 2017); and
- increased use of renewable energy (wind and solar) to reduce the water demand for electricity generation (DOE, 2014; IChemE, 2015).

- Increasing supplies
  - Food
    - closing the yield gap of crop productivity between low- and middle income countries and higher-income countries while minimizing environmental, social, and other impacts ("sustainable intensification");
    - producing genetically modified (GM) crops that are insect- and herbicide-resistant and tolerant to drought;
    - producing GM livestock to change the fat content in milk;
    - producing in vitro meat that does not involve raising livestock (cell-cultured meat) or plant protein–based meat; and
    - developing hydroponics- and aquaponics-based technologies (D'Odorico et al., 2018; Scanlon et al., 2017).
  - Water
    - further developing desalination options for brackish groundwater or seawater so that desalination expands beyond specific uses that deal with small volumes of water (e.g., drinking water) and population groups that can accommodate the higher costs;
    - capturing stormwater;
    - developing advanced materials for removing chemical and biological contaminants from water (e.g., removing lead contamination from drinking water to address contamination scenarios such as those faced by residents of Flint, Michigan); and
    - treating municipal wastewater (D'Odorico et al., 2018).
  - Energy
    - developing second- (i.e., from agricultural waste) and third- (i.e., from nonedible biomass) generation biofuels,[1]
    - developing advanced energy crops,
    - using waste for energy production,

[1] Today's ethanol from fermentation of sugars and biodiesel from plants are considered first-generation biofuels, and in both cases, the feedstocks can be used for food as well.

- improving the energy efficiency of production of chemicals and products, and
- developing technology for cost-effective recovery of dissipated energy from electricity generation that could also help with carbon capture and storage (DOE, 2014).

- Managing storage and transport
  - developing technologies that enhance food preservation and storage (D'Odorico et al., 2018), and
  - improving battery and other energy storage technologies (IChemE, 2015).

Interdisciplinary research that integrates the physical, agroecological, and social sciences, as well as economics, is needed, along with the involvement of academia, industry, and government. While chemical engineering's role in the energy sector is discussed in Chapter 3, the ties among food, water, air, the environment, and energy are apparent and are key to the future of both the discipline and society writ large.

MOLECULAR SCIENCE AND ENGINEERING OF WATER SOLUTIONS

Chemical engineers have a leading role in molecular science and engineering that requires a fundamental understanding of water structure, dynamics, and interactions, as well as in the development of new complex separation processes. However, issues of water scarcity, preservation, and purification are on a global scale, and solutions to these complex issues therefore require an unprecedented ability to think across scales ranging from the atomic to the geologic. The ability of chemical engineers to solve problems from the molecular to the system-design level will be critical to meeting these challenges, but they will also have to learn from and interact with civil and environmental engineers and to understand system boundaries that go far beyond a unit operation, process, or plant scale.

Water Purification

The purification of water involves the removal of a variety of chemicals or solid materials using a range of processes (Figure 4-4). Several water sources and associated opportunities for chemical engineering are highlighted below.

Desalination

The conversion of ocean or brackish water to drinking water is one of the great engineering advances of recent decades, and this technology is applied at scale in numerous locations (e.g., the Ashkelon Seawater Reverse Osmosis Plant in Israel[2]). The two main technologies applied to this problem are distillation, in several forms, and reverse

---

[2] See https://www.water-technology.net/projects/israel.

osmosis (RO) membrane systems, the latter of which account for 65 percent of global capacity (Abdelkareem et al., 2018; Bhojwani et al., 2019). Modern RO plants operate near the thermodynamic limit but are still very large consumers of energy. Opportunities to improve the overall energy efficiency of a seawater RO plant involve energy and waste-heat management as much, if not more, than the development of new membrane materials or processes. Many RO plants are located in regions where renewable power sources (wind or solar) can be contributors, or in remote regions where nuclear power is a viable alternative. In the latter case, it is possible to use an arrangement whereby the power plant runs at full capacity, with electricity fed into the grid, during periods of high demand, and is otherwise used to purify water, which is much easier to store than electricity.

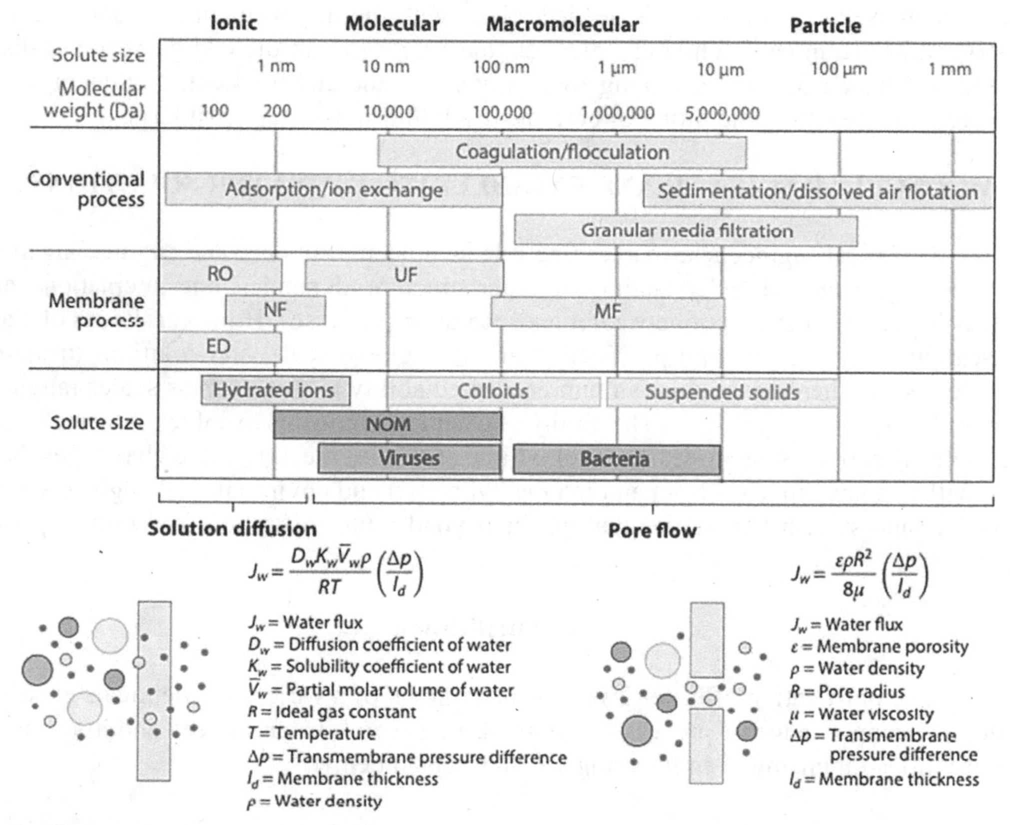

FIGURE 4-4 Comparison of several conventional and membrane processes for water purification and the subsequent mechanisms of water transport through membranes based on solute size and molecular weight. NOTE: ED = electrodialysis; MF = microfiltration; NF = nanofiltration; NOM = natural organic matter; RO = reverse osmosis; UF = ultrafiltration. SOURCE: Landsman et al. (2020).

Produced Water

While sea- or brackish-water desalination is well developed, current and future challenges relate to dealing with water contaminated in different ways. For example, oil production yields oil along with so-called produced water, which in many cases is larger in volume than the oil. This water contains suspended oil, additives, and solids, and in many cases, heavy metals or radioactive elements. If hydraulic fracturing has been used, at least some of the produced water will contain polymeric or surfactant-based "fracking" fluids (this water, sometimes called flowback, was, ironically, initially formulated with freshwater). The oil and water may be produced as an emulsion (either water-in-oil or oil-in-water, or as a multiple emulsion), which if stabilized by asphaltenes or resins can be quite stable. Economical treatment and reuse are very difficult, so this water is often reinjected into the formation. The development of technology for cost-effective treatment of this water, particularly on site, presents a substantial opportunity for chemical engineers.

Similar challenges continue to exist in, for example, managing the water produced during coal and other mining; the iron-ore treatment in the taconite process; and multicomponent radioactive liquid wastes such as those at the Hanford, Washington, nuclear reactor site (where 45 years of operation resulted in an estimated 440 billion gallons of wastewater [Washington State Department of Ecology, 2021]). Other challenges include the removal of boron and other neutral materials, as well as the removal of environmentally persistent perfluorocarbon molecules.

In the case of oil-field or mining water, the separations need to be carried out in steps to accomplish the sequential removal of colloidal and larger particles, then oils, and finally salts. Each of these steps could be considered a unit operation. The science of coagulation and flocculation is well known and practiced in municipal water treatment plants, but the implementation of new coagulants or flocculates is often too costly. Nonetheless, there are substantial opportunities to develop new polymer or surfactant chemistries that can address these challenges and to leverage the principles of self-assembly to treat these waste streams.

After solids and oils have been removed, membrane processes can be applied. The membranes used for water treatment are usually porous polymeric films with pore structures and sizes designed for the application at hand. The flux of water and solutes across the membrane always depends on a balance between permeability and selectivity, as outlined by Landsman and colleagues (2020). In this process, the design parameters are the pore size and distribution; solute size and shape; solute concentration; osmotic pressure; and, importantly, interactions between the solute and the membrane, which can be electrostatic, chemical, or biospecific (Landsman et al., 2020). Each of these parameters can be exploited to design separations for water contaminated in various ways.

A range of research-based approaches have been suggested for new membranes. Molecularly heterogeneous polymers (multiblock copolymers) can be designed to form bicontinuous solids, and then treated to make solids of controlled porosity or synthesized with carbon nanotube channels. Such membranes could be functionalized to be electroactive or biomolecularly specific. They could incorporate elements of ion-specific exchange, or be catalytic and react with and separate target molecules simultaneously. This is a rich field for chemical engineers with an interest in transport, catalytic and reaction chemistry, thermodynamics, self-assembly, and materials science. To make an impact on real-world applications, research in this area needs to focus on the factors that typically limit existing technology solutions (e.g., biofouling, durability, cost). Moving forward, an important consideration will be management of the life cycle of such membranes and incorporation of recycle or upcycle characteristics into their design. For example, the production of polymeric membranes often relies on polar aprotic solvents, such as N,N-dimethylformamide (DMF), or 1,4-dioxane and tetrahydrofuran (THF). These solvents pose considerable environmental challenges, and it would be preferable to replace them with environmentally responsible alternatives.

Trace-Element Recovery

There is an increasing need to recover recyclable and reusable nutrients and minor or trace elements—particularly phosphorous, rare earth elements, and energy-related elements—from waste streams. In many cases, natural deposits of their associated minerals are limited within the United States (Jyothi et al., 2020). In addition, challenges remain for the extraction and/or destruction of both emerging and legacy trace contaminants that threaten human and ecological health. The low concentrations of trace elements make removal challenging, especially in waters containing high salinity or highly complex organic matrices. A wide variety of conventional separations have been employed (e.g., chemical precipitation, coagulation-flocculation, flotation, solvent extraction, ion exchange, adsorption, membrane processes, filtration, reverse osmosis, and electrochemical techniques), with varying levels of technoeconomic feasibility depending on the contaminant properties, background water composition, and treatment goals (Naidu et al., 2019; Pereao et al., 2018). Two of the more promising approaches, especially for recovery of trace contaminants, use sorption (including adsorption, absorption, ion exchange, and precipitation) and/or membrane processes. For many trace elements, however, significant advances in these technologies are needed to expand recovery and reuse and reduce treatment costs (Elbashier, et al., 2021; Li et al., 2021b). Several promising new techniques employing novel sorbents have emerged, including electrospun nanofibers with a highly specific surface for adsorbent applications (e.g., Sharma, 2013) and ionic imprinted polymers (e.g., Branger et al., 2013; Luo et al., 2015). In the case of membrane technologies, novel materials targeted for recovery of rare earth elements include metal organic frameworks and liquid membranes (Smith et al., 2019; Tursi et al., 2021).

The recovery of phosphorus from water presents formidable challenges. Reserves of phosphate rock are rapidly being depleted, and the current cost of recovering phosphorus from wastewater or agricultural runoff exceeds the cost of separating conventional phosphate from rock. Therefore, new and cost-effective technologies for recovering phosphorous from wastewater are needed. To this end, as Peng and colleagues (2018) indicate, more than 50 such technologies, including biological, chemical, and physical processes, have been developed. One of the most common end products for phosphorus recovery is precipitated struvite or vivianite. Fluidized bed reactors that achieve upwards of 80 percent recovery from wastewater have been reported (Nelson et al., 2017), but the energy consumption, cost, and footprint of available processes are all too high to encourage their widespread adoption. Newer alternatives based on adsorption technologies, ranging from alginates to peptide-based materials, as well as biobased systems, are rapidly being explored, with encouraging results (e.g., Jama-Rodzenska et al., 2021; Su et al., 2020; Zhang et al., 2020a). Alternative approaches, such as Donnan dialysis with ion-exchange membranes, are also being investigated for concentrated water containing high concentrations of ions, organic matter, and suspended solids in concentrated waste streams (Shashvatt et al., 2021). However, much research is still needed to bring such technologies into practice.

The extraction of lithium or uranium from water streams poses similar challenges, including extraction from seawater and water produced during oil and gas operations. In the case of lithium, concentrations range from about 5 mg/L to about 500 mg/L, and removal can be accomplished using solar evaporation, adsorbents, membrane-based processes, and electrolysis-based systems (Kumar et al., 2019). Current challenges include accelerating concentration processes and dealing with lower lithium concentrations. In the case of oil and gas wastewater, metal oxide adsorbents and membrane technologies are promising. However, current membrane materials lack sufficient lithium-ion selectivity, and novel membrane approaches, such as 12-Crown-4-functionalized polymer membranes (Warnock et al., 2021) and Cu-m-phenylenediamine (MPD) membranes (Wang et al., 2021a), require further research.

Finally, concerns resulting from the widespread contamination of drinking water by legacy contaminants, such as perfluorinated compounds and lead released from water distribution systems, highlight the need for innovative treatment technologies that address recalcitrant organics and metals with known human health risks at trace concentrations. Per- and polyfluoroalkyl substances (PFAS) have no known half-life and have been found in more than 2,000 locations within the Unites States. The toxicity of the more than 4,000 PFAS compounds varies widely; state-level water quality regulations for some PFAS compounds, such as perfluorooctanoic acid (PFOA), are at the part-per-trillion level. Significant research into treatment strategies includes technologies focused on contaminant destruction (e.g., advanced oxidation and reduction, photolysis, electrochemical oxidation, and incineration) and those aimed at separating the compounds from water (e.g., activated carbon adsorption, polymeric adsorbents, ion exchange, ozofractionation, and membrane separation; Ross et al., 2018). To date, however, no single technology has been identified that is cost-effective, not energy intensive, and universally effective at treating the range of PFAS compounds (e.g., varying functionality and chain length) within the complex matrices of source waters (Crone et al., 2019; Gagliano et al., 2020). The development of treatment technologies for these persistent chemicals is a major opportunity for chemical engineers.

Lead contamination of drinking water resulting from dissolution of lead solder, fixtures, and piping is another concern requiring innovative solutions, including consideration of point-of-use treatment technologies. Maintenance of lead scales within distribution systems is the typical control mechanism for ensuring water quality; however, changes in water quality can dramatically affect lead release and compromise drinking water quality either across the distribution system or within premise plumbing (Wahman et al., 2021). Thus, the development of effective point-of-use technologies (e.g., reverse osmosis, adsorbents) that can remove both dissolved and particulate lead is needed (Brown et al., 2017; Verhougstraete et al., 2019). Nanoenabled technologies are also useful for lead and other metal contaminants, such as copper, Cu(II); chromium, Cr(VI); and arsenic, As(V). For example, Greenstein and colleagues (2019) have examined polymer-iron oxide nanofiber composites and iron oxide–coated polyacrylonitrile fibers for removal of lead, Pb(II), and these other metal ions, as well as suspended solids. In these cases and others, solutions to such separation problems depend on fundamental chemical engineering principles, such as thermodynamics, transport phenomena, chemical kinetics, engineering of nanoscale materials, and process design.

Removal of Microplastics

Chemical engineers have opportunities to address the challenges of remedying the damage done by pollution in large bodies of water. A significant challenge is the remediation of microplastics in oceans and large lakes. An estimated 8 million metric tons of plastic enters waterways annually (NOAA, 2021b), and a significant portion of that total is in the form of so-called microplastics, which are less than 5 mm in size. Microplastics enter the water directly as waste from consumer products, such as cosmetics, but are also produced by attrition of larger pieces of plastic or incomplete incineration of plastics. The Great Pacific Garbage Patch (GPGP) covers approximately 1.6 million square kilometers in the North Pacific (Lebreton et al., 2018). By comparison, the *Deepwater Horizon* oil spill covered about 150,000 square kilometers in the Gulf of Mexico, or 10 percent of the area of the GPGP (NOAA, 2021a). Microplastics pose a risk to larval fish and thus can potentially impact a significant source of food protein (Gove et al., 2019).

Remediation of such a large area of open sea is challenging. Some microbial treatments have been proposed (Auta et al., 2017), but appear to be applied more effectively in treatment plants. Open-sea cleanup of microplastics will likely have to rely on their physical removal. Such harvesting will require the concentration of microplastics into flocs, which may be promoted by highly effective flocculating agents (Roh et al., 2019). Such flocs could then also be subjected to bioremediation processes.

Removal of microplastics from water and other media presents opportunities for chemical engineers. Removal methods include physical sorption and filtration, biological removal and ingestion, and chemical treatments (Iyare et al., 2020; Padervand et al., 2020). Physical sorption methods include adsorption on green algae (depending on the microparticles' surface charge), removal using membrane technology, and removal using filtration technology (e.g., ultrafiltration, disc filter, rapid sand filtration, and dissolved air flotation [Talvitie et al., 2017]). Chemical removal methods include coagulation, agglomeration, and settling processes (Lapointe et al., 2020) using Fe^{3+}- and Al^{3+}-based salts and other coagulants (e.g., polyacrylamide [PAM]); electrocoagulation; sol-gel reactions; and photocatalytic degradation. Biological removal and ingestion methods include biological degradation using microorganisms (Padervand et al., 2020).

Fundamental Properties of Water

The technological advances needed to secure a global freshwater supply, as described above, rely on fundamental breakthroughs in understanding the structure of water and its dynamics (Debenedetti and Klein, 2017). For example, the structuring of water near an interface plays a critical role in the fouling of that surface, whether it is on the outside of a naval vessel or the inside of a pipe in a water treatment plant. Similarly, whether ions, for example, are depleted or concentrated at the air–water interface is of considerable importance in the development of new water-purification technologies or of water transport models for agriculture or climate prediction. Chemical engineers, primarily in academia, are heavily involved in experimental and theoretical studies of water both with and without other additives. This section reviews and summarizes some of the opportunities and challenges therein.

Considerable advances have been made over the past two decades regarding the structure and dynamics of water under a wide range of conditions and environments. Much of that progress has been fueled by theoretical and computational advances, coupled with the development of experimental techniques capable of probing structure and dynamics over a wide range of length and time scales (Figure 4-5).

Structure of Pure Water

Some of the more intriguing developments in understanding the structure of pure water have been enabled by developments in synchrotron light sources and vibrational spectroscopy, coupled with progress in models and computation that has permitted a deeper interpretation of such measurements (Bjorneholm et al., 2016). Various types of vibrational spectroscopy, as well as elastic and inelastic neutron scattering, have gradually been refined to provide detailed insights into the structure and dynamic processes in bulk water and at interfaces, particularly when coupled with selective deuteration, at time scales ranging from fractions of a picosecond to tens of nanoseconds. These time scales encompass a wide range of characteristic molecular processes, from O–H bond vibrations to the motion of water and water-bound molecules, such as lipids in bilayer membranes or polymers in solution. Two-dimensional infrared spectroscopy, for example, continues to push its limits of sensitivity and applicability and has provided previously inaccessible information about how molecules, such as amyloid proteins, self-assemble in aqueous solution (Middleton et al., 2012; Shim et al., 2009). Sum frequency generation (SFG) is providing much-needed insights into chemical reactions at interfaces, including those occurring at electrodes for energy generation (Neri et al., 2017). In situ experiments, such as inelastic neutron scattering measurements, can now provide critical information about industrially relevant reactions as they occur in real-world processes, such as the formation of NaOH and Na_2SiO_3 activated in low-CO_2 cements (Gong et al., 2019). Experiments at light sources (synchrotrons) have for decades been important for structural characterization, such as crystallographic studies of water, hydrates, and aqueous solutions of biomolecules, including proteins. Synchrotron light sources continue to be upgraded, and with

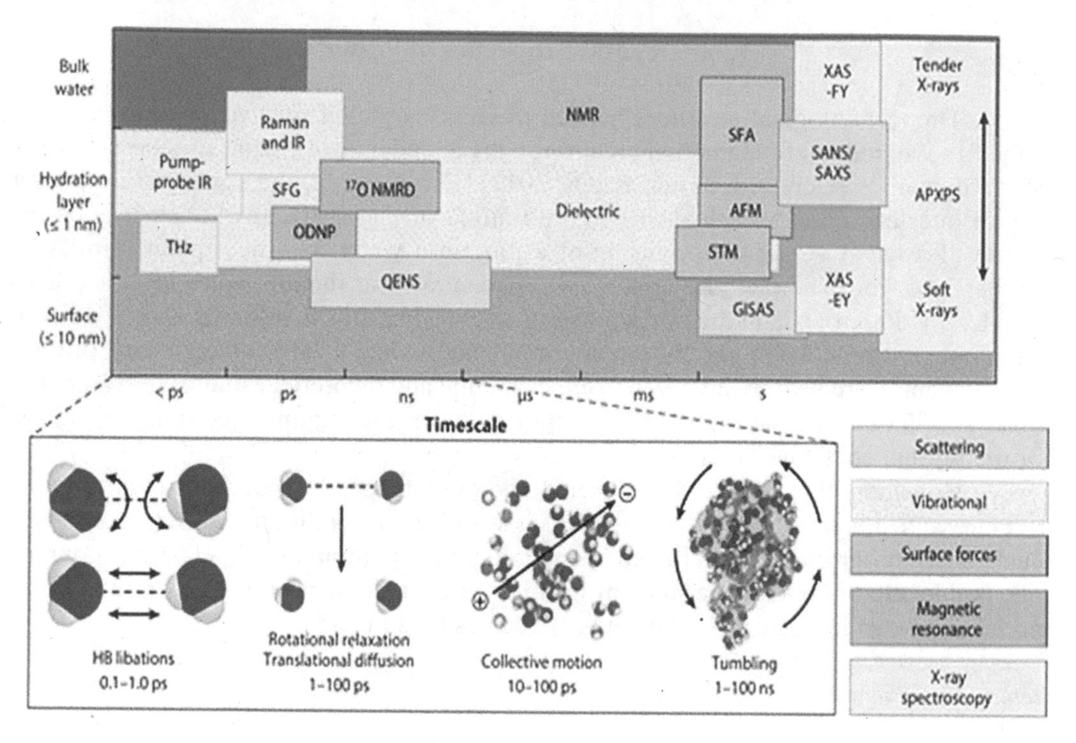

FIGURE 4-5 Illustration showing the structure and dynamics of water when probed by various spectroscopies at multiple length and time scales. NOTE: AFM = atomic force microscope; APXPS = ambient-pressure X-ray photoelectron spectroscopy; GISAS = grazing-incidence small-angle scattering; HB = hydrogen bond; IR = infrared; SANS/SAXS = small-angle neutron scattering/small-angle X-ray scattering; SFA = smooth factor analysis; SFG = sum frequency generation; STM = scanning tunnel microscopy; ODNP = Overhauser dynamic nuclear polarization; O NMRD = oxygen nuclear magnetic resonance dispersion; QENS = quasielastic neutron scattering; XAS-FY = X-ray absorption spectroscopy—fluorescence yield. SOURCE: Monroe et al. (2020).

those upgrades, new methods are providing an unprecedented view into the structure of water in a wide range of scenarios, including as ultrathin films or in samples undergoing unusual phase transitions (Byrne et al., 2021).

Over the past 30 years, the issue of water polymorphism in pure water and aqueous solutions has gradually been fleshed out (Bachler et al., 2019; Debenedetti, 2003; Gallo et al., 2016; Handle et al., 2017). The phase diagram of water has been expanded to recognize the existence of several forms of amorphous ice, low-density amorphous (LDA) ice, and high-density amorphous (HDA) ice. LDA ice can be formed by rapid vapor deposition of water onto ultracold substrates (below 120 K); it has a density of 0.94 g/cm^3, but its viscosity is higher than that of water. In contrast, HDA ice has a density of 1.17 g/cm^3 and can be formed by pressurizing a particular phase of ice (ice I_h) at temperatures in the vicinity of 140 K. A third form of amorphous ice, referred to as very-high-density amorphous ice, has a density of 1.26 g/cm^3 and can be formed by compressing HDA to pressures above 1 GPa. The existence and origin of these newly found forms of ice have been understood largely based on sophisticated molecular simulations, many of which have been carried out by chemical engineers (Palmer et al., 2018). The discovery of these ices has helped explain several astrophysical observations, as they arise, for example, in comets or icy moons and in some of the coldest regions of the Earth's atmosphere. Amorphous ice is also relevant for engineering applications and could be used for the chemical and structural stabilization of biological molecules at low temperatures (as needed, for example, in cryo-transmission electron microscopy) over extended periods. The discovery of LDA ice formed by vapor deposition has inspired the creation by chemists and chemical engineers of other classes of emerging engineering materials, such as ultrastable vapor-deposited glasses, which offer unusual mechanical characteristics and important advantages for applications in electronics, such as organic light-emitting diodes and polymorphic metallic films that respond to light.

Structure of Water at Interfaces

Considerable progress has also been made in understanding the structure of water in inhomogeneous environments—for example, water at interfaces and under extreme confinement. Consensus has gradually emerged, primarily from various types of infrared measurements and molecular models, that an individual —OH group from the molecule is freed from the hydrogen-bonding network and points toward the vapor phase. Water molecules near the interfacial "layers" are believed to be slightly more disorganized than bulk, fully hydrogen-bonded water, and considerable debate continues around the acid/base characteristics of interfacial water. This last issue is of particular significance to electrochemistry and, more generally, to chemical reactions at interfaces, and presents exciting opportunities for chemical engineering. It is also important for the interpretation and design of adsorption processes at aqueous interfaces. Clathrate formation, for example, with both methane and CO_2, is believed to be nucleated at the water interface (Li et al., 2020a; Liang et al., 2019). A better understanding of such interfaces would therefore enable the discovery and development of clathrate formation inhibitors or stabilizers.

Beyond pure water, the structure of aqueous solutions at interfaces presents important challenges (Bjorneholm et al., 2016). The issue of whether ions, for example, are depleted or concentrated at the air–water interface is still being debated. Current thinking is that large, more polarizable halide anions are depleted from the air–water interface to a lesser extent relative to smaller and less polarizable ions (Ghosal, 2005; Jungwirth and Tobias, 2006; Tong et al., 2018). Iodide, however, is believed to be enriched at that interface. These trends are strongly influenced by the presence of molecular solutes, and depending on the polarity and size of such solutes, it is difficult to predict the structure of an aqueous solution at an interface, posing considerable challenges for the design of adsorption operations, for example.

Theory and simulation have the potential to provide the tools needed to describe such systems, but molecular models of water and ions are currently unable to describe the solubility of ions in water or to predict the segregation or enrichment effects observed in experiments (Mester and Panagiotopoulos, 2015). Such models have been developed for bulk pure water, and the effects of ions have been included as an afterthought, thereby restricting the models' ability to describe solutions or mixtures and interfaces. Attempts to circumvent this problem by relying on quantum mechanical methods have been hindered by the computational demands of such calculations, which continue to exceed available resources. Recent approaches involving various combinations of advanced sampling concepts from statistical mechanics, quantum mechanical calculations of intermolecular interactions, and emerging concepts from machine learning offer considerable promise for reducing the description of water and aqueous solutions and their interfaces to a tractable problem. As chemical engineers strive to conceive and design chemical processes involving aqueous interfaces, it will be important for such predictive models to be developed and brought to bear on the design and optimization of modern water-based technologies.

The challenges and gaps in understanding that arise at the air–water interface are exacerbated at solid interfaces with metals; oxides; or organic matter, including biomolecules. Water–metal interfaces are central to catalysis and electrochemistry, and probes capable of directly reporting the structure of interfacial water are severely limited. Importantly, existing measurements require interpretation based on molecular models, and such models continue to suffer from the same shortcomings highlighted in the context of air–water interfaces. Much work remains to be done not only in understanding the structure of water at such interfaces but also in manipulating that structure to control reactivity and transport (Ruiz-Lopez et al., 2020).

Considerable effort has been focused on understanding both hydrophobicity and hydrophobic surfaces. Experimental evidence from SFG experiments has now shown that, as with the air–water interface, water molecules have an individual "dangling" —OH group at a hydrophobic surface (Bjorneholm et al., 2016). Results of experiments on individual surfaces also indicate that, beyond the first monolayer of water molecules at such an interface, water rapidly adopts a bulk, isotropic structure. Those findings are in conflict with measurements of the forces between two hydrophobic surfaces, which indicate that an attractive force can already be felt at separations on the order of 20 nm (Kekicheff et al., 2018), suggesting that some level of structural influence is felt at relatively large distances. Simulations of water structure and forces between hydrophobic surfaces have added some clarity to this ongoing debate, with some calculations finding evidence of long-range attractions and others finding only short-range interactions, depending on the type of surface and water model adopted in the calculations (Altabet and Debenedetti, 2017; Eun and Berkowitz, 2011; Hua et al., 2009).

Even less is known about how these interactions might change in the presence of ionic species, including ions or charged molecules. Recently, however, chemists and chemical engineers have made important advances by relying on synthesis of carefully conceived organic molecules and surfaces that present hydrophobic groups and charges in precise arrangements, coupled with atomic- or surface-force apparatus measurements (Ma et al., 2015). Those experiments have revealed that chemical heterogeneity at the nanometer level plays a significant role in the structure of water and the resulting hydrophobic interactions. A key advance is the realization that hydrophobicity is not an inherent characteristic of nonpolar domains but is instead controlled by functional groups separated by nanometer scales. Cleverly designed simulations have helped explain and exploit some of these observations (Kim et al., 2015). From an engineering perspective, these findings suggest that judicious positioning of charges next to hydrophobic domains represents a strategy for tuning hydrophobic forces, with a wide range of implications, from the design of water-repellent coatings, antifreeze proteins, and self-assembling systems to the stabilization of colloidal suspensions.

Dynamics of Water

For chemical engineering, the relevance and excitement of many of these new discoveries surrounding the structure of water are matched only by new observations pertaining to the dynamics of water under confinement or near interfaces. Early experiments with membranes consisting of carbon nanotube pores (less than 2 nm in diameter) revealed that flow rates through such pores are roughly three orders of magnitude faster than anticipated by continuum flow calculations (Holt et al., 2006). Fast transport of gases in nanopores had been anticipated on the basis of simulations by chemical engineers (Skoulidas et al., 2002), and the experimental observations were explained by invoking the formation of a frictionless interface at the nanotube wall. It has also been reported that such a phenomenon occurs with both organic solvents (e.g., alkanes) and hydrogen-bonding solvents, and that in the case of water, the flow rate gradually drops over time, presumably as a result of some level of ordering in the pore. Subsequent experiments and simulations have shown that such ordering can in fact lead to even faster transport of water—as fast as 1 meter per second, and at small dimensions, in the range of a nanometer (Marbach et al., 2018; Striolo, 2006; Tunuguntla et al., 2017).

Chemical engineers have been particularly adept at drawing inspiration from biology to develop solutions for problems ranging from reengineering proteins to co-opting bacteria for water reuse. Research on technologies for water transport through protein pores, such as ion pumps or aquaporins, has been used in the design of the high-flux systems mentioned above or selective pores for ion transport. Fundamental theoretical and computational research has paved the way for the design of intriguing separation or energy-generation processes, including recent discoveries about the transport of aqueous ionic solutions through Janus nanopores—designer pores consisting of adjacent sections with different diameters and surface charges—that could potentially become important sources of energy from simple salinity gradients (Yang et al., 2018; Zhang et al., 2017; Zhu et al., 2018). More work is needed to scale up and optimize observations that have thus far been limited to laboratory-scale observations, but these and other "tailored-pore" approaches offer considerable promise.

Similarly, additional research on aqueous electrocatalysis, photocatalysis, and photoelectrocatalysis (Li et al., 2020b; McMichael et al., 2021; Ochedi et al., 2020)—for which the basic idea is to use electrons and/or photons to transform readily available substances, such as water and CO_2, into sustainable chemical fuels such as hydrogen and methanol—is likely in the near future to see application in commercial projects that could shape the industrial landscape for generations. In the face of rapidly changing climate patterns, fundamental research on topics related to ice nucleation, clathrate formation or destabilization, and the fate of aerosols can inform technologies aimed at mitigating the impacts of climate change. As water sources change and conservation is accelerated, monitoring will be of critical importance at a global scale. For example, "dating" efforts to gauge the age of aquifers using atom trap trace analysis of krypton and noble gases are only a few years old and are starting to yield intriguing data, but are as yet not widely known or utilized (Gerber et al., 2017).

FEEDING A GROWING POPULATION

The advent of farming-based agriculture was an essential enabler of human civilization. Innovative strides in agriculture in the 20th century, many of which were enabled by chemical engineers or researchers working in what would ultimately become the chemical engineering discipline, allowed substantial improvements in food yields per land area, the ability to efficiently prolong the life of foods, and the ability to deliver food to consumers over long distances. These innovations are exemplified by the development and scale-up of the Haber–Bosch process for ammonia production, which rivals global biological nitrogen fixation in magnitude (Galloway et al., 2008). The development of refrigeration, today a standard lesson in chemical engineering's thermodynamics courses, is another critical example of the technological march toward improved efficiency in food production.

Today, agricultural pursuits use an astounding 38 percent of all land on the planet outside of Antarctica (FAO, 2020), and the raising of livestock is responsible for 14.5 percent of global GHG emissions (FAO, 2021). Researchers estimate that besides decarbonizing the energy sector (as discussed in Chapter 3), better management of food production and agriculture represents the other primary way to reduce global GHG emissions (Mbow et al., 2019; see Figures 4-6 and 4-7). As a result of continued population growth and anthropogenic climate change, humankind will need to rethink and reinvent agricultural and food production practices toward more sustainable land and resource use (Tilman et al., 2011).

These global-scale challenges present significant opportunities for chemical engineers, especially in the application of systems-level approaches to evaluate and optimize products and processes. Examples of specific challenges are mentioned below, but are by no means exhaustive. Generally, opportunities for innovation can be viewed in terms of increasing efficiency in agricultural practices and developing "leapfrog" technologies that are more sustainable than current agricultural methods.

Improving Agricultural Efficiency

The last three decades have seen significant technological innovations in agriculture, resulting in more precise application of resources, such as fertilizer; the advent of modern biomolecular techniques for engineering improvements in plant- and animal-based products; and the use of automated machinery and, more recently, robotics to replace manual labor. These and other agricultural advances will inevitably continue.

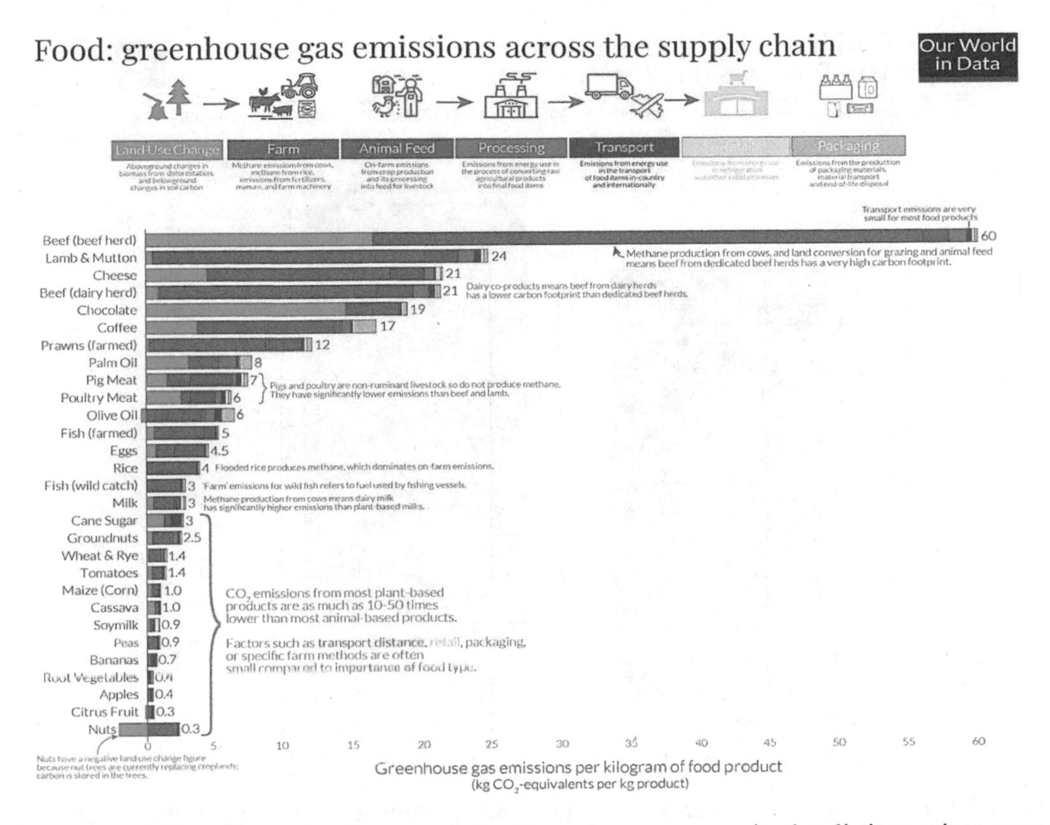

FIGURE 4-6 Chart illustrating the magnitude of greenhouse gases emitted at distinct points across the supply chain for several food products. NOTE: Greenhouse gas emissions are given as global average values based on data across 38,700 commercially viable farms in 119 countries. SOURCE: Our World in Data (2021b).

Since the emergence of metabolic engineering and synthetic biology, tools of modern molecular biology have become commonplace in chemical engineering. Tremendous opportunities exist to combine these tools with systems-level approaches to improve per-land-area crop yields; enhance the efficiency of water and nutrient use; and increase resistance to fungal infections, insect infestations, drought, and other adverse climate/weather and biological effects. While the public has generally viewed genetic modifications of plants with mistrust, techniques based on CRISPR (clustered regularly interspaced short palindromic repeats) technology may ultimately enable precise genomic editing without being considered genetic modification, representing a major boon for application in food crops. The use of genome-wide association studies, especially in the broad gene pools that exist in undomesticated crops, and translation to domesticated food crops present an opportunity to apply computational approaches common to chemical engineering curricula in concert with analysis-driven research.

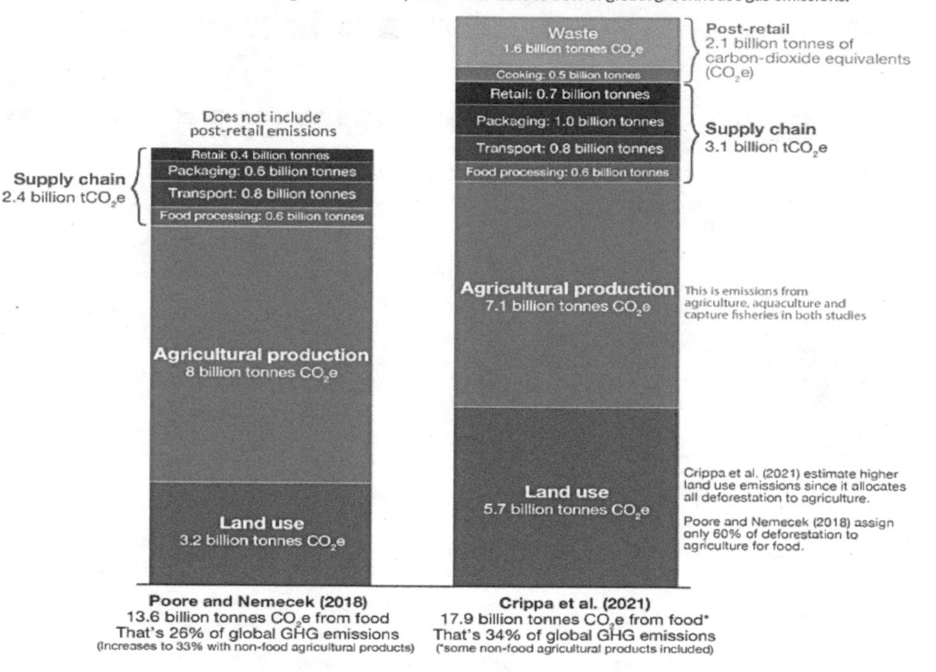

FIGURE 4-7 A side-by-side comparison of two leading estimates of global greenhouse gas emissions from the global food system. NOTE: Crippa and colleagues (2021) include emissions from a number of nonfood agricultural products, including wool, leather, rubber, textiles, and some biofuels. Poore and Nemecek (2018) do not include nonfood products in their estimate of $13.6 billion metric tons of CO_2e. This may explain some of the difference. SOURCE: Our World in Data (2021b).

Protein-rich products—including dairy, eggs, fish, meat, and poultry—account for an increasing amount of the human diet, with demand growing rapidly in low- and middle-income countries. In higher-income countries, protein-rich foods are often available in excess and make up a substantial fraction of food-related consumer waste (Spiker et al., 2017). Reducing net consumption of animal-based protein is likely to be an important contributor to improving sustainability and curbing methane emissions. Achieving this reduction will require, among many other efforts, both the development of substitutes for animal-based proteins and, in cases in which animals are still harvested for meat products, understanding of methods for mitigating related emissions. Exciting avenues of research in the latter area include understanding of the microbiome interaction of agricultural animals with feed. Examples include recent research wherein dairy cows were fed small amounts of seaweed, which reduced overall per-animal methane emissions (Roque et al., 2021).

Beyond the development of improved agricultural practices for plants and animals, improved water usage is a core element of the role of chemical engineering in the future of agriculture. Anthropogenic climate change, among other forces, is already shifting the global and regional balances of water resources. Systems-level approaches, including life-cycle assessment, will play a critical role in meeting this local, regional, and global challenge. Technological approaches to the use of water in agriculture are a prime example for which modular processes will be vital, as deployment is likely to require many units operating in a localized and autonomous manner rather than large, centralized facilities. Many of the same technical challenges apply to reducing the energy use associated with agriculture.

The judicious application and effective recycling of reactive nutrients, traditionally exemplified by ammonia and urea as fertilizer but also including phosphorous and potassium, is another area presenting multiple chemical engineering challenges. The dead zone in the Gulf of Mexico (and in bodies of water around the globe) due to fertilizer runoff is a legacy issue presenting an opportunity for chemical engineers to contribute to environmental protection through targeted applications of nutrients to enhance nitrogen management. Chemical engineers can also pioneer more efficient production of ammonia (beyond the Haber–Bosch process), which has reemerged as a grand challenge and has been the focus of intense research in the past decade (e.g., Boerner, 2019; Garnier, 2014; Smith et al., 2020). Any potential technology for this process needs to ultimately produce ammonia in very large quantities to be globally relevant. Addressing challenges in catalysis, reaction engineering, and process development overall will be critical to this end. Some nutrients beyond ammonia/reactive nitrogen are finite relative to the expected growth in agriculture, and their availability today still relies on mining, as was the case with ammonia in the late 19th century. These finite nutrients include phosphorous and potassium, which, along with ammonia, are commonly applied to feed and food crops. Meeting challenges in their economical and efficient recovery along multiple supply chains will be critical as agricultural demands increase.

Modern agriculture and food processing produce massive amounts of industrial and postconsumer waste. For example, nut processing results in the production of remarkable amounts of waste lignocellulose. These waste streams offer a direct and immediate pathway for the needed development of biofuels, biochemicals, and biomaterials—areas in which chemical engineers already play a leadership role. Going beyond anaerobic digestion or simple combustion of organic waste to produce products of higher value than methane (often used for combustion) or enable direct production of power is an area ripe for immediate contributions from chemical engineers.

The above discussion includes but a sampling of ways to improve modern agricultural practices in which chemical engineers can play pivotal roles. There are undoubtedly many others. The problem of food production is inherently global in nature, and systems-level thinking at multiple scales—a hallmark of the chemical engineering profession—will be critical to enable positive change toward a more sustainable agricultural system.

Reinventing Agriculture

Today's agricultural practices are inherently land, water, and nutrient intensive. Almost universally, agriculture today is practiced in two dimensions and with batch processes. The use of land means that agricultural productivity scales with surface area, whereas most chemical engineering manufacturing processes, as described further in Chapter 6, scale volumetrically. Moreover, crops and animal products are typically not harvested continuously but in a manner that depends on the time of year, with many external factors (e.g., weather, climate, disease, fire, drought) affecting the process conversion efficiency, yield, and rate. For many generations, these observations appeared to be intrinsic to the production of food and therefore barely worth stating. However, now is the time to consider thoughtfully how immutable these notions are. Examples of this shift in thinking can already be seen with the surge of companies focused on the production of non–animal-based meat, dairy, egg, and oil substitutes. In various ways, these companies are exploring the concept of transitioning agriculture from an areal farming practice to a volumetric, continuous industrial manufacturing process. The outcomes of this approach will likely have effects across the entire food and agriculture supply chain.

One example—among many—of plant-based products used today is lipids in oil-producing crops. The production of palm oil, for example, is quite energy intensive, leading to substantial land-use change effects, especially in tropical regions. The application of synthetic biology, metabolic engineering, bioprocess development, and analysis-driven research to produce oil products that can displace plant-based oils in a cost-effective and sustainable manner will likely have positive effects on reforestation and preservation of the natural environment (Parsons et al., 2020). Producing food in a continuous, industrial manufacturing setting could also vastly increase the number of feedstocks available for food production beyond CO_2, NH_3, and sunlight. As discussed in the section on feedstock flexibility in Chapter 6, the use of waste-based feedstocks for valuable products, especially for food production enabled by biological and catalytic transformations pioneered by the chemical engineering community, offers a clear path toward a more circular carbon and nitrogen economy that is more sustainable than today's agricultural practices, and chemical engineers will play a critical role in this much-needed transition.

Food Engineering and Processing and Storage

Supplying the world's population with food that is nutritious, affordable, and sustainable is a global challenge. Societal-scale pressures associated with climate change and population growth and distribution will demand substantial changes in the world's food sources in the coming decades. Despite the large number of chemical engineers working in the food industry, food science has historically not been an area of strong emphasis in U.S. chemical engineering research. Multiple opportunities exist for chemical engineers to play a critical role in the future of food engineering and processing. These topics share several features with more traditional chemical manufacturing, including the need to operate at very large scales at low cost and with stringent safety, sustainability, and quality standards. Any proposed solution that fails to meet any of these criteria will be ineffective in leading to real change in the world's food system. Of critical importance is to evaluate the cost and sustainability of new concepts using systems-level life-cycle assessment while taking account of the role of local environmental conditions and traditions.

Food Engineering

Chemical engineers have led in adapting the tools of molecular and systems biology for applications in diverse areas; however, the application of these methods to food engineering is in its early stages. Although genetic modifications can be controversial, molecular biology has the potential to enable enormous advances in crop science. An example is golden rice, which could eliminate vitamin A deficiencies in large populations. The pursuit of research and development (R&D) in chemical engineering with plants rather than single-celled organisms, such as yeast, poses both technical and nontechnical challenges. The cell walls in plants make extracting molecules from plant cells more challenging relative to animal or microbial cells, and the longer life cycles of plants compared with single-celled organisms or model organisms such as *C. elegans* create logistical challenges for research. Nevertheless, the potential for chemical engineers to combine their skills in systems thinking with a rich set of biomolecular tools to improve food-related plants is great, particularly if they make concerted efforts to work in concert with related fields of biological and agricultural science.

In addition to improvements in food at the molecular level, the coming decades are likely to see significant changes in the macroscopic sources of food, particularly protein and lipid sources. Meat-free protein alternatives are already becoming mainstream in the United States, but considerable advances will be required if these kinds of foods are to be used on a global scale. Scaling up the volume of products that involve processing solids and liquids at low cost and with high reliability has been a core focus of chemical engineering since the field's inception, and chemical engineers will have a major impact if they can adapt their existing expertise in chemical engineering and biotechnology to the challenge of producing new foods at scale. Valuable opportunities exist for collaboration between chemical engineers and researchers who are pioneering initial demonstrations of "lab-grown" foods on small scales and for diversification in such areas as self-assembly into food-relevant applications.

Although the impact of new food sources will accelerate, the scale of primary crops and agriculture will remain vast. Chemical engineers can play a role in continuing developments in precision agriculture, defined as the precise delivery of fertilizer, pesticides, and herbicides to minimize environmental impact. In the past, only a limited connection has existed in this area between the ambitions and needs of industry and farmers and the scope of academic research in the U.S. chemical engineering community.

Food Processing and Storage

While producing food in better ways is critical to meeting global food needs, reducing food waste could also have an enormous impact. The challenges associated with food waste differ around the world: broadly, in higher-income countries, food waste is dominated by high-quality food that is not completely used by consumers, whereas in low- and middle-income countries, food waste more often results from a shortfall in the delivery of food from its original source to consumers. Consequentially, although significant advances in packaging and food treatment to prolong shelf life have the potential to make a major impact in both low- and middle-income countries and higher-income countries, there can be no single solution to the overall challenge of reducing food waste. Even so, chemical engineers have the potential to enable important advances in this globally important area. To ensure impact in low- and middle-income countries, researchers need to focus on ultra–low-cost solutions and appropriate technologies that account for local context and cultural traditions. In higher-income countries, the sustainability of food production and packaging is a major concern that can be addressed through the use of non-food containers to hold food during shipping or storage and edible coatings that can be applied to food to prolong shelf life. Chemical engineers are well suited to the challenging task of applying systems-level life-cycle assessment to the development of solutions in these areas.

UNDERSTANDING AND IMPROVING AIR QUALITY

Even when air pollutants are at very low levels—and despite the progress that has been made in this area since 1970—air pollution continues to harm the environment, affect the climate, and harm human health (Figure 4-8; EPA, 2021a). On a global scale, the World Health Organization (WHO) estimates that about 4.2 million people died prematurely worldwide in 2016 because of outdoor air pollution in cities and rural areas, and 91 percent of those deaths occurred in low- and middle-income countries (WHO, 2018). The causes of death were heart disease, stroke, chronic obstructive pulmonary disease, and lung cancer. Additionally, the WHO estimates that of the 3 billion people who still use solid fuels (such as wood, crop wastes, charcoal, coal, and dung) and kerosene for cooking, about 3.8 million died prematurely in 2016 from diseases attributable to indoor air pollution caused by these types of cooking practices (WHO, 2016). Globally, air pollution is the fourth-highest contributing risk factor for death, behind high blood pressure, smoking, and high blood sugar (Figure 4-8). Indoor air quality[3] is far less well studied and not as uniformly regulated as outdoor, yet people in the United States and other high-income countries spend more than 85 percent of their time indoors (EPA, 2018). Efforts to address indoor air pollution are complicated by the highly heterogeneous building and ventilation types that exist across even relatively small regions. Building codes related to ventilation have focused thus far on energy efficiency and less on air quality, and in many cases, improvements related to energy efficiency have led to less air exchange and thus poorer indoor air quality (EPA, 2020).

[3] The National Academies' Committee on Emerging Science on Indoor Chemistry is currently examining the links among chemical exposure, air quality, and human health in indoor environments.

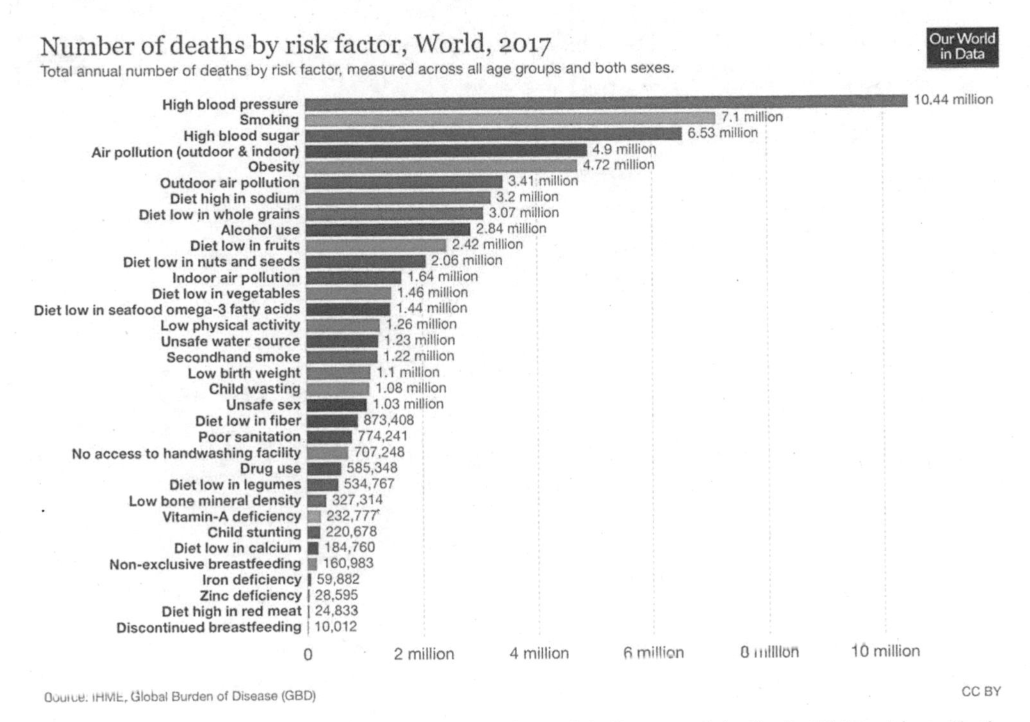

FIGURE 4-8 Number of deaths caused by various risk factors globally in 2017. Air pollution is the fourth-largest risk factor, contributing to 4.9 million deaths annually. SOURCE: Our World in Data (2019).

The six most common outdoor air pollutants are particulate matter (PM), ground-level ozone (O_3), sulfur dioxide (SO_2), nitrogen dioxide (NO_2), airborne lead, and carbon monoxide (CO). While many of these pollutants are still present in the air, their concentrations in the United States have declined significantly since 1970 (Figure 4-9). In addition, 187 toxic air pollutants that can cause cancer and birth defects—including such chemicals as acetaldehyde, asbestos, cadmium and chromium compounds, benzene, dioxin, epichlorohydrin, and methanol—are listed in the U.S. Clean Air Act (42 U.S.C. § 85 [1955]); some of these chemical species form atmospheric aerosols, while the rest remain in the gaseous phase. Another air pollution challenge is to protect stratospheric ozone, which was achieved to a great extent by banning chlorofluorocarbons (CFCs), hydrochlorofluorocarbons (HCFCs), and other harmful compounds (UNEP, 2020).

Still another challenge is posed by acid rain, caused by the reaction of atmospheric SO_2 and NO_x with water, oxygen, and other chemicals to form nitric and sulfuric acids. Acid rain can turn lakes and streams acidic, harming fish and wildlife, and flowing through soil can leach aluminum and remove minerals and nutrients from the soil. The impact of acid rain in the United States and Europe has largely been reduced by the regulation of sulfur and nitrogen emissions. Other air pollutants include CO_2 and other GHGs, which are discussed further in the context of energy (Chapter 3) and the circular economy (Chapter 6).

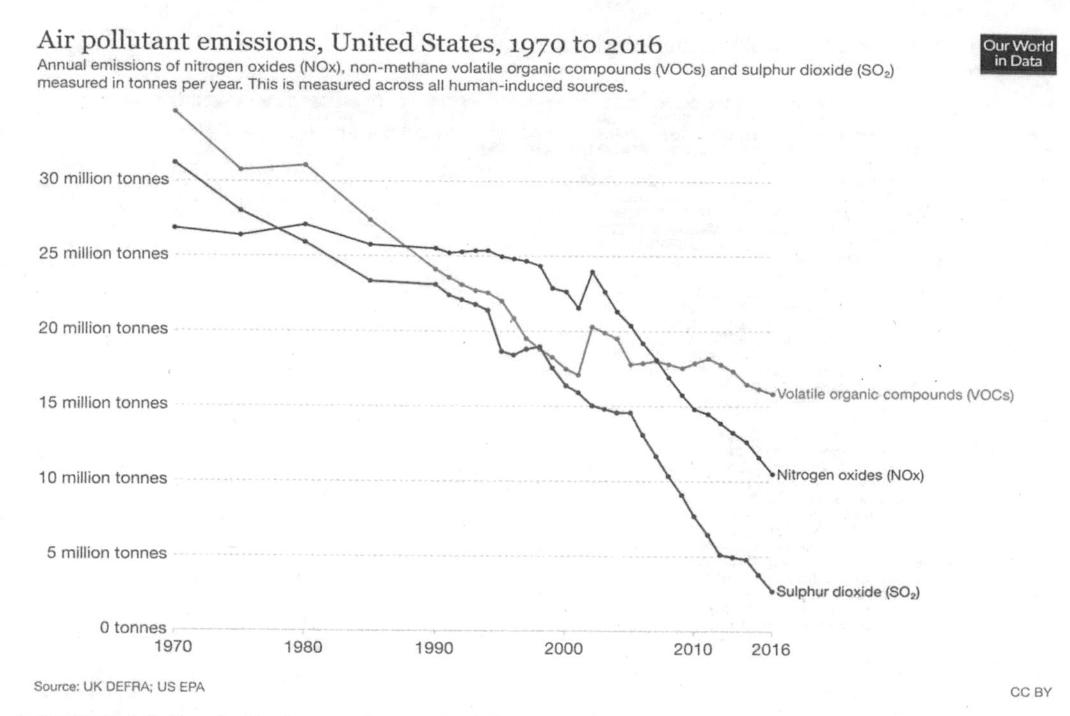

FIGURE 4-9 Graph displaying the gradual decrease in the emissions of three major air pollutants in the United States over time. SOURCE: Our World in Data (2021a).

Atmospheric aerosols are suspensions of microscopic and inhalable solid or liquid (typically, aqueous) particles in the air. About 90 percent of aerosols, by mass, occur naturally (e.g., from volcanic eruptions, ocean phytoplankton emissions, sea and salt sprays, forest and grassland fires, or dust aerosols from wind erosion of arid land; Voiland and Simmon, 2010). Anthropogenic aerosols include those from fossil fuels (e.g., those composed of sulfates) and agricultural waste burning, road dust, cooling towers, industrial processes, and deforestation fires (which are composed primarily of carbon). Aerosol concentrations can be as high as 10^6–10^7 particles per cm^3. The typical size of aerosol particles is 1 nm to 10 μm; they are classified as PM_{10} when their diameter is less than 10 μm and $PM_{2.5}$ when it is less than 2.5 μm. Aerosols greater than 10 μm in diameter settle to the ground in a matter of hours because of gravity, but those less than 1 μm in diameter remain in the atmosphere for weeks and are ultimately removed by impacting the surface of the Earth (dry deposition) or via precipitation (wet deposition). Aerosols are characterized as either primary (i.e., emitted as particles at the source) or secondary (i.e., formed from gaseous precursors; Haywood, 2016). This class of pollutants has garnered considerable attention with respect to indoor air quality during the COVID-19 pandemic because of the ability of exhaled virus-laden aerosols to cause infection.

Aerosols have both direct and indirect effects on climate. Models estimate that direct cooling effects have counteracted about half of the warming caused by GHGs since the 1880s (Voiland and Simmon, 2010). Indirect effects include nucleating the formation of clouds and controlling their numbers and lifetimes. For humans, inhalation of PM and its subsequent deposition in the lungs and even entrainment into the bloodstream are the main causes of aerosol-related health problems. Specifically, $PM_{2.5}$ can cause cardiovascular, mental, dermatological, and reproductive health problems (McNeill, 2020).

A major challenge for improving air quality is bridging the molecular scale of a chemical reaction and the massive scale of an atmospheric model. Some of the foundations of chemical engineering education (e.g., transport phenomena, thermodynamics, and chemical reaction engineering), along with a systems approach focused on complexity and scale, can be applied directly to solving environmental engineering problems. In many respects, the atmosphere (more generally, the environment) functions as a large chemical reactor in which pollutants are transported, mixed, and transformed while being distributed across the atmosphere, land, and water. More specifically, the extremely wide ranges of the temporal and spatial scales of the atmosphere are well suited to a chemical engineering modeling approach. Spatial scales span about 15 orders of magnitude, from the nanometer range of molecular dimensions to thousands of kilometers, and temporal scales span about 12 orders of magnitude, from the millisecond range of fast chemical reactions to the thousand-year scale of waste disintegration.

The chemical engineering profession has made key contributions to improving air quality, including the development of catalytic converters for vehicles (Box 4-1), cleaner-burning fuels, flue gas desulfurization and selective catalytic reduction systems for NO_x conversion to nitrogen gas and water, wet flue gas scrubbing methods, and better coal gasification technologies (Haywood, 2016). Furthermore, most chemical industries are required to control their aerosol emissions, and the principal equipment for that control includes cyclonic separators, fabric filter collectors (baghouses), electrostatic filters and precipitators, and wet scrubbers, all of which were developed with the help of chemical engineers.

Moving forward, chemical engineers have an opportunity to minimize and eliminate the formation of air pollutants at the source. When "benign by design" is not possible, chemical engineers can play a role in minimizing or eliminating emissions of these pollutants into the environment or even apply treatments to make the emissions safe for the environment. Ultimately, it may even be possible to treat the environment to remove air pollutants (AIChE, 2017; Sánchez, 2019). Specific opportunities for chemical engineering include the following:

- reduction of emissions from smokestack and exhaust tailpipes through improvements in engines and industrial plants (e.g., engines with better computer control and fuel economy, and multivalve engines), advanced monitoring and diagnostics, enhanced maintenance, development of catalytic pathways to destroy pollutants, use of cleaner fuels (e.g., low-sulfur and renewable fuels) and filters (e.g., diesel filters), and development of electric and fuel cell vehicles;
- better management of waste, such as the use of environmentally friendly solvents, and equipment for capture of methane gas emitted from waste sites and its use as biogas (e.g., Cao et al., 2018b);

BOX 4-1
Control of Nitrogen Oxide Emissions from Diesel Engines

Energy-efficient diesel-powered engines have enabled the growth and harvesting of the food supply, the mobility of people and goods over large distances, and the construction of modern infrastructure. The higher air-to-fuel ratio in diesel engines leads to lower greenhouse gas emissions relative to gasoline-powered engines, but removing NO_x and soot from diesel exhaust poses formidable challenges. Modern exhaust treatment systems integrate several catalysts and filters to trap carbon particulates, oxidize CO and organics, and reduce NO_x. For gasoline engines, three-way precious metal catalysts have succeeded spectacularly in mitigating emissions, representing one of the most meaningful contributions of catalysis and chemical engineering to human health. But these catalysts are ineffective in removing NO_x from air-rich diesel effluent. Particulate and NO_x emission limits required the urgent development of alternative after-treatment strategies for diesel engines. A timely combination of advances in materials design, mechanistic insights, and systems control and optimization led to the successful deployment of Cu-exchanged chabazite (Cu-CHA) zeolites as catalysts for urea-based selective catalytic reduction (SCR) of NO_x.

These breakthroughs required effective, robust, and regenerable materials expected to function over broad ranges of temperature and effluent flow rates and compositions, including brief excursions to about 900 °C, and devices able to meet emission targets for 100,000–400,000 miles under harsh hydrothermal environments containing strong catalyst poisons (S, P) derived from fuels and lubricants. The relevant reaction networks are complex (with more than 20 parallel and sequential steps). They include urea decomposition to NH_3, which then selectively reduces NO_x to N_2 and H_2O. This reaction must occur without parallel reactions of NH_3- with O_2, which is about a thousand times more concentrated than NO_x. Research led to the development of catalysts with isolated Cu cations grafted onto zeolite exchange sites. The microporous zeolite scaffold confines and protects Cu active sites from contact with fuel components that form carbonaceous residues, while at the same time providing acid sites for additional synergistic catalytic functions. These developments were enabled by engineering concepts in chemical kinetics and spectroscopy, and were leveraged by using theoretical methods to show how the unique reactivity of the Cu centers reflects their solvation and mobilization by NH_3 to form the Cu dimers required for the O_2 activation steps in the catalytic redox cycles. CHA zeolites also proved to be remarkably robust in the harsh hydrothermal environments that degrade most other zeolites. These innovations in catalyst design led to the deployment of Cu-CHA catalysts for urea-SCR, representing the first large-scale implementation of an exhaust after-treatment strategy using catalysts containing earth-abundant elements.

In practice, these SCR catalysts are deployed in compact devices that seamlessly integrate them with catalytic particulate filters, the CO/hydrocarbon oxidation catalysts, and other components to form a diesel after-treatment system (see Figure 4-1-1). Robust process models and control schemes are essential for the system to function, and they require accurate rate and selectivity data, together with thermodynamic, hydrodynamic, and kinetic descriptions. Today, these modular chemical plants are deployed in millions of diesel-powered vehicles. The vehicles' efficient engines have helped decrease CO_2 emissions, and their exhaust streams no longer contaminate the environment with NO_x and particulates. This achievement represents the result of a combination of novel materials and core chemical engineering concepts, made to work through the systems-based approach that characterizes chemical engineering as a discipline.

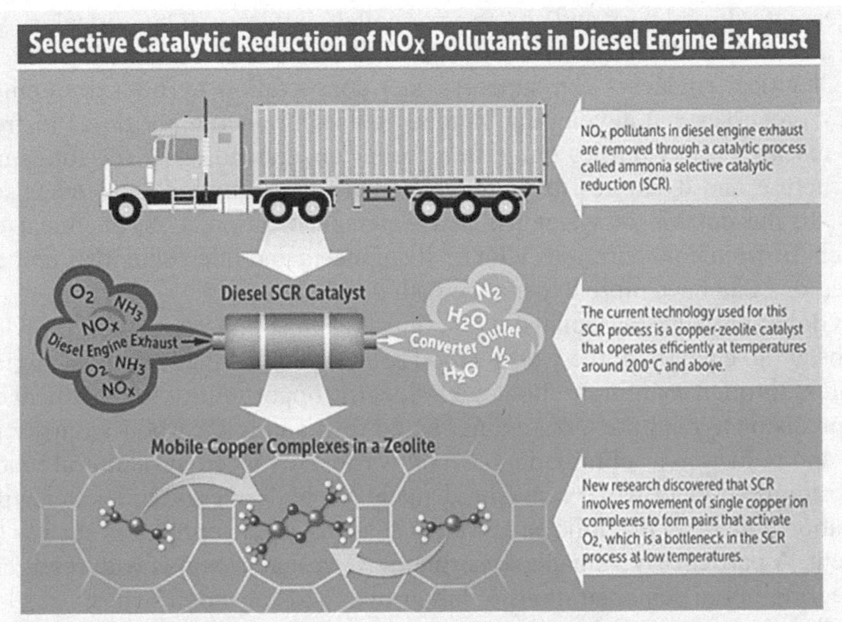

FIGURE 4-1-1 Emissions of nitrogen oxides (NO_x) from diesel vehicles are converted to environmentally benign products using urea-selective catalytic reduction (SCR) technology based on copper-exchanged zeolite catalysts that have high reactivity at low temperatures and are durable over long lifetimes. SOURCE: Purdue University/Mo Lifton.

- use of clean energy solutions for power generation, cooking, heating, and lighting;
- use of strict emissions control in waste incineration sites; and
- capture of CO_2 emissions.

In addition to these mitigating actions, chemical engineers can contribute to the fundamental understanding of aerosol formation, aerosol dynamics in the atmosphere, and their chemical characterization (e.g., Seinfeld, 1991; Seinfeld and Pandis, 2016). Sensor technologies that would allow better monitoring, chemical characterization, and knowledge of the spatial distributions of aerosols would also help address air pollution. Finally, the use of data science and multiscale models to illustrate atmospheric chemistry and the incorporation of process modeling to bridge the gap between observations and theory are opportune areas for the engagement of chemical engineers.

CHALLENGES AND OPPORTUNITIES

Food, energy, and water are highly interconnected, and solutions in this complex system need to be both environmentally sustainable and economically viable. A continued increase in the global population will lead to increased resource demands, posing a broad set of challenges for chemical engineers to address in the coming decades. Across all these challenges, interdisciplinary and cross-sector collaborations will be critical, as will coordination across the federal agencies with responsibilities in these areas.

Chemical engineers can support water conservation by both designing higher-efficiency processes and developing methods for using alternative fluids to freshwater. Specific research opportunities range from better understanding of the fundamentals of water structure and dynamics to the development of membranes and other separation methods. In the domain of water use and water purification, U.S. chemical engineers would benefit from collaborations with civil engineers and other scientists and engineers in arid regions that have more experience with desalination.

Global pressures associated with climate change and population growth will require substantial changes in the world's food sources, a need that chemical engineers can help address through enabling technologies. Specific opportunities for chemical engineers include precision agriculture, non–animal-based food and low-carbon-intensity food production, and reduction or elimination of food waste. Advanced agricultural practices designed to improve yield while reducing demand for both energy and water will require collaboration with other disciplines, as well as systems-level approaches such as life-cycle assessment. A particularly valuable opportunity for collaboration is with researchers who are pioneering initial demonstrations of lab-grown foods on small scales.

Chemical engineers have an opportunity to help in improving air quality by advancing understanding of the nature and physics of aerosol particles and applying separation technologies, as well as the molecular- and systems-level understanding that will be necessary to address this global challenge. Atmospheric science is already an interdisciplinary field that includes chemistry, physics, meteorology, and climatology, making it a promising area in which chemical engineering can contribute through increased collaboration.

Recommendation 4-1: Federal research funding should be directed to both basic and applied research to advance fundamental understanding of the structure and dynamics of water and develop the advanced separation technologies necessary to remove and recover increasingly challenging contaminants.

Recommendation 4-2: To minimize the land, water, and nutrient demands of agriculture and food production, researchers in academic and government laboratories and industry practitioners should form interdisciplinary, cross-sector collaborations focused on the scale-up of innovations in metabolic engineering, bioprocess development, precision agriculture, and lab-grown foods, as well as the development of sustainable technologies for improved food preservation, storage, and packaging.

5
Engineering Targeted and Accessible Medicine

- Chemical engineers have been involved for at least a century in reactor design and separations and more recently in cell engineering, formulations, and other aspects of drug manufacturing, and they have the potential to make many more contributions to health and medicine at scales ranging from the molecular to manufacturing facilities.
- Since the first attempts to isolate small molecules from biological organisms and control and reengineer cell behavior, the development of biologically derived products has increased, with major advances resulting from recombinant DNA technology, the sequencing of genomes, the development of polymerase chain reaction, the discovery of induced pluripotent stem cells, and the discovery and implementation of gene editing.
- The development of disease treatments is a multidisciplinary enterprise, and chemical engineers can contribute to many aspects of medicine by applying systems biology to physiology, the discovery and development of molecules and materials, and process development and scale-up.
- Many health disparities are a result of systemic issues that will require larger social changes to address, but chemical engineers can develop engineering solutions that help address disparities requiring more focused efforts.

Spending for health care in the United States is enormous (Figure 5-1) and has grown over the years, from 5 percent of gross domestic product (GDP) in 1960 to nearly 18 percent in 2019. The largest fractions of this total are for hospital care (31 percent) and physician and clinical services (20 percent). The costs of prescription drugs have skyrocketed as well, despite a decrease in retail drug prices, and represent a growing share of health care costs (e.g., 3.8 percent in 2018, compared with 5.7 percent in 2019; CMS, 2020; Martin et al., 2021). On the other hand, revenue from the biotechnology sector, including drugs and agricultural and industrial products, is a significant contributor to the national economy, accounting for more than 2 percent of GDP in 2012 (Carlson, 2016).

Therapeutic drug molecules are either small molecules (molecular weight [MW] below 1000 Da) synthesized chemically, or biologic molecules of any MW produced by biological (cell-based) methods. Drug development today includes both therapeutic classes, although small molecules now account for roughly 70 percent of U.S. Food and Drug Administration (FDA) drug approvals and about 60 percent of U.S. market share (Makurvet, 2021). This is the case in part because, for small molecules, the cost of synthesis is lower and the delivery of therapeutic material is simpler. Biologics have the advantages

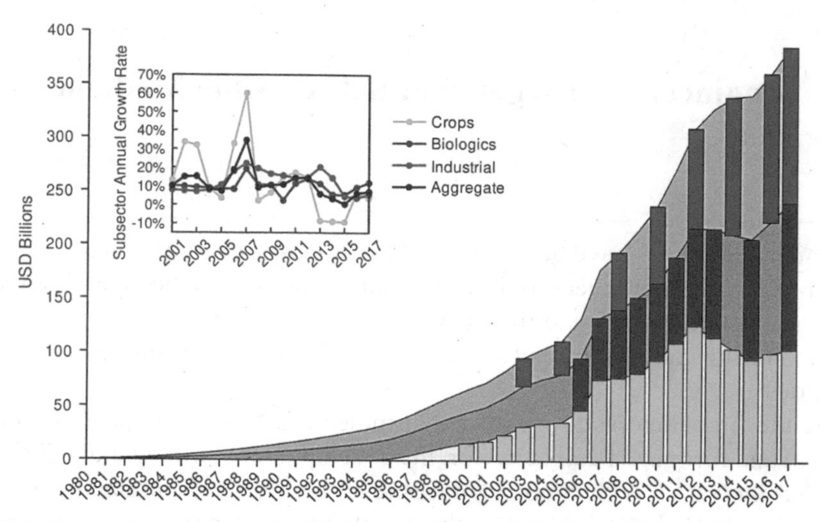

FIGURE 5-1 Estimated total annual U.S. biotechnology revenues, 1980–2017. Bars show data, while shaded areas are a numerical model based on annual growth rates. The inset shows annual growth rates for subsectors (crops, biologics, industrial, aggregate) between 2000 and 2017. SOURCE: Bioeconomy Capital (2018).

of offering highly selective targeting and, from an intellectual property perspective, being more difficult to copy. However, they can be much less stable; are unable to cross many biological barriers, including the blood–brain barrier; and are typically inactive when administered orally. Both small (chemical) and biologic molecules are likely to continue sharing the medical discovery and therapeutic markets, with the choice between the two depending on the application and disease at hand.

This chapter describes the role of chemical and biochemical engineering in human health applications, with a particular focus on biomolecular engineering. Following an overview of the role of biomolecular engineering in health and medicine, the chapter describes opportunities for chemical engineers to advance personalized medicine; improve therapeutics; understand the microbiome; design materials, devices, and delivery mechanisms; and develop hygiene technologies. Finally, the chapter examines how chemical engineers can contribute to addressing health disparities that result from societal inequity and reducing the costs of therapeutic treatments.

THE ROLE OF BIOMOLECULAR ENGINEERING IN HEALTH AND MEDICINE

Biomolecular engineers drive the development of computational and experimental techniques for identifying drug targets using multiscale modeling tools and data science approaches (see Chapter 8), and then employ organ-on-a-chip and/or tissue model development to replicate in vivo behavior. Small-molecule drug development and manufacturing still depend on chemical engineers who understand colloid science, particle dissolution, and multicomponent phase behavior to design formulations. For small-molecule discovery, automated, high-throughput screening of potential drugs will continue to integrate chemical synthesis, modeling of binding surfaces, and automation. Scale-up and manufacturing of both small molecules and biologics also rely on traditional process development skills of biomolecular engineers in addition to novel approaches, such as the integration of production and purification.

The origins of biomolecular engineering as applied to therapeutics can be traced back centuries to the use of such organisms as baker's yeast (*Saccharomyces cerevisiae*) to ferment grape juice and leaven bread. The first formal application of engineering principles to biological processes is often recognized as the production of penicillin in the 1930s and 1940s. The bacteriologist Sir Alexander Fleming is credited with discovering in 1928 the ability of an accidental mold growth to inhibit the growth of staphylococcus colonies, naming the active component of the mold *penicillin.* After significant work by a collaborative team of scientists and engineers to produce the drug at scale, this serendipitous discovery led to a major medical advance in the treatment of sepsis. Other similar compounds from molds led to the field of antibiotics, which of course remains critical in medicine to this day.

Despite the eventually profound impact of the discovery of penicillin, it was more than a decade later (in the late 1930s, as the world was poised to enter World War II) when Ernst Chain, Howard Florey, and Norman Heatley at Oxford identified the potential agent from *Penicillium notatum* (Florey, 1949). They worked to produce it economically by extracting it from the broth of the growing mold. Because of the intense bombings that began in the United Kingdom, Florey and Heatley visited the United States to try to address the needs for large-scale production; the collaborative transatlantic effort of biochemists, bacteriologists, physicians, and chemical engineers (and more) made it possible to produce penicillin biosynthetically, as chemical synthesis appeared to have little chance of success. Even biological routes were difficult—hundreds of scientists and engineers worked to create enough penicillin to treat all allied troops wounded in the D-Day invasion of Europe (~250,000 patients/month). Notably, Margaret Hutchinson Rousseau—the first woman to earn a PhD in chemical engineering and the first female member of the American Institute of Chemical Engineers (AIChE)—designed a deep-tank fermentation process for growing the surface-growing mold to enable penicillin production on a large scale. Among other significant advances were media development, strain improvement, and purification and recovery innovations that resulted in a 5,000-fold improvement in yields to 50 g/L (ACS and RSC, 1999; Gaynes, 2017). These efforts illustrate how classic chemical engineering concepts were applied to human health as early as the mid-20th century, ultimately contributing to the broader convergence of human health and engineering.

Since the first attempts to produce small molecules from biological organisms, the development of biologically derived products has increased. As described in Box 2-2 in Chapter 2, biochemical and biomolecular engineering and production of medicines have continued to grow, with major advances resulting from the development of recombinant DNA technology, the sequencing of genomes, the development of polymerase chain reaction (PCR), the discovery of induced pluripotent stem cells, and the discovery and implementation of CRISPR (clustered regularly interspaced short palindromic repeats) (see Figure 5-2). The first recombinant protein therapeutic approved by the FDA was insulin (5.8 kDa) in 1982, derived via production in *Escherichia coli* (*E. coli*) with subsequent purification and refolding (Carter, 2011; Leader et al., 2008). Since the 1980s, many more proteins produced in *E. coli* have been approved, and they are still produced in this simple bacterium. However, mammalian cells, particularly Chinese hamster ovary (CHO) cells, have been used to produce around 70 percent of all recombinant therapeutics since the late 2010s (Jayapal et al., 2007; Wells and Robinson, 2017).

Today, monoclonal antibodies (mAbs) represent the highest volume of sales and a major focus of drug development. These therapeutics can bind to a specific antigen and have been developed as highly specific treatments for such diseases as cancer, asthma, arthritis, Crohn's disease, migraines, and infectious diseases (Lu et al., 2020; Wells and Robinson, 2017). The first therapeutic mAb was muromonab-CD3, approved by the FDA in 1986, which was utilized in the treatment of transplant rejection until 2011. Humanizing antibodies, and later producing fully human mAbs as part of a library for protein engineering, rapidly accelerated the development of therapeutic antibodies (Jones et al., 1986; Tsurushita et al., 2005). Over the last 25 years, many antibodies have entered the market (Figure 5-3), with 61 mAbs being approved by the FDA for clinical use at the end of 2017 and an additional 18 new antibodies being approved from 2018 to 2019 (Lu et al., 2020).

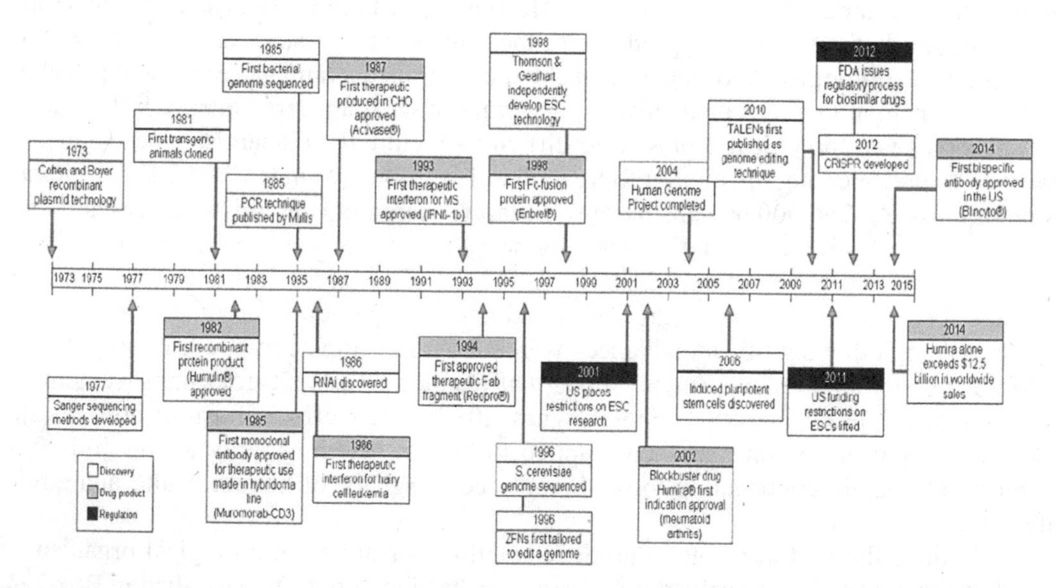

FIGURE 5-2 Biotechnology discoveries and notable drug approvals from 1973 to 2014. NOTE: CHO = Chinese hamster ovary; CRISPR = clustered regularly interspaced short palindromic repeats; ESC = embryonic stem cell; FDA = U.S. Food and Drug Administration; MS = multiple sclerosis; PCR = polymerase chain reaction; RNAi = RNA interference; TALENs = transcription activator-like effector nucleases; ZFNs = zinc-finger nucleases. SOURCE: Wells and Robinson (2017).

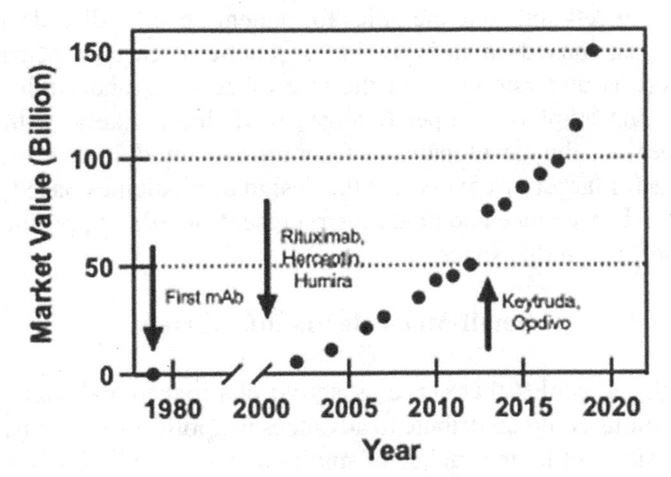

FIGURE 5-3 Trademarked antibody therapies contribute to a large fraction of the drug market, with major advances resulting from humanizing antibodies in the late 1990s. As indicated by the arrows, the several anticancer antibodies introduced in the early 2000s represent 5 of the top 10 bestselling antibody drugs in 2018. NOTE: mAb = monoclonal antibody. SOURCE: Lu et al. (2020).

Modern biomolecular engineering is very much at the intersection of chemical engineering and molecular biology, biochemistry, materials science, and medicine. The role of chemical engineers in addressing problems in health and medicine continues to lie in process and scale-up, as well as in discovery and development of molecules and materials with pharmaceutical and medical applications. For example, the development of the small-molecule therapeutics Remdesivir and Molnupiravir highlights the value of biomolecular engineering in combatting the COVID-19 pandemic. However, the repercussions of the pandemic also serve as a reminder of the importance of the availability and transport of raw materials, as well as the critical importance of scale-up of manufacturing where medicines are needed and the fact that even in higher-income countries, health disparities can result in needless deaths.

The next 20 years of chemical and biomolecular engineering will feature opportunities in personalized medicine; advances in the engineering of biologic molecules, including proteins, nucleic acids, and such entities as viruses and cells; growth at the interface between materials and devices and health; the use of tools from systems and synthetic biology to understand biological networks and their intersections with data science and machine learning; development of the next steps in manufacturing; and the use of engineering approaches to address health equity and access to health care.

PERSONALIZED MEDICINE

An important area in which chemical engineering can contribute to the future of medicine is personalized medicine, which denotes the tailoring of treatments for individual health, including cell and gene therapies for patient-specific disorders, as well as preventive care regimens based on an individual's genetic-disease propensity or biomarker presence. As more is understood about the role of genetics and environment in future disease, a continuing emphasis on personalized medicine is likely, with a clear role for chemical engineers in the development of appropriate models at scales ranging from atomic to systems for target discovery and the design of medicines based on those targets. The application of data science and modeling represents another opportunity for chemical engineers to contribute in this space.

Small-Molecule Manufacturing

For small molecules that serve as active pharmaceutical ingredients (APIs), a number of innovations could contribute to advances in applications during manufacturing. A detailed discussion of the importance of small molecules and related research opportunities for chemical engineers can be found in the recent report of the National Academies *Innovations in Pharmaceutical Manufacturing on the Horizon: Technical Challenges, Regulatory Issues, and Recommendations* (NASEM, 2021b). This section focuses on the application of additive manufacturing applied to personalized medicine. One such application is polypharmacy, which entails product formation through the use of 3-D printing. This method is of growing interest because in the United States, more than 40 percent of those over age 65 use five or more medicines, and 5 percent use more than ten (Kantor et al., 2015), creating complexity and confusion for patients. Polypharmacy is also common in Europe and Australia (Schöttker et al., 2017). Beyond the pharmaceutical implications of understanding how these medicines may interact in treating disease, the ability to deliver medicines appropriately in a safe and efficacious way and to ensure that the dosage and route of delivery are optimal is a need that chemical engineers and others can address through both systems biology and advances in manufacturing. In addition, the ability to combine multiple medicines into one delivery on a patient-demand basis would likely improve health care outcomes.

The use of 3-D manufacturing for APIs poses many technical challenges, including physics-based modeling of the fluid mechanics of drop formation and extrusion for liquid and powder solidification, as well as challenges involved in the extrusion process itself (NASEM, 2021b). Improving and simplifying the delivery of multiple drugs is a key opportunity for chemical engineers.

Additional opportunities for improvements in small-molecule manufacturing lie in predictive analytics and in the emerging field of translational medicine bioinformatics. Pharmaceutical companies use predictive analytics to develop algorithms that predict optimal conditions for robust production by leveraging large historical datasets. For low-dose or limited-duration drugs for which limited scale-up data are available, Bayesian statistics are used to estimate acceptable parameter ranges and product specifications. Translational medicine bioinformatics is an emerging field that draws on the fundamentals of chemical engineering and has the potential to transform personalized medicine. By analyzing raw data from DNA sequencing, patterns can be detected for specific mutations or genes to help predict which small-molecule drugs a patient might respond to. This field holds the potential to turn biomarkers into clinical diagnoses.

An additional need is the implementation of plug-and-play analytics, allowing rapid shifts from one production process to another (e.g., for small-batch production), and more uniformity in data formats for analytical devices across vendors. As supply chains continue to be stretched, the widespread adoption and implementation of digital twinning, or simulated process models, to predict and plan for supply chain issues that affect downstream challenges (e.g., the availability of specific vials or packaging in a given number of months) will also be beneficial for biotechnology and pharmaceutical production.

Cell-Based Therapies

Another aspect of personalized medicine is cell-based therapies—most important, CAR (chimeric antigen receptor) T cell immunotherapy. Cancer cells typically carry a ligand that blocks a specific receptor (PD-1) on the cell surface of T lymphocytes, rendering this protective part of the human immune system ineffective, and thereby enabling cancer cells to evade the body's natural defenses. CAR T cells are specially engineered to express CAR with a high affinity for tumor antigens. Immunotherapy—often termed immuno-oncology—has now become a mainstay of cancer treatment, with cell therapies representing the largest growth area and immunomodulation a close second. To some degree, the currently available therapies are influenced by what is known to be successful, with 22 cell and gene therapies being approved by the FDA in 2021 (FDA, 2021) and more than 460 targets in the current global pipeline (Xin Yu et al., 2019).

Therapeutic application of CAR T cells relies on retrieval of cells from a patient's blood, isolation of these cells followed by engineering in vitro, and then reintroduction to the patient. This approach has been quite successful, particularly for leukemias; it is highly individualized, allowing high specificity. In 2017, Kymriah® became the first FDA-approved immunotherapy, wherein a patient's blood is collected, and then individualized therapy is created by introducing a virus to the T cells to allow them to express a 4-1BB costimulatory domain to enhance cellular responses and express the CD-19 receptor. However, the cost of CAR T therapy is high for single-patient treatments, and the product's stability and success can be patient specific. At present, a large number of immunotherapies are in clinical trials, and the annual growth rate is expected to be 15 percent for cell therapies and nearly 30 percent for gene therapies (Mullin, 2021).

New directions for improving reliability and lowering costs are an ongoing focus in the fields of biomolecular engineering and immunology. In particular, there is interest in scaling down manufacturing practices that were developed for large-scale culture to reduce the need for highly skilled workers (compared with the present process of collection of blood by clinical staff and its shipment to manufacturing sites). Also, to improve the reliability and efficacy of treatment, key variables or biomarkers that lead to successfully engineered cells need to be identified, and simple analysis methods developed (Wang et al., 2021b; Whitehead et al., 2020). The use of induced pluripotent stem cells has the potential to reduce patient-specific cell collection, but also requires improvements in both modulating and measuring differentiation and manufacturing of these surface-dependent cells. Finally, T cell engineering requires viral vectors or particles that are manufactured with consistent quality and functionality.

Computational Tools and Modeling to Improve Personalized Medicine

Systems biology applied to physiology is an additional avenue for chemical engineers to contribute to personalized medicine. As early as the 1960s, Yeats and Urquhart (1962) noted that biological systems and their responses to physical stimuli can be understood using the engineering principles of process control. In fact, the impressive ability of biological systems to show resilience in the face of perturbations is possible with biological control systems that have components akin to their engineering counterparts, such as sensors, controllers, and actuators. Diseases can be seen as a malfunction or failure of a component in the control system, and quantitative approaches to understanding this behavior can benefit physicians' ability to diagnose and treat these failures (Figure 5-4).

Chemical engineers have several opportunities to contribute to systems engineering of biological systems, such as sensor design and analysis, identification of fault detection (using physiological, cellular, metabolic, or other data to identify changes in function), and process modeling to represent complex relationships for biological systems and predict behavior (Ogunnaike, 2019). There are various approaches to developing these models (Janes et al., 2017); as they mature, the models will play a role in the health care industry in predicting and avoiding catastrophic health consequences for patients. Systems biology encompasses both the approach to framing questions about function and understanding of feedback, adaptation, and dynamics to enable the development of methods for treating or resolving illness (Janes et al., 2017). A key feature of systems biology is examination of the whole system, deconstruction or identification pathways and networks of biomolecules, and re-creation and integration of the information (Figure 5-5). As one might expect, the use of large, accessible data repositories (so-called big data) is a complementary approach to understanding biological behavior that can be used to shape information gained from the models (see Chapter 8).

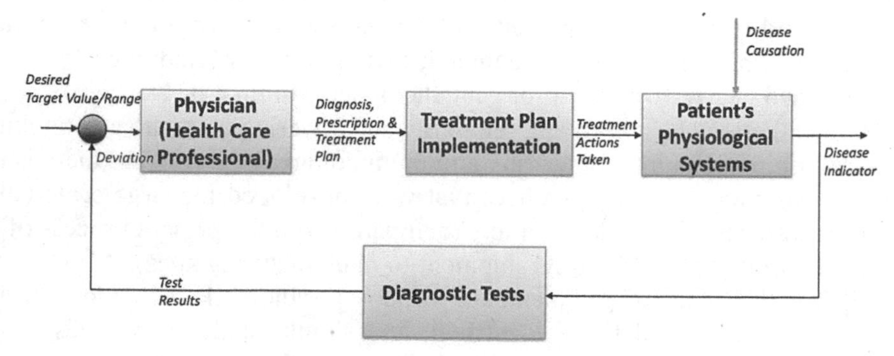

FIGURE 5-4 Feedback control paradigm for implementing personalized medicine. SOURCE: Ogunnaike (2019).

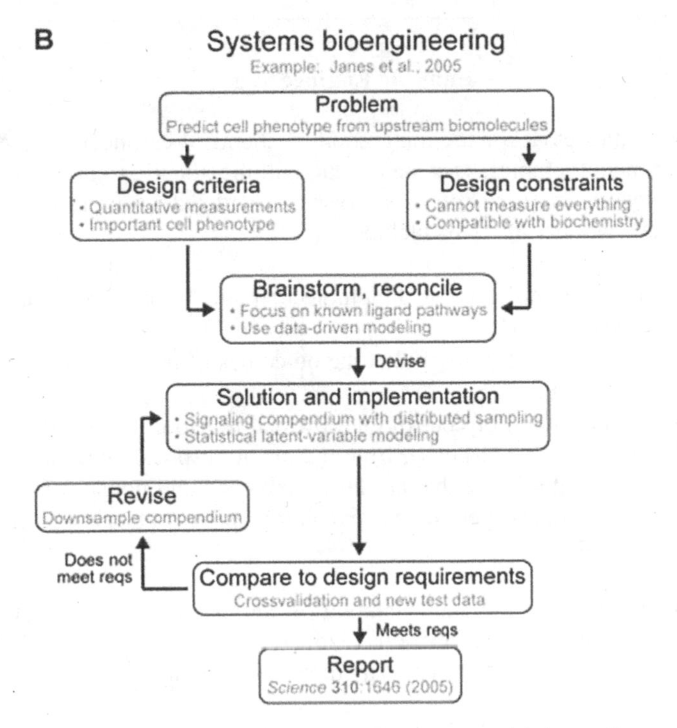

FIGURE 5-5 Engineering design approach applied to systems biology. Gray text provides an example case study. SOURCE: Modified from Janes et al. (2017).

The continuous assessment of patient health—particularly for immunocompromised individuals—would improve understanding of the spread of disease and enable early-stage prevention and treatment. The potential ability to determine the status of a patient with autoimmune or other health conditions is highly attractive, yet this is currently a daunting task that requires a simple and well-designed method for identifying specific biomarkers, collecting leukocytes from the skin or blood, and evaluating the presence of inflammation or immune activation (Box 5-1). Other means of patient-state monitoring might include the design of nanoparticle or microparticle probes targeting specific cell types so they can be tracked and monitored over extended time periods.

ENGINEERING APPROACHES TO IMPROVING THERAPEUTICS

Vaccines as Biomolecular Therapeutic Agents

Although drug discovery is key to the development of new types of drugs, discovery is only the beginning of the development phase; significant challenges arise in the large-scale manufacture of drugs, including industrial-level product generation, purification, and formulation. The COVID-19 pandemic has highlighted some of the challenges

BOX 5-1
Immune Engineering

Several opportunities exist for the integration of chemical engineering and immunology, including cancer immunotherapies, vaccine design, and therapeutic treatments, particularly in addressing challenges pertinent to infectious diseases and autoimmune disorders. Relevant quantitative chemical engineering skills include

- the ability to design molecular and biomolecular agents that can stimulate or modulate the immune system;
- the use of systems biology to predict the outcomes of manipulations of complex biological networks; and
- the ability to understand or control the flow and transport of key signaling molecules involved in the immune response, from the innate response that yields the generation of cytokines and interferons that are fairly universal and nonspecific, to the adaptive immune response that is specific to a given target or set of target agents via engineered T cells or NK cells.

Broad areas of interest include understanding of the physical and biological mechanisms underlying how the immune system functions, applied virology, and efforts that leverage this knowledge and engineering design to develop therapies and vaccines capable of being translated to the clinic.

Chemical engineers are involved in the development of new vaccination concepts and the molecular design of new delivery approaches for vaccines and biologic therapies such as antibodies, including rapid drug development and commercial scaling of proteins or protein components. Manufacturing methods that enable increased global availability or reliability of vaccines and therapies are vital for new technologies being introduced, from molecular to cellular systems. Diagnostic approaches are critical to enable rapid and reliable detection of pathogens and monitoring of biomarkers of immunity. Also of interest are systems biology, machine learning, physics-based modeling, and engineering approaches directed toward understanding innate and adaptive immunity, with the goal of preventing and curing disease.

of bringing drugs to market. Key to the success of the rollout of COVID-19 vaccines, messenger RNA (mRNA) vaccines from Pfizer-BioNTech and Moderna, was the generation of production lines to accommodate large-scale production and formulation. Fortunately, Moderna had already gained experience in this area and built a manufacturing facility for the development of other vaccines that were approaching early clinical trials.

Protein-based biologic drugs have been manufactured using cell-based bioreactors to generate proteins for therapeutics, and a significant amount of biotechnology know-how is based on the engineering of bioreactors that require management of titers from microbes or mammalian cells, management of cell viability and oxygen and nutrient levels, and extensive removal of cellular debris and waste products as part of the purification process. Unlike these processes, the production of mRNA uses a cell-free enzymatic reaction; RNA polymerase is utilized in the presence of a DNA plasmid and an RNA promoter that accelerates amplification of the DNA sequence. Additional enzymatic steps may also be used to modify the end groups of the mRNA. Because the process does not use cells, the concentration of the product can be much higher than is possible in a bioreactor for which cell density is limited and high concentrations of cells can lead to cell death and toxicity. And because the reactions are enzymatic, the reaction rates, yield, and purity are based on the nature of the promoter and the frequency of enzymatic errors that lead to short oligonucleotide impurities, double-stranded RNA, and similar by-products that need to be removed to create a pristine product free from potentially dangerous constituents. RNA self-replication machinery can be encoded in the sequence, thus increasing rates of generation of mRNA. Purification steps have involved high-performance liquid chromatography or other column-based chromatography. In general, the manufacture of mRNA therapeutics is a very new area, with a very different set of reagents, kinetics, products, and processes.

At the time that the SARS-CoV-2 virus appeared, Moderna had 10 other mRNA-based drugs approved by the FDA as Investigational New Drugs or for clinical trials, and had already launched a manufacturing facility for mRNA vaccines that were in various phases of clinical trial. Because the manufacturing machinery was already in place, the company was able to take full advantage of the highly versatile and modular nature of an mRNA vaccine, along with a significant funding stimulus from the U.S. government. Once the SARS-CoV-2 sequence had been revealed publicly, the U.S. National Institute of Allergy and Infectious Diseases collaborated with Moderna to determine a most probable target sequence from the spike protein, and Moderna generated DNA templates and launched production of a new mRNA encoding for the COVID-19 antigen within days, and using its platform, it took just 3 weeks to go from sequence to vaccine. By contrast, the turnaround time for a standard vaccine—typically derived from a known protein antigen or a deactivated form of the virus—can range from several months to years. The speed at which the mRNA vaccine was developed, tested in laboratory animals, and produced at levels sufficient for clinical trial was record breaking (Box 5-2).

Traditional vaccines use attenuated virus or virus capsids, but a great deal of work is currently being done on subunit vaccines—vaccines that rely on the use of a protein antigen that presents a key and broadly recognized epitope of the original infectious agent that is able to help initiate an immune response. The many advantages of this method include the ability to generate and mass produce a well-defined protein or polysaccharide biomolecule product instead of dealing with isolated viruses, typically using classic bioreactor synthesis; increased safety of the resulting vaccine for patients, including those who are immunocompromised; and the increased physical and thermal stability of the vaccine. Critical to the success of such vaccines is the ability to identify key sequences on the original infectious agent that can elicit the formation of neutralizing antibodies by B cells and/or initiate a T cell–mediated response. The nature of this response and its efficacy are highly dependent on the selection of the correct antigenic sequence or structure; ideally, this selected region would be a part of the virus that does not undergo many genetic mutations, so as to ensure efficacy against disease even in the presence of variants. Furthermore, subunit vaccines generally are not sufficiently immunogenic on their own

BOX 5-2
mRNA Vaccine Development

The discovery and development of disease treatments have been multidisciplinary enterprises for many years. One of the most important recent examples of the integration of engineering and biology to address medical needs is the rapid development of a COVID-19 vaccine. In less than a year, vaccines were developed, tested in preclinical animal and clinical human trials, and produced at large scale. This box summarizes a discussion with Moderna's chairman of the board, Noubar Afeyan, that addressed the vaccine technology; the role of chemical engineers in advancing that technology; and the nature of drug discovery, formulation, and expedited biomanufacturing.

Why mRNA? RNA is an information molecule that is modular and specific. In typical biologics, each protein drug is a unique molecule that requires significant adaptation of formulation. In contrast, messenger RNA (mRNA) is a coded molecule—when the drug is changed, only the RNA sequence is changed, while the physical and chemical characteristics of the drug molecule remain the same. Thus, mRNA is especially ideal for vaccines because the immune target is also a code, and the goal is to replicate the antigenic molecular sequence—thus mRNA is a perfect fit for vaccines. Additionally, because it is a platform technology, when the drug needs to be adjusted or changed, the manufacturing process remains the same.

The role of chemical engineers. The idea to use mRNA as a vaccine molecule grew in part from conversations between two chemical engineers—Bob Langer and Noubar Afeyan—who launched Moderna to explore the possibility of delivering mRNA to enable cells within the body to produce the protein antigens for vaccines. Chemical engineers are able to take scientific advances, such as the successful protecting groups that stabilize mRNA in the bloodstream, and think more broadly about how to generate a therapeutic. Systems-level thinking aided in considering the factors needed to get from the molecular and cellular levels to the efficient delivery and distribution of RNA macromolecules within the body.

Vaccine development. A key aspect of developing a viable mRNA vaccine is the ability to systemically deliver mRNA, a highly negatively charged macromolecule, to cells that will generate proteins that are actively accessed by immune cells. The generation of an appropriate formulation using cationic lipid nanoparticles was a problem suited for chemical engineers, and finding formulations that were effective in enabling both mRNA delivery to the cell cytoplasm and simultaneous upregulation of immune cells as an adjuvant was key to the formation of a successful, effective, and commercially viable vaccine.

Frontiers in immune engineering. Chemical engineers can contribute to advances in such frontiers as guide RNAs that can alter the expression of multiple genes at once, drugs generated within red blood cells, and the engineering of biological systems. These kinds of developments will ultimately lead to fewer small-molecule drugs—the body operates on proteins derived from RNA, and relatively few of a cell's most important regulating molecules are small molecules. Ultimately, for many biomedical/pharmaceutical drugs, bioreactors may become irrelevant. It is likely that in 50 years, all protein drugs will be mRNA drugs.

Clinical relevance of emerging technologies. CRISPR (clustered regularly interspaced short palindromic repeats), synthetic biology, chimeric antigen receptor (CAR) T, and other cell therapies are the future of medicine. The prospect of manufacturing these types of therapeutics is challenging, but many similar challenges have proven to be solvable. There is no mystery here—good engineers can solve even highly complex problems.

to induce a strong humoral or T cell response; therefore, for subunit vaccines that utilize much smaller portions of the protein antigen, adjuvants need to be added to enhance the body's innate immune response and upregulate immune-cell activation. The future of vaccine development will include the use of genetic data mining and machine learning with biomolecular computation to enable a predictive approach to vaccine subunit selection, the ability to rapidly examine the immunogenic response of multiple protein subunits, and the establishment of protein scaffolds that facilitate such rapid assay approaches.

Chemical engineers will play a significant role in the development of the protein engineering and computational and biomolecular design space for the discovery of new antigens. The design of antigens, including the selection of specific sequences, and the use of different kinds of molecular adjuvants can affect the immunological response by directing different cellular pathways. For example, for certain viral infections, such as HIV, it is desirable to elicit a strong intracellular T cell response to prevent further viral replication; in other cases, a humoral response with highly effective neutralizing antibodies might be sought. Recent efforts have been focused on tissue-resident T cells and other memory T cells that can play critical roles in advancing immunity. The selection of the correct epitope to elicit a desired response and the engineering of the means of delivery of a vaccine to maximize that response introduce additional complexities that chemical engineering is well equipped to address. Aspects of vaccine delivery include the mode of delivery, control and impact of pharmacokinetics of antigen exposure, route of delivery, and degree and nature of uptake and presentation of antigens by antigen-presenting cells. Such materials as degradable polymers or other forms of delivery devices, such as microneedles for transdermal delivery, may be used to help control these factors or influence which cells are accessed upon release of the vaccine (Caudill et al., 2018; Prausnitz, 2017).

As important as the generation of antigenic molecules is, it is equally important to design their solubility and biocompatibility for ease of manufacture and downstream generation of purified vaccines. Because subunit vaccines are smaller biomolecular units, they also provide a more accessible route to creating manufacturing platforms that allow rapid vaccine development and scalable on-site manufacturing. An interesting biological engineering consideration is the selection of biological hosts for the manufacturing process. As an example, replacing mammalian CHO cells with rapidly dividing yeast cells for the production of new subunit vaccines can cut by a factor of four the time for both generation of clinical test compounds and, ultimately, manufacturing-scale generation (Brady and Love, 2021). The ability to use stable, easily transfected, and rapidly dividing cell lines can also enable the manufacture of key vaccines at low cost and high yield in underserved parts of the world (Matthews et al., 2017). These kinds of innovations are indicators of the potential impact of chemical engineers in enabling rapid response to diseases and a much more accessible and equitable approach to human health.

Engineered Proteins as Therapeutics

Engineered antibodies are now perhaps the largest and most expansive area of protein engineering. Since the first monoclonal antibody therapies were introduced in the

1980s, the domain of antibody therapies has exploded; by the end of December 2019, nearly 80 antibody-based therapies were on the market, and eight of the top ten drugs on the market were biologics (Lu et al., 2020). The design, development, and manufacturing of antibodies have been critical in the development of treatments of several disorders. Antibodies work via a highly specific binding and blocking mechanism that allows them to be highly effective inhibitors of biomolecules and specific cell-membrane receptors, and to bind to antigens to fight viral and other infectious agents.

Several challenges remain in the generation of antibodies on a commercial scale. Chief among them is the time and scale needed for antibody production. The cellular machinery needed to make complex antibody molecules is available in mammalian cells such as CHO; however, these kinds of cells typically exhibit low yields and slow rates of growth, and they can require significant maintenance because of their sensitivity to temperature and other environmental factors. Yeast and bacteria cells are much more productive, have rapid doubling times, and are readily manipulated to generate different protein sequences. Unfortunately, however, these systems need to produce proteins intracellularly, which requires cell lysis and complex separation steps. The design of the molecules can impact bioavailability, solubility during manufacture, and blood stability. Chemical engineers will play a key role in determining the best routes for antibody production.

The molecular design of antibodies is modular and provides a significant design space; the Fc (fragment crystallizable) determines the cell effector function—the specific cell type targeted and the nature of cellular impact. Recently, antibody fragments and engineered fusion proteins have also entered the market, utilizing variants or portions of the original antibody while introducing a range of advantages over full antibodies for a number of applications. Each of the Fab (fragment antigen-binding) components contains an inner binding region, and the single chain fragment (scFv) contains the recognition and binding sequence from the Fab. These forms allow for more direct incorporation of a much smaller binding molecule, thus providing molecules that do not require the complex manufacturing steps needed to produce whole antibodies. Protein engineering of these molecules has led to exciting therapeutic opportunities—the combination of Fab components from two different antibodies can yield bispecific antibodies with dual-binding capability, and it is also possible to fuse an antibody with a targeting protein to create a therapy that binds to specific cell types for targeted therapies. Furthermore, additional modifications can be made in the degree or type of glycosylation, the presentation of charge and charge heterogeneity, and control of the hydrophobic/hydrophilic nature of the molecule. These options represent myriad possibilities for therapies that have not yet been explored. Chemical engineers can manipulate these systems on the molecular level using rapid assay methods and computation in combination with studies of manufacturing properties to achieve high yields and lower overall cost.

Biomolecular engineering has had an enormous impact on therapeutic designs. It has come in a variety of modes, including computational peptide designs; display libraries of phage, yeast, and bacteria for target discovery; and protein engineering of antibodies and antibody fragments. The design of phage, bacteria, and yeast libraries has allowed the generation of numerous diverse peptide motifs to identify target binding agents (Arnold, 2018; Wittrup, 2001). Identification of specific ligands has been a key driver of biologics discovery. Chemical engineers have made prominent contributions to this field through the introduction of quantitative analyses and optimization. These display tools have been extended to in vivo application, enabling direct in vivo identification of target-specific binding sequences for organs, which can enable targeted drug delivery—the so-called vascular zip code approach (Teesalu et al., 2012). Viral libraries have also been used to identify target-specific gene therapies. For example, adeno-associated virus libraries can be evolved to facilitate in vitro and in vivo gene delivery (Bartel et al., 2012). Such directed evolution approaches have contributed to efficient optimization of agents in the rather complex biological landscape.

Keeping proteins stable under physiological conditions for prolonged times is also a significant hurdle. Engineering of agents that promote protein stability continues to be an area of opportunity in chemical engineering. Biologics now represent a major class of therapeutics, monoclonal antibodies have emerged as the dominant therapeutic modality, and nucleic acid–based therapeutics are poised to increase their role in the therapeutic landscape. However, these delicate biologics require stringent storage conditions. Maintenance of a "cold chain" for frozen formulations represents a serious challenge in the future use of biologics, especially in resource-limited communities and geographic regions. Strategies for biologics stabilization have their foundation in protein biophysics, an area that has benefited from the active participation of chemical engineers. Fundamental studies founded in the thermodynamics of water structure, influenced heavily by chemical engineers, have laid the foundation for strategies for protein stabilization. Solutions for future challenges posed by the need for nucleic acid stability can potentially leverage the knowledge generated about protein stability.

MODELING AND UNDERSTANDING THE MICROBIOME

The complexity of the microbiome—which includes multiple cell types involved in both the microbial community of a given organ and the different host cells that support and respond to that community—presents a challenge for modeling of host–microbiome interactions to develop predictive solutions for human health (Box 5-3). The past 20 years have seen several advances in the toolsets available to chemical and biomolecular engineers for creating predictive models for cellular systems; these capabilities are enhanced when combined with access to large banks of genomic data. Genetic information about a microorganism, along with the location of key genes, can be determined, and this information, in combination with understanding of the biochemical and metabolic activities of the cell (effectively the reactome), makes it possible to determine the fully reconstructed signaling networks of cells. This information can then be converted to a mathematical representation and analysis so that computational approaches can be applied to enable predictive capabilities of specific network signaling pathways.

These genome-scale metabolic models (GEMs) have progressed and expanded in large part as a result of the constraints-based reconstruction and analysis (COBRA) approach (O'Brien et al., 2015). This approach narrows large numbers of possible pathways in optimizations by applying key constraints, such as nutrient availability and rate of uptake by cells. Experimental data can be used to enable cell-type and conditions-based constraints that enable more specific models. More recent advances have made it possible for computational approaches to take into account dynamic genomics due to evolution, as well as changes in cell states. In the previous decade, the expression matrix (E-matrix) was introduced for computationally predicting the expression of genes in an organism through knowledge of its transcriptional and translational activity. Metabolic networks and E-matrices can now be used together for modeling of cellular membrane composition and thus of the impact of such environmental factors as reactive oxygen and hypoxic pH shifts (Chowdhury and Fong, 2020).

BOX 5-3
The Human Microbiome and its Impact on Human Health

One of the areas of biomedicine that remains a relative frontier is the development of treatments for human health based on the microbiome—the collection of naturally occurring microbes that reside in coexistence with the body, facilitating or complementing the biological function of the human host. The most notable regions of localization of such microbiota include the gastrointestinal (GI) tract, cervix, lung, ovary, skin, and oral mucosa, along with the mucosa of other organs. Biologists have found a remarkable connection between the different microbes present in the body and such factors as GI health and function, neurological processes, and even the immune system. These commensal bacteria live and grow in the microenvironments created within the body, utilizing nutrients generated by the body. They emit biomolecules that engage signaling networks to modulate the behavior of other surrounding microbes and the human host.

Work over the past several decades has led to a better understanding of the role played by these native bacteria in maintaining and enabling human health. Chemical engineers can make a vital contribution toward understanding the complex signaling and resultant biological responses of the body, and in applying this knowledge toward human health treatments. Systems biology approaches and the use of genomics can be deployed by engineers to describe more completely the molecular pathways and genomic networks involved in regulation of the microbiome and human host. Ultimately, this kind of understanding will uncover the role of bacterial communities in affecting health and indicate ways to translate that knowledge into therapies. These treatments vary, and there are many possibilities in the design of therapies that transfer or transplant healthy human microbiota to correct the existing bacterial systems and help restore the original microbiota function. Furthermore, synthetic biology tools have made it possible to directly modify bacteria to create new therapeutic approaches for addressing a range of chronic disorders or enabling settings with greater resistance to infections or susceptibility to disease.

The advancement of these methods in metabolic engineering in general has opened up many opportunities to apply them to understanding the microbiome. With the toolsets provided by GEMs, it is possible to connect microbiome and host-cell interactions with final biological function (Chowdhury and Fong, 2020). These efforts are greatly facilitated by significant cross-institutional efforts, including the National Institutes of Health's (NIH's) Human Microbiome Project (HMP), a massive NIH Common Fund effort involving public and private institutions that led to identification and sequencing of the microorganisms in multiple human organs and the generation of tools for connecting microorganism genomic data to disease state and function. A second phase of HMP—HMP2—led to a series of publications on the role of the gut microbiome in inflammatory bowel disease (IBD), preterm births, and diabetes (*Nature*, 2019). The gut microbiome is the most studied microbiome of the human body, and significant progress has been made toward building databases that will inform GEM models, as well as other computational approaches, particularly machine learning and artificial intelligence methods that require large datasets. An example of a comprehensive GEM model is AGORA (assembly of gut organisms through reconstruction and analysis; Magnusdottir et al., 2017), an approach that uses gut metabolic and genomic data to semiautomatically generate reconstructions of metabolic pathways for 773 independent gut microorganisms. This accomplishment is of great significance in illustrating the power of such models to assist in determining the functional behavior of myriad combinations of cells in consortia, thus enabling such capabilities as determination of predictors of disease identification of targets for treating disorders. In 2020, AGORA2 was introduced (Heinken et al., 2020). The number of microbes included increased 10-fold; additional factors, such as microbial drug degradation, were introduced; and both singular and pairwise bacteria members are considered. These kinds of models can lead to better understanding of disease (e.g., understanding the role of microbial consortia as opposed to individual bacteria in regulating key processes) and exhibit predictive capabilities (e.g., predicting patient susceptibility to IBD or Crohn's disease based on nutritive deficiencies).

Cell Transfer Approaches to Medical Applications for the Microbiome

Increased understanding that the gut microbiome of healthy humans can be extremely beneficial for the management of infection, inflammation, and metabolism has led to the use of native human microbial consortia—the collections of different naturally occurring bacteria species that reside within the body—as a means of regulating disease (Colman and Rubin, 2014). The approach uses fecal matter transplants (FMTs) to transfer the healthy and established microbiome from donor to patient. The efficacy of this and related approaches, such as the use of oral pills formed from purified fecal matter, has been demonstrated to show efficacy in restoring the ability to fight gut-related infections (e.g., aggressive forms of colitis, severe sepsis, diarrhea) and are in clinical trials for many related conditions (Wischmeyer et al., 2016)—more than 900 clinical trials for these therapies were listed in clinicaltrials.gov in 2020 (Wang et al., 2021b). Chemical engineers can help advance these kinds of native microbiota-based therapies to a new level by addressing their expansion, cost, and scalability—for example, by investigating isolation and purification of native microbiota to enable more consistent therapeutic products and approaches that can be readily replicated. Much is to be gained as well from a deeper understanding of the impacts and effects of such therapies. Machine learning, in combination with information gained from genomics and systems-biology models, may enable more effective predictive methods and more personalized approaches to treating patients and pairing these therapies with given characterized FMT systems.

The use of native microbial consortia, or communities of microbes of different types and function, demonstrates the power of the synergism that exists among the many different cells within the human microbiome. Cellular communities consist of independent cells that work together to share the labor of generating different metabolites, maintaining different roles in managing energy, producing molecules, and enabling breakdown or processing of different nutrients. The various bacterial components create supportive environments that enable the fully functional microbiome to provide dramatic gains in immunological and anti-infective properties. Together, these communities are more resistant to disease, and exhibit reduced metabolic burden and a broader range of function than individual microbe types. Therefore, the next step in microbiome therapeutic approaches is the engineering of microbes and their combination to form similarly synergized consortia with potentially greatly enhanced capabilities (McCarty and Ledesma-Amaro, 2019). Given the complexity of the interacting cellular systems and their various functions, a systems approach will be required to replicate important interdependencies while incorporating new or enhanced functionalities.

Synthetic Biology as a Tool for Engineering Microbes to Improve Microbiome Health

Bacteria have unique qualities that can be manipulated to enable the generation of synthetic microbiota with programmed function. In conjunction with gene editing tools, this work can yield a wide range of functions that depend on communication and interaction with other cell types. These functions include intercellular signaling, among them chemical, biophysical, and adhesion molecule–based interactions between cells, as well as ionic and electrical channels. Quorum-sensing capabilities in bacteria populations, based on the sensitivity of cells to an autoinducer molecule that increases concentration as cell density increases, are an example of a microbial behavior system that can be manipulated to advantage in synthetic communities. It is possible to engineer quorum-sensing mechanisms that are independent and thus completely orthogonal to native commensal bacteria, thus enabling unique or independent "programs" instituted on the basis of specific cell types that will be responsive to a given quorum-sensing signal. Other mechanisms of cell–cell communications include N-acyl-L-homoserine lactones (NHL) and autoinducing peptides that can cause bacterial-cell circuits, programmed into given microbial species, to turn on or off based on their concentrations (Hwang et al., 2018). These tools can be combined with additional means of triggering response; for example, bacteria can be designed to respond to exogenous signals. As another example, inducer molecules can be used to trigger the expression of transgenes that activate a promoter. Synthetic biology enables the design of independent gene regulatory systems, and this orthogonality makes it possible to generate a range of independent controls that can be used to manage complex systems with multiple organisms.

Synthetic biology tools such as CRISPR and the assembly of DNA parts or "circuits" that can control genetic expression on demand can be used to modify a microorganism by introducing the ability to generate a given molecular signal, gene regulator, or indicator in the presence of a given disease condition. Examples of singular commensal or symbiotic bacteria that have been engineered include

- *E. coli* that can generate glycotransferases that yield molecular inhibitors to dangerous enterotoxins that cause diarrhea,
- lactobacillus engineered to prevent infection through the generation of bacteria-killing lactoferrins,
- cholera-controlling microbes that express a regulator of virulence genes, and
- *E. coli* microbes engineered to secrete antimicrobial peptides that can reduce the effects of *Salmonella* (Tan et al., 2020b).

As shown in Figure 5-6, harmless commensal bacteria can be modified synthetically and then reintroduced to the microbiome in the gut using oral delivery or transplant approaches. These newly introduced bacteria are then able to incorporate into the native microbiota and generate molecules that can address infection or greatly lower the impact of toxins released by undesired bacterial infections.

Along with modifying susceptibilities to infection, microbes can be engineered to release regulators of the immune system. This ability is particularly relevant for the treatment of inflammatory disorders such as IBD; *Lactis* bacteria (a probiotic used in production of cheese and yogurt) can be modified to produce anti-inflammatory cytokines such as IL-10 to treat IBD effectively. This and related approaches have generated great interest in the use of engineered microbes for therapeutic applications for conditions ranging from infection and inflammation to metabolic disorders and potentially even endocrine and autoimmune disorders, such as diabetes and obesity (Tan et al., 2020b). Given recent findings regarding the gut microbiome–neurological connection and the likely advances in biological understanding of both the gut and other human microbiome communities with respect to disease and overall human health, this area of technology is likely to expand significantly in the next decade.

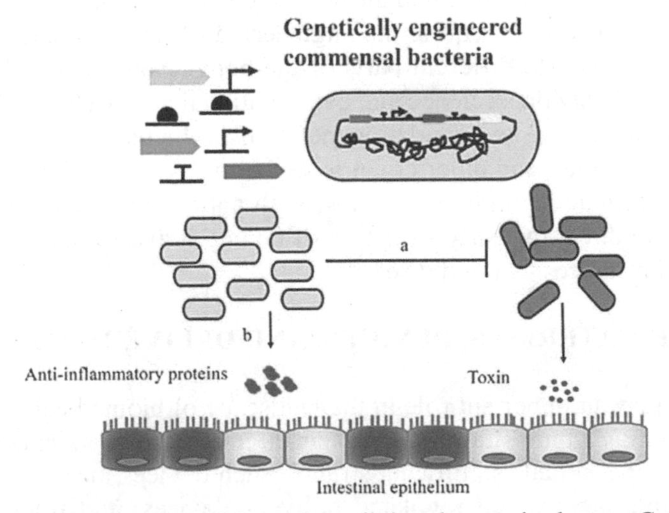

FIGURE 5-6 Engineered commensal bacteria as living therapy in the gut. Commensal bacteria can be engineered to (a) inhibit pathogenic bacteria or toxins or (b) reduce inflammation through secretion of anti-inflammatory proteins. SOURCE: Tan et al. (2020b).

Activating molecules can include toxins, antibiotics, or specifically determined inducer molecules. Syntrophic dependencies, in which a molecule produced by one microorganism synergizes the activity or metabolism of another, further increase the complexity of these systems and enable the kinds of metabolic interdependencies found in nature (Harimoto and Danino, 2019; Hwang et al., 2018). It is clear that using these kinds of engineering tools to better design therapies based on engineered consortia will require applying understanding of complex signaling behaviors, modeling, and the manipulation of synthetic biology tools.

Chemical engineering, which has had a traditional focus on metabolic engineering, is a discipline particularly well poised to enable advances in these areas. In the late 20th century, the field of biochemical engineering brought forth the concept of the cell as a reactor and the use of cellular pathways to create molecules of interest with a high and controlled yield. Early developments in this field led to work on metabolic manipulation of cells to direct and control certain cellular reaction pathways over others. The microbiome is in some sense a highly tuned and sensitive bioreactor or series of bioreactors within the human body. The challenges for chemical engineering lie in designing cells for manipulating not only the generation of desired compounds but also the rate of generation, the selectivity for a given target, and the longer-term stability of the reacting system. Chemical engineers have embraced synthetic biology as a critical tool in addressing such problems and are well equipped to advance this work.

The unique skillsets of chemical engineers can contribute to gaining knowledge, enabling discovery, addressing disease, and enhancing human health through understanding and regulation of the human microbiome. Key challenges and opportunities include further advancement of synthetic biology tools that incorporate environmental and conditional responses, regulation of the reactome across multiple species, and engineering of cellular consortia to achieve patient-specific outcomes. The expansion of engineered therapies to other microbiomes will result in the need to determine effective modes of delivery to specific organs, which may require the engineering of new biomaterials that support the transfer of microbiota to different parts of the body. Along with advances in GEM methodologies, additional data science and computational approaches will likely become more important in predicting and regulating metabolism. Increased availability of data on the gut, skin, vaginal, oral, and other organ microbiomes will enable the use of artificial intelligence and machine learning algorithms, with anticipated increased use of models and computation to direct the early detection of potential disease and the application of preventive health measures to avoid disease.

DESIGN OF MATERIALS, DEVICES, AND DELIVERY MECHANISMS

Devices play an important role in the landscape of biomedical technologies. Devices that control drug administration are of special interest to chemical engineers and an area in which they have made significant strides. Such devices include external infusion pumps, implantable pumps, self-regulated delivery devices, and transdermal patches, among others (Anselmo and Mitragotri, 2014; Vargason et al., 2021). Implantable pumps offer a patient-compliant means for long-term disease management and have had a transformative impact on diabetes, pain management, and neurological disorders.

Patients with type 1 diabetes are on lifelong treatment with insulin, and the need for frequent insulin injections represents a significant limitation in diabetes management. Implantable pumps offer an excellent alternative, especially for the delivery of basal insulin (Cescon et al., 2021; Shi et al., 2019; Wolkowicz et al., 2021). Several challenges, including those related to insulin stability and device biocompatibility, had to be overcome for these devices to reach the clinic. Another major challenge has been managing the compatibility of implanted pumps with the human body (Kleiner et al., 2014; Park and Park, 1996). Foreign devices, including implanted pumps, evoke a foreign-body response that includes the arrival of immune cells, leading ultimately to the formation of fibrous capsules, whose properties play an important role in determining the durability of the device. Chemical engineers have made important contributions to this field through understanding of the foreign-body response and the development of coatings to minimize fibrous capsule formation.

Control strategies are especially relevant for insulin pumps, for which an active control of insulin release is necessary to maintain euglycemia. A healthy pancreas adopts a complex control algorithm to control insulin release such that glucose levels remain within a tight window. Accordingly, chemical engineers have led extensive efforts to develop model-based and adaptive control strategies that enable communication between glucose sensors and insulin pumps. As glucose sensors have advanced to the point of providing continuous measurement over several days in patients, effective control algorithms have supported the development of closed-loop insulin delivery devices—the so-called artificial pancreas (Doyle et al., 2014). The paradigm of self-regulated pumps can be extended to other indications for which continuous and regulated delivery is essential. The development of such devices requires novel formulation strategies, pump designs, control algorithms, and continuous sensors, challenges for which chemical engineers are well suited.

Another exciting frontier for chemical engineering is the development of devices for completely noninvasive methods of drug delivery. Such devices can be placed on the skin, where they can offer a needle-free method of continuous and regulated drug administration. These transdermal patches provide a natural means for the sustained release of drugs (Prausnitz and Langer, 2008), providing the key benefit of active patient control and ease of termination when needed. Several transdermal patches are currently available, allowing easy delivery of nicotine and fentanyl, for example. However, these patches are limited to the delivery of small-molecule drugs because of the transport limitations of human skin, leaving many emerging drugs beyond their scope.

Several advances have been made in improving drug delivery across the skin to expand the use of transdermal patches to biologics. One such advance, and an area in which chemical engineers have already made an impact (Hao et al., 2017; Prausnitz, 2017), is microneedles, which penetrate the skin to a depth sufficient for delivering drugs but not for inducing pain. Microneedles are showing high potential, particularly for delivering vaccines, skin being an excellent organ for vaccine delivery because of the presence of immune cells. Further, microneedles carry vaccines in solid form, thus improving their storage stability. Such solid, stable, self-administered vaccines could be particularly suitable for use during the current COVID-19 pandemic. Strategies have also been developed for enhancing noninvasive delivery of biologics through oral or inhalation routes, with enabling contributions from chemical engineers (Brown et al., 2020; Matthews et al., 2020; Morishita and Peppas, 2006).

The human body is a connected system of various organs, each designed for a unique function. While the macroscopic behavior of each organ has historically been studied in medicine, quantitative understanding and control of transport properties in these systems are limited. In the absence of this control, access to many organs is severely limited. At a fundamental level, transport is limited by the body's natural metabolic processes and transport barriers. These biological barriers, while serving the important purpose of regulating the body's metabolic functions, also limit how drugs can be delivered in the body (Figure 5-7). Accordingly, many potential drugs fail to reach their destination in the body, and thus most molecules never become drugs. This limitation ultimately reflects the high cost and lengthy timeframes of drug development.

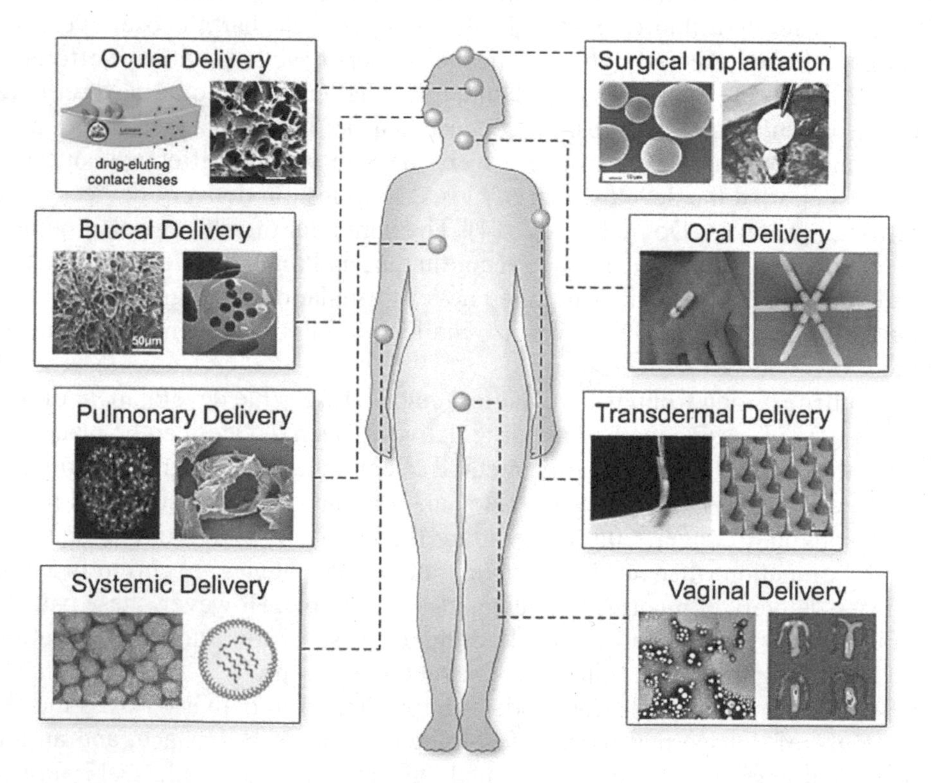

FIGURE 5-7 Examples of biomaterials and their routes of administration for in vivo use. In addition to pills and injections, biomaterials have been developed to administer drugs successfully in a variety of other ways. SOURCE: Fenton et al. (2018).

The early contributions of chemical engineers to small-molecule drug delivery were focused on developing encapsulation strategies to control the release of drugs. Macroscopic and microscopic depots have been designed to encapsulate small and large drugs. The most commonly used microsphere-based drug delivery systems consist of a suspension of polymer microspheres that can be delivered by subcutaneous injections. Fundamental studies describing the diffusion and degradation kinetics and mechanisms of diffusion in the polymer matrix, led by chemical engineers, have played an important role in the establishment of these systems (Ritger and Peppas, 1987). Triggered release of drugs from these polymeric depots, mediated by such external triggers as light, ultrasound, and magnetic fields, has allowed better control over the drug release (Sun et al., 2020). Chemical engineers have numerous opportunities to advance sustained-release depots, especially with respect to extending the release duration, which can further improve patient compliance. Such advances are especially relevant for such applications as ocular delivery, given the strong motivation to minimize the number of injections in the eye. Long-term compatibility of depot systems with the immune system also represents a continued challenge.

At a nanoscale level, the same polymeric systems have provided benefit for targeted drug delivery, a concept first introduced for the delivery of chemotherapeutic agents. Numerous efforts have been made to improve the targeting ability of nanoparticles by virtue of their surface modification of polyethylene glycol (PEG), which reduces surface protein adsorption and subsequent hepatic sequestration. Nanoparticles can be further modified by targeting ligands, including peptides, aptamers, and antibodies, to improve target accumulation (biomaterials are discussed in more detail in Chapter 7). A number of nanoparticles, including liposomes, polymeric nanoparticles, and inorganic nanoparticles, are already commercially available, and several are available in the clinic (Karabasz et al., 2020). Lipid nanoparticles have played a central role in packaging of mRNA to enable its stabilization and intracellular delivery (Mitchell et al., 2021). While a large number of materials, targeting moieties, and design strategies have been explored, achieving exquisite targeting remains an unmet need.

Moving forward, opportunities for chemical engineers in the field of drug delivery include developing a better understanding of transport processes and leveraging this understanding to accomplish better targeting. One of the key challenges is the limited spatial and temporal resolution of drug imaging in the body. Model systems (e.g., organs-on-chips) can provide a deeper understanding of such transport processes (Bhatia and Ingber, 2014; Ghaemmaghami et al., 2012). Drug delivery methods also need to be patient-centric; that is, they need to improve patient compliance, including such considerations as cost, simplicity, and convenience.

Building Sensors and Diagnostic Tools

Rapid, frequent, and inexpensive diagnosis is the foundation of successful health care. Blood-based diagnostic methods, including those in the routine medical checkup and in focused disease diagnosis, have been the cornerstone of past and even current diagnostic infrastructure. With perhaps the notable exception of glucose monitoring, blood-based diagnosis is typically done in the hospital setting, requiring clinical supervision.

Extensive efforts have recently been made to shift away from this inconvenient, discrete operation of blood sampling to a new paradigm focused on continuous measurements of physiological markers. Most such advances have been made in glucose monitoring for diabetes (Lipani et al., 2018). Continuous, wearable devices now available for patient use measure glucose for days through a single wearable device. In addition to providing multiple measurements, continuous sensors offer an educational tool enabling patients to understand how their glucose levels respond to various metabolic and external factors (e.g., Brown et al., 2019). Efforts have also been made to measure glucose concentrations in a completely noninvasive manner, making use of spectroscopic methods, noninvasive collection of tissue fluid through the skin, or the use of sweat-based sensors. The latter devices measure various analytes present in the sweat and use these measurements to derive a physiological assessment (Chung et al., 2019). A variety of analytes are present in sweat, including small molecules, such as glucose and hormones, and large molecules, such as proteins. Improvements to sensitivity, accuracy, and modeling based on these measurements are needed if these devices are to achieve broad health care application.

In addition to molecule-based diagnosis, wearable sensors can perform a variety of other diagnostic measurements, including measurements of temperature, blood oxygenation, pulse, and even biopsy alternatives (Gao et al., 2019; Kim et al., 2019). The ability to derive these key measurements through a wearable sensor has already transformed the collection of human physiology data. Traditionally, these measurements were possible only using large, bulky sensors, often connected to stationary electronic processors and displays. However, the ability to perform these measurements in a truly nonintrusive manner has enabled the collection of information about human responses in real-life situations.

Extrapolating these trends into the future, wearable sensors are expected to collect massive amounts of information about human behavior in healthy and diseased conditions. Some of the technical challenges in this field include the development of sensors to accurately collect and measure the small amounts of analytes that are available in sweat. Another challenge lies in accurately correlating these measurements with blood concentrations, a challenge that pertains specifically to the transport of analytes from the blood into the interstitial fluid, and relating this information to individual patient health. Can models be developed using these data to predict catastrophic physiological events through such analysis? Can drug reactions be better understood or predicted through analysis of such massive amounts of data? Opportunities exist for chemical engineers to develop large-scale network models with which to understand the dynamics and connectivity of adverse events. By virtue of their training in understanding and appreciating multiscale dynamics, chemical engineers are especially well suited to undertaking this challenge. Personalization of drug therapies is an exciting opportunity to reduce adverse events without compromising therapeutic efficacy. However, data need to be available to support such personalization at the design stage and reduction of adverse events at the follow-up stage.

Cell-, Organ-, Organism-on-a-Chip to Model Biological Functions

Drug development is typically a lengthy and expensive process, with 10–15 years and approximately $2.6 billion dollars typically being required for development of a new drug (DiMasi et al., 2016). A large fraction of this time and expense is associated with late preclinical and clinical development. Acceleration of lead identification enabled by high-throughput screening, rational drug design, or computational platforms has greatly enhanced the availability of early lead candidates. However, the transition of these leads to preclinical and clinical programs is limited by challenges associated with the unknowns involved in evaluating the interactions of the leads in the body. In fact, the overall probability that a drug entering clinical trials will be approved by the FDA is about 12 percent (DiMasi et al., 2016). Currently, despite advances in computational systems biology and in vitro models, more than 60 percent of these failures are due to lack of efficacy and another 30 percent to toxicity (Waring et al., 2015).

Another key limitation is the limited relevance of animal to human pathophysiology. Small animals, especially mice, are commonly used in preclinical studies, their use being driven largely by the existence of disease models in these rodent systems. Specifically, several genetically engineered murine models exist with which to assess some of the key biological aspects of a drug, including phenotypic disease manifestation, target specificity, and improved survival. However, translation of these benefits to the engineering aspects of drug development is severely limited. For example, some of the major biological barriers to drug transport in the body, such as skin and mucous membranes, are substantially different in mice and humans. Some key aspects of these barriers that dictate diffusion differ greatly between mice and humans, in some cases primarily because of the differences in body mass and in others because of the innate differences between these species. Beyond mice, the universe of large-animal models is highly fragmented, and their use depends on the disease in question. Ultimately, the nonhuman primate model, which is often a key preclinical model, is ethically, logistically, and financially challenging. In addition to limiting the speed of drug development, these hurdles bias the landscape of drug developers because the required resources and time are often a luxury available only to large companies. Tools to minimize the burden of preclinical drug development will not only reduce the cost and time of development and provide preclinical information that is more relevant to clinical programs, but also level the playing field for drug developers.

Several advances have been and continue to be achieved in the development of scaled-down microphysiological systems (the so-called organ-on-a-chip or human-body-on-a-chip) that will provide an output at least as predictive as animal testing (Low et al., 2021). This field was pioneered by the cosmetic industry with the goal of eliminating animal testing for its products. That strategic decision led to the development of in vitro human epidermis models that can provide meaningful information about the safety of cosmetic products (Faller et al., 2002). The last two decades have seen similar devices able to mimic the function of other vital organs, including the liver, brain, and lungs, among others (Low et al., 2021). Such systems require considerations not only of the biology—that is, incorporation of different cells (e.g., lung epithelial cells, astrocytes, hepatocytes) in the system—but also of the engineering aspects of flow and interfaces. For example, the liver is the most well-perfused organ in the body, and its function depends on that. Hence, a liver-on-a-chip would need to account for the intricacies of vascular flow (Schepers et al., 2016). Such devices can play a role in assessing drug toxicity. Hepatotoxicity is a key safety concern for many drugs, and a means of screening for it at an earlier stage could substantially accelerate drug development. Liver chips could also help screen nanomedicines, which are actively cleared by the liver after their administration to the body. Similarly, a lung-on-a-chip would need to incorporate the key features of an air–water interface in the presence of lung surfactant and mucus. Such lung chips could aid in assessing the interactions of drugs with the lung microenvironment, a topic that has become critically important during the current COVID-19 pandemic (Saygili et al., 2021).

Systems that capture the key elements of the immune system in vitro are also expected to support the design of future therapeutic products. The immune system is central to many major health concerns, including infections, cancer, and autoimmune diseases. The immune system functions through extensive orchestration of many cell types, including dendritic cells, T cells, and others, to stimulate a holistic response to drugs and vaccines. Systems that capture the key events of such orchestration in vitro (e.g., lymphatics-on-a-chip) would provide insights into the interaction of therapeutics with the immune system.

Moving forward, such organs-on-a-chip will require active incorporation of biological and engineering aspects into the design (see Figure 5-8). From the biological perspective, incorporating all essential cell types in the system is critical. More is not necessarily better because the model systems need to capture the essential complexity of the organ in the human body while being as simple as possible. Critical as well will be maintaining the cells in a correct phenotypic state, and ensuring cellular communication in these systems is also critical. Such key attributes include the barrier function—for example, diffusive properties of tissues and permeation across tight junctions or dynamics arising from vascular flow. Such systems need to leverage advances in microfabrication to capture key structural complexities in organs and advances in biological tools, such as gene editing, to control cells, and they require the means to incorporate and address complexity in a tissue microenvironment.

From an engineering perspective, most nonorganoid-tissue chips have very low throughput. Thus only a few replicates can be performed at any one time, which limits the ability to screen thousands of potential hits during drug discovery. More automated and miniaturized systems will be needed for commercial use. In addition, most systems are fabricated in-house in academic laboratories, so reliability and reproducibility become limiting. Clear approaches to quality control are needed, including physiological validation (sensitivity, specificity, and precision). Particular opportunities of interest to chemical engineers include understanding the connectivity of various organs-on-chips and assessing the role of flow in cells and organs. Both development and validation of these systems are research areas that chemical engineers are well suited to address.

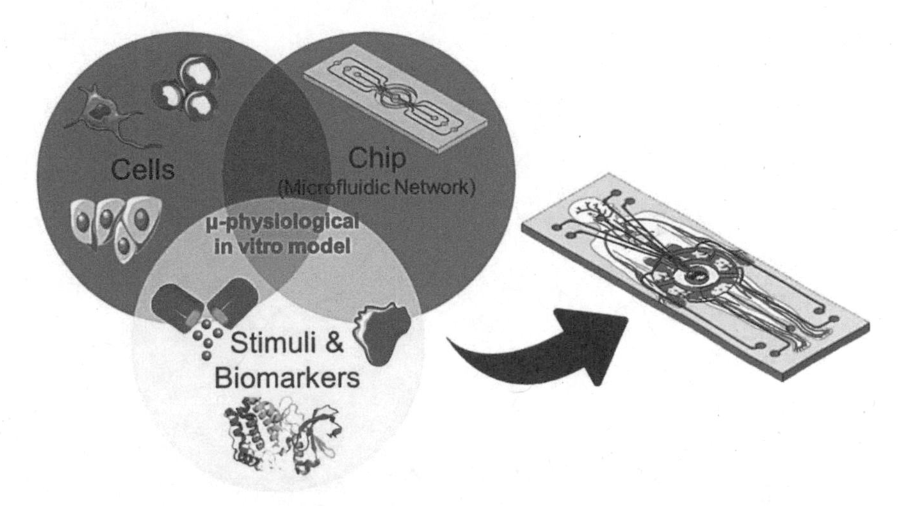

FIGURE 5-8 Advances in microphysiological systems rely on the development of three main components: the cell source, chip technology, and biomarker discovery. SOURCE: Ramadan and Zourob (2020).

HYGIENE AND THE ROLE OF CHEMICAL ENGINEERING

Historically, engineers have played a major role in the development of hygiene and the sanitary infrastructure, developments that have led to an increase in life expectancy in much of the world. Indeed, diarrhea (endemic cholera) is no longer among the leading causes of death in world statistics (see Figure 4-8 in Chapter 4). Today in the era of COVID-19, chemical engineers, especially those collaborating with other scientists and engineers in environmental sciences and technology, have opportunities to contribute to societal health and well-being and help narrow the disparities between low- and middle income countries and higher-income countries.

Recent years have seen great strides in better understanding the link between indoor air quality and health,[1] an area of growing concern in which chemical engineers can be expected to make major contributions. Within months of the onset of the COVID-19 pandemic, for example, long-lived aerosols (airborne particulates or droplets) were identified as the primary source of human-to-human transmission of the virus (Edwards et al., 2021; Prather et al., 2020) and a focal point for mitigation strategies, including the use of social distancing, face masks, and high-efficiency air filters. Figure 5-9 depicts the mechanistic framework for such superspreading events as the choir practice of the Skagit Valley Chorale, showing an aerosol size distribution from Bazant and Bush (2021). From their fundamental understanding of fluid mechanics, chemical engineers can readily confirm the time scale (many hours) for airborne suspension of aerosols in the 1-micron- to submicron-length scales in the figure, which indeed is the time scale described in the early COVID-19 literature (Prather et al., 2020).

[1] The National Academies' Committee on Emerging Science on Indoor Chemistry is currently examining the state of science regarding chemicals in indoor air.

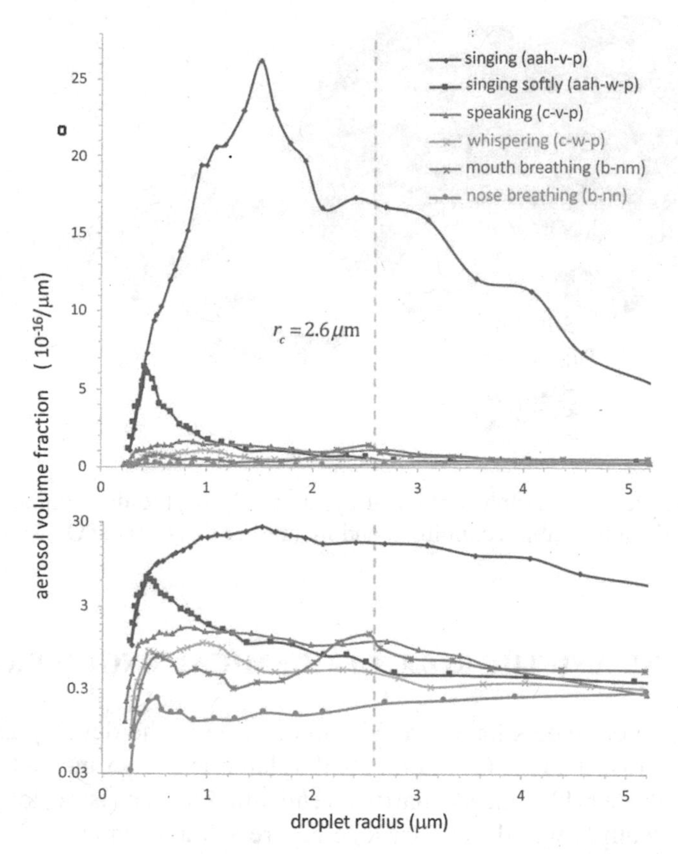

FIGURE 5-9 Model predictions for the steady-state, droplet radius–resolved aerosol volume fraction (linear scale—top panel; logarithmic scale—bottom panel) produced by a single infectious person in a well-mixed room. The model accounts for the effects of ventilation, pathogen deactivation, and droplet settling for several different types of respiration (singing, singing softly, speaking, whispering, mouth breathing, nose breathing) in the absence of face masks. The ambient conditions are taken to be those of the Skagit Valley Chorale superspreading incident. SOURCE: Bazant and Bush (2021).

Chemical engineering and aerosol experts Edwards and colleagues (2021) took the analysis of aerosol size distribution in another useful direction—variation within the population as a function of several factors, including COVID-19 infection, age, and body mass index. They showed that all three of these factors can influence aerosol sizes by three orders of magnitude, and thus serve as a useful starting point for estimating of spreading distances and infection transmission rates. Ultimately, such insights on transmission rates can help guide policies (Samet et al., 2021) and mitigation efforts, such as the use of masks and air filters.

Hand sanitizers and related consumer/cleanser products are another aspect of hygiene that has risen to prominence during the COVID-19 pandemic, and an area in which chemical engineering has an important role to play, especially with respect to the balance between antimicrobial/antiviral efficacy and product safety and environmental impact.[2] The unprecedented high volume of use of hand sanitizers to combat COVID-19 has raised concerns regarding potential toxicity and/or negative impact on the environment (Mahmood et al., 2020) and rekindled interest in the development of new materials and formulations that can address these concerns.

Historically, the chemical engineer's understanding of particulate and aerosol filtration has been based on mathematical models for capture by a single fiber (Spielman, 1977), with additional enhancements to build in geometric considerations, such as tortuosity and porosity. Such modeling approaches are common to other areas of chemical engineering that also analyze flow through porous media (e.g., modeling of oil reservoirs and mass transport in porous catalysts). Recent advances in coating materials, additive manufacturing, and systems engineering (Bezek et al., 2021; Christopherson et al., 2020) highlight further opportunities for filtration technologies for face masks and indoor filters. As attention moves past the current pandemic, and other potential challenges to indoor air quality and possible perturbants are better understood, chemical engineering will have growing opportunities to help develop materials that can change color or otherwise indicate health challenges so they can be addressed by physical barriers or treatment.

ENGINEERING SOLUTIONS FOR ACCESSIBILITY AND EQUITY IN HEALTH CARE

NIH defines health disparities as differences among specific population groups in the attainment of full health potential that can be measured by differences in incidence, prevalence, mortality, burden of disease, and other adverse health indicators. While the term "disparities" is often used or interpreted to reflect differences among racial or ethnic groups, disparities can exist across many other dimensions as well, such as gender, sexual orientation, age, disability status, socioeconomic status, and geographic location (NASEM, 2017). Despite overall improvements in population health over time, many such disparities have persisted, and in some cases, widened.

There is a persistent lack of awareness of and engagement with health disparities within science and engineering in both research and educational activities (Vazquez, 2018); indeed, one barrier to the greater involvement of engineers in health disparities research is their lack of knowledge in this area (McCullough and Williams, 2018). Although many health disparities result from systemic issues that can be addressed only through larger social changes, chemical engineers can still play a role in helping to resolve these issues. If they are to do so, however, these issues need to be introduced in the classroom and the workplace, imparting an understanding of the complexities and implications of health disparities, including the associated public health concerns and the social context of differential medical treatment based on race, gender, sexual orientation, age, disability

---

[2] One aspect of this balance can be seen on the FDA webpage "FDA updates on hand sanitizers that consumers should not use," with the following hazards (including product recalls): contains methanol or 1-propanol; contains microbial contamination; is subpotent, containing insufficient levels of ethyl alcohol, isopropyl alcohol, or benzalkonium chloride; is packaged in a container that resembles a food/beverage container, increasing risk of accidental ingestion.

status, socioeconomic status, and geographic location (Barabino, 2021). Addressing these disparities in chemical engineering education can even help attract students from diverse backgrounds (the importance of which is discussed in Chapter 9), students who are more likely to engage in exploring these issues as they go forward in the field (Thoman et al., 2015).

One opportunity for applying engineering solutions to reducing disparities is in low-resource settings. With respect to COVID-19, for example, chemical engineers can play a role in developing affordable vaccines for low-income populations who may not have access to traditional vaccine supply chains, distribution facilities, or cold-storage options (e.g., Frueh, 2020). Generally, creating appropriate technologies for low-resource settings requires that engineers consider not only the scientific rigor and effectiveness of the technologies but also their adaptability to local needs and ability to be maintained using resources the community can afford (TARSC, 2015). Today, while information is easily available to anyone with internet access, scientific tools and health care devices still prove to be expensive and inaccessible for many communities. As an example of addressing this problem, chemical engineers played a role in creating an appropriate and low-cost technology for diagnosing sickle cell disease in rural sub-Saharan Africa. What followed was a variety of new, low-cost, point-of-care testing devices that enable individuals to seek out sickle cell treatment in time (McGann and Hoppe, 2017; Nnodu et al., 2019; Oluwole et al., 2020).

Ethics, empathy, and attitudes are interrelated and important for eliminating disparities and building pathways to health equity. Conversations surrounding the ethics of engineering typically emphasize the integrity of the procedural steps of the scientific process. When creating and implementing new technologies, chemical engineers need to incorporate a broader set of ethical considerations and cultural competencies. Essential ethical considerations include the impact of new technologies or processes on low-resource communities and marginalized populations who experience greater health disparities, including how new treatments or technologies will be received among different cultures and populations and the impact on the environment. Chemical engineers have an opportunity to help reduce health disparities when they explicitly incorporate human-centered design into technologies to make them more accessible, equitable, and culturally sensitive. Collaboration across disciplines—engineers joining with clinicians, social scientists, policy makers, and members of the communities being served—will help address multilevel determinants of disparities and lead to better and more equitable interventions.

Another aspect of increasing accessibility to reduce disparities is lowering the cost of therapeutic interventions. The U.S. demand for monoclonal antibodies (mAbs) and a number of other biological compounds continues to grow, in part because of the increasing average age of people in the United States (Figure 5-10) and the diseases associated with aging, including cancer and cardiovascular and respiratory diseases. The success of mAbs in addressing those diseases drives demand. Unfortunately, the cost of producing biologics and the subsequent costs to consumers create pressure to improve flexibility and reduce costs so as to increase health care equity while maintaining reliability and stability during manufacturing and distribution.

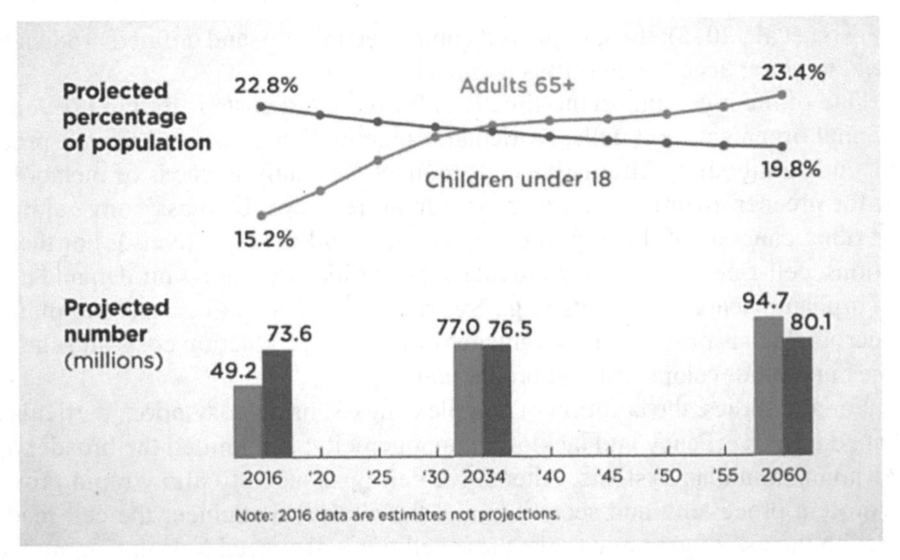

FIGURE 5-10 Projected population share of children under 18 and adults aged 65+ up to 2060. Adults aged 65+ are expected to outnumber children under 18 by 2034. SOURCE: U.S. Census Bureau (2018).

Approaches to reducing costs include streamlining workflows and increasing plant size, as well as developing strategies for producing higher protein yields per unit volume both in bioreactors and during purification and storage. One of the most obvious ways to produce high protein yields is to use continuous bioreactors, or perfusion culture. As in a traditional continuous reactor, this approach entails keeping catalyst (in this case the cells) within the reactor while adding fresh reactants (media) and removing product and spent media. Industry has long avoided this approach, as product yields have reached 2–10 g/L in fed-batch culture through media and process-based optimization. However, increased application of single-use technologies has opened up the possibility of broader application of perfusion technology, which can lead to three-fold increases in volumetric productivity (Bausch et al., 2019). In addition, expanding manufacturing space quickly is much more challenging with traditional plant designs, with new facilities costing more than $400 million and taking 5 years to build (Jagschies, 2020).

Some avenues toward improving perfusion culture processes include better scaled-down models to enable early development and screening of cells and optimization of media, which during traditional industrial development can take 3–6 months. Blending of concentrated media stocks by in-line dilution is also needed; even four-fold concentrates can save time and resources. Needed as well is the ability to combine these blending and perfusion runs with adaptive control technology driven by models of nutrient consumption, as well as improved cell-line development to ensure that the cells provide robust product (yield and quality) over the length of time expected for growth (2–3 months). In addition, supply chain issues, such as quality and consistency of raw material, become even more important with a perfusion approach to minimize product variance. Regulatory issues become more significant as well in that the product may change composition with time (Allison et al., 2015); thus, improved control technology and defined product batches could lead to better acceptance of this approach.

One of the most important directions for reducing costs is moving beyond traditional model organisms and cells, particularly mammalian cells used for the production of therapeutic antibodies. Alternative cells to meet the catalytic needs of metabolic engineering for greener chemistries, wastewater treatment, and biomass conversion are the focus of other chapters of this report (Chapters 3, 4, and 6, respectively). For therapeutic applications, cell-free systems may provide opportunities to address on-demand therapeutics and orphan disease treatments (e.g., Swartz, 2012). As an alternative to mammalian cells, microbial hosts provide the advantages of reduced production costs and shorter process times in both development and production.

For antibodies, the addition of complex sugars, or glycosylation, particularly of a type that addresses efficacy and lack of immunogenicity, has limited the broader applicability of nonmammalian systems. Alternative cell types need to allow rapid growth; authentic protein processing and secretion; and the ability to engineer the cell readily, including with the use of genomic tools; these cell types also need to lack mammalian viral infectivity (Matthews et al., 2017). A number of chemical engineers have already made compelling cases for and shown successes in developing and implementing genetic tools for the use of yeasts, from *Pichia pastoris* to *K. lactis, Y. lipolytica,* and *K. phaffi* (Hamilton et al., 2006; Jiang et al., 2019; Miller and Alper, 2019; Panuwatsuk and Da Silva, 2003); *S. cervisiae* and *P. pastoris* are already FDA-approved production organisms for vaccines and cytokines. To enable widespread application, continued improvements in volumetric productivities are needed, as are improved genetic tools and models for metabolism, as well as a better understanding of why some products are not well expressed.

More broadly, developing technologies or tools that replicate the capabilities of their expensive counterparts—so-called frugal science—enables out-of-the-box solutions for global science and health care. For example, considering the different parts of key devices and creating alternative low-cost parts resulted in the creation of a $1 hearing aid (Sinha et al., 2020) and a hand-powered paper centrifuge (<$0.25) that enables blood separation in resource-poor settings (Bhamla et al., 2017).

CHALLENGES AND OPPORTUNITIES

Modern biomolecular engineering is very much at the intersection of chemical engineering and molecular biology, biochemistry, materials science, and medicine. Current challenges for applications in health and medicine include advancing personalized medicine and the engineering of biological molecules, including proteins, nucleic acids, and other entities such as viruses and cells; bridging the interface between materials and devices and health; improving the use of tools from systems and synthetic biology to understand biological networks and the intersections with data science and machine learning; developing the next steps in manufacturing; and using engineering approaches to address equity and access to health care. All of these challenges present opportunities for chemical engineers to apply systems-level approaches and their ability to work across disciplines. Specific opportunities include the application of systems engineering to biological systems in such areas as

- sensor design and analysis;
- identification of fault detection (using physiological, cellular, metabolic, or other data to identify changes to function);
- process modeling to represent complex relationships for biological systems and predict behavior; and
- understanding and modification of molecular pathways and genomic networks involved in the regulation of normal physiology, as well as disease.

Opportunities to apply quantitative chemical engineering skills to immunology include cancer immunotherapies, vaccine design, and therapeutic treatments for infectious diseases and autoimmune disorders. The development of completely noninvasive methods for drug delivery represents an exciting frontier of device- and materials-based strategies. Chemical engineers are also well positioned to advance work on sustained-release depots and targeted delivery of therapeutics.

The demand for mAbs, therapeutic proteins, and mRNA therapeutics continues to grow, in part because of the increasing average age of people in the United States. Unfortunately, the costs to produce biologics and the subsequent costs to the consumer create pressure to improve flexibility, reduce costs, and increase health care equity while maintaining reliability and stability during manufacturing and distribution. This challenge provides an opportunity for chemical engineers to develop novel bioprocess and cell-based improvements through collaborations with biologists and biochemists.

Across all these areas, interdisciplinary and cross-sector collaborations will be critical, as will coordination across the federal agencies with responsibilities in these areas. An additional major challenge for chemical engineers is their lack of awareness of health disparities. Introducing these issues in the classroom will enable future chemical engineers to play an active role in reducing health inequities. Furthermore, increased diversity of both students and instructors in the classroom will provide a broader perspective on the challenges requiring engineering solutions.

Recommendation 5-1: Federal research investments in biomolecular engineering should be directed to fundamental research to

- **advance personalized medicine and the engineering of biological molecules, including proteins, nucleic acids, and other entities such as viruses and cells;**
- **bridge the interface between materials and devices and health;**
- **improve the use of tools from systems and synthetic biology to understand biological networks and the intersections with data science and computational approaches; and**
- **develop engineering approaches to reduce costs and improve equity and access to health care.**

Recommendation 5-2: Researchers in academic and government laboratories and industry practitioners should form interdisciplinary, cross-sector collaborations to develop pilot- and demonstration-scale projects in advanced pharmaceutical manufacturing processes.

6
Flexible Manufacturing and the Circular Economy

- Chemical engineering as a discipline was founded in the need to deal with heterogeneous feedstocks, especially petroleum, and this need will be amplified in the transition to more sustainable feedstocks.
- The production and manufacturing of useful materials and molecules enabled by chemical engineers are now creating problems that must be solved at scale.
- Chemical engineers play a critical role in manufacturing and can thus contribute to more sustainable manufacturing through efficiency, nimbleness, and process intensification.
- The principles of green chemistry and green engineering will be important in the shift from molecular to larger system scales, and to more sustainable manufacturing and a circular economy.

The chemical engineering discipline is broadly concerned with enabling realistic, cost-effective, efficient, and safe physical and chemical transformations of matter into more useful molecules or materials. In the last century, the discipline of chemical engineering enabled transformations of the entire landscape of both modern society and the planet. This chapter focuses on some key examples of the challenges and opportunities for chemical engineering in moving toward more flexible manufacturing and a circular economy.

Manufacturing is generally defined as the synthesis and formulation of useful products. In the last century, chemical engineers revolutionized manufacturing across all sectors of the economy, including agrochemicals and fertilizers, cement, consumer goods, flavors and fragrances, food and feed, fuels, paints and coatings, paper and pulp, pharmaceuticals and biologics, polymers, semiconductors, and many others. To give some idea of scale, for plastics alone—a prominent example discussed later in this chapter—the mass of these synthetic materials manufactured in just the last 70 years now exceeds the mass of all living animals on the planet (Elhacham et al., 2020). Additionally, the chemicals and materials manufactured at scale today contain substantial embodied energy and produce significant greenhouse gas (GHG) emissions (Figure 6-1). These two metrics are important benchmarks, along with issues of environmental and social justice and supply chain resilience, for calibrating new technology development that can mitigate anthropogenic climate change.

Given their critical role in manufacturing, chemical engineers have many opportunities to increase its environmental sustainability. This chapter provides an overview of the intersection of manufacturing and chemical engineering, followed by a discussion of feedstock flexibility, distributed manufacturing and process intensification, and the importance of transitioning from a linear to a circular economy.

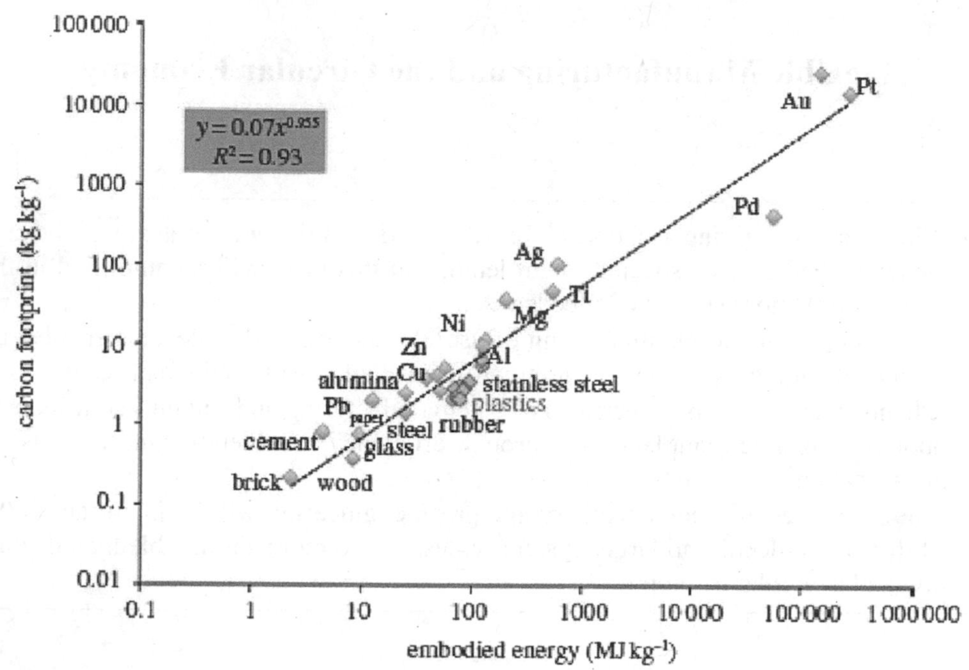

FIGURE 6-1 CO_2 emissions, in kg CO_2/kg material, and embodied energy, in MJ/kg material, for various materials. SOURCE: Gutowski et al. (2013).

INTERSECTION OF MANUFACTURING AND CHEMICAL ENGINEERING

The intersection of manufacturing and chemical engineering is founded on the systems-level, quantitative approach intrinsic to the profession (Peters et al., 2002; Turton et al., 2012). This approach includes the ability to conduct rigorous mass and energy balances, coupled with appropriate technoeconomic assessment (TEA) and life-cycle assessment (LCA). The tools of TEA and LCA enable detailed analysis of developed processes to identify potential efficiency gains from reducing cost, energy use, or material inputs. These efficiency gains are accomplished by an improved ability to recycle materials or by the transition from batch to continuous processes, among other changes. TEA and LCA tools are also critical in manufacturing to identify "leapfrog" processes that can displace current methodologies (TEA and LCA are described in more detail in Chapter 8). More recently, consideration of environmental and social justice has become increasingly important for chemical engineers in these analyses.

In addition to TEA and LCA, the overall principles of green chemistry (Anastas and Warner, 1998) and green engineering (Anastas and Zimmerman, 2003), highlighted in Box 6-1, provide qualitative guidelines that chemical engineers can place in a quantitative, objective context with systems-level approaches. This capability ultimately enables more efficient and responsible manufacturing practices and better decision making regarding trade-offs.

BOX 6-1
Principles of Green Chemistry and Green Engineering

Prevention—it is better to prevent waste than to treat or clean up waste after it has been created.

Atom Economy—synthetic methods should be designed to maximize the incorporation of all materials used in the process into the final product.

Less Hazardous Chemical Syntheses—wherever practicable, synthetic methods should be designed to use and generate substances that possess little or no toxicity to human health and the environment.

Design Safer Chemicals—chemical products should be designed to effect their desired function while minimizing their toxicity.

Safer Solvents and Auxiliaries—the use of auxiliary substances (e.g., solvents, separation agents) should be made unnecessary wherever possible and innocuous when used.

Design for Energy Efficiency—energy requirements of chemical processes should be recognized for their environmental and economic impacts and should be minimized. If possible, synthetic methods should be conducted at ambient temperature and pressure.

Use of Renewable Feedstocks—a raw material or feedstock should be renewable rather than depleting whenever technically and economically practicable.

Reduce Derivatives—unnecessary derivatization (use of blocking groups, protection/deprotection, temporary modification of physical/chemical processes) should be minimized or avoided if possible, because such steps require additional reagents and can generate waste.

Catalysis—catalytic reagents (as selective as possible) are superior to stoichiometric reagents.

Design for Degradation—chemical products should be designed so that at the end of their function they break down into innocuous degradation products and do not persist in the environment.

Real-time Analysis for Pollution Prevention—analytical methodologies need to be further developed to allow for real-time, in-process monitoring and control prior to the formation of hazardous substances.

Inherently Safer Chemistry for Accident Prevention—substances and the form of a substance used in a chemical process should be chosen to minimize the potential for chemical accidents, including releases, explosions, and fires.

Inherent Rather than Circumstantial—designers need to strive to ensure that all materials and energy inputs and outputs are as inherently nonhazardous as possible.

Design for Separation—separation and purification operations should be designed to minimize energy consumption and materials use.

Maximize Efficiency—products, processes, and systems should be designed to maximize mass, energy, space, and time efficiency.

Output-Pulled Versus Input-Pushed Products—process and systems should be "output pulled" rather than "input pushed" through the use of energy and materials.

Conserve Complexity—embedded entropy and complexity must be viewed as an investment when making design choices on recycle, reuse, or beneficial disposition.

continued

BOX 6-1 continued

Durability Rather than Immortality—targeted durability, not immortality, should be a design goal.

Meet Need, Minimize Excess—design for unnecessary capacity or capability (e.g., "one-size-fits-all" solutions) should be considered a design flaw.

Minimize Material Diversity—material diversity in multicomponent products should be minimized to promote disassembly and value retention.

Integrate Material and Energy Flows—design of products, processes, and systems must include integration and interconnectivity with available energy and material flows.

Design for Commercial Afterlife—products, processes, and systems should be designed for performance in a commercial "afterlife."

Renewable Rather Than Depleting—material and energy inputs should be renewable rather than depleting.

SOURCES: *Green Chemistry* (Anastas and Warner, 1998); *Green Engineering* (Anastas and Zimmerman, 2003).

Safety and safe operations are the most critical responsibility of the chemical engineering field. Safety is more important than reaching the goals of improving efficiency, increasing cost-effectiveness, and lowering environmental impacts of manufacturing processes. Simply put, many of the centralized industrial manufacturing facilities that chemical engineers have enabled over the last century handle gases, liquids, and solids at such scales and under such operating conditions that an accident can harm people and damage local and regional environments. Strict adherence to safety, including its inclusion in chemical engineering education, is essential to the discipline and needs to be rigorously maintained from the laboratory to the refinery.

Most conventional manufacturing processes in a chemical engineering context are operated at extremely large scale and in capital- and operating-intensive centralized facilities to harness economies of scale. However, advances in the valorization of flexible feedstocks, electrification of the manufacturing sector (Schiffer and Manthiram, 2017), and the concepts of scale-out and distributed manufacturing will play a key role for chemical engineering going forward. Some of these concepts are likely to play major roles in the deployment of manufacturing to low- and middle-income countries and to economically depressed regions of higher-income countries, as well as in various efforts at reshoring of manufacturing through new technologies. Indeed, industrial manufacturing has the potential in this century to at least partially transform physically from the scale of the petrochemical complexes studied by today's undergraduate chemical engineers to more heterogeneous intensified and distributed manufacturing sites, including those with electrically driven power sources. Notably, flexible and distributed manufacturing have already been a major focus of research and development among the international community, but substantial opportunities remain for the U.S. chemical engineering community to contribute meaningfully in these areas.

Lastly, the concept of the circular economy is commonplace in today's vernacular, including across many STEM (science, technology, engineering, and mathematics) professions. The Ellen MacArthur Foundation defines the circular economy as a systemic approach to economic development designed to benefit businesses, society, and the environment; in contrast to the "take-make-waste" linear model, a circular economy is regenerative by design and aims to gradually decouple growth from the consumption of finite resources (EMF, 2021). This broad concept has a figurative home squarely in the chemical engineering approach to manufacturing and green engineering principles because the ideas of materials recycling and, more generally, the efficient use of matter and energy are central to the field's systems-level thinking. Transitioning manufacturing from a linear to a circular economy is a key opportunity for chemical engineers.

FEEDSTOCK FLEXIBILITY FOR MANUFACTURING OF EXISTING AND ADVANTAGED PRODUCTS

The chemical engineering profession emerged in large part to confront the urgent challenges faced more than a century ago in the then-burgeoning petroleum refining industry. Petroleum is a highly heterogeneous organic resource. The global-scale petrochemical refining industry, in concert with the chemical engineering profession, was thus born out of the ability to characterize, fractionate, and ultimately valorize feed streams that are highly diverse in chemistry and that change as a function of time and source. Indeed, many of the original fractionated streams that could be derived from petroleum processing as a consequence of producing the original target fuels were considered waste. The ingenuity of chemists and chemical engineers, however, gave rise to uses for these waste compounds, including such diverse applications as asphalt, building blocks for synthetic polymers, and such formulated products as lubricants and processing fluids. These and myriad other high-value chemical applications today form the highly profitable chemicals backbone of the petrochemical business. The scale of the petrochemical industry worldwide is staggering: in 2019 its annual production volumes were 5–5.5 billion metric tons (100.69 million barrels per day) of crude oil, 4.1 trillion cubic meters (3.6 billion tons of oil equivalent) of natural gas, and 7.96 billion metric tons of coal (IEA, 2021c,d; EIA, 2021d).

The continued drive toward more efficient, environmentally friendly, and cost-effective manufacturing processes will likely benefit greatly from a much wider range of available feedstocks for use as building blocks to produce the chemicals and materials demanded by modern society (Figure 6-2). This concept of *feedstock flexibility* can be broadly defined as the input of diverse feedstocks into a transformation process that is able to process various starting compounds or mixtures of compounds to produce the target product. Petrochemical refineries already do this today. Importantly, today's petroleum feedstock requires oxidative chemistry to produce oxygenated molecules from hydrocarbons. For a different feedstock that is already oxygenated, such as lignocellulosic biomass, reductive chemistry would be needed to manufacture such products as hydrocarbons.

The use of a flexible feedstock also influences ideas around distributed manufacturing, as discussed below. For example, the availability of stranded natural gas resources, along with the potential harm of leakage of GHGs from those resources, makes conversion of these streams via chemical, biological, electrochemical, or other means a key opportunity for chemical engineers. The use of nonthermal approaches requiring minimal utility infrastructure may be critical for the ultimate feasibility of small-scale and distributed harnessing of such feedstocks as stranded natural gas; industrial waste gases; and industrial, commercial, or municipal wastewater (Khalilpour and Karimi, 2012; Tuck et al., 2012). This work also includes, as shown in Figure 6-2, the use of selective transformations afforded by biological and chemocatalytic transformations. Indeed, biological transformations of conventional feedstocks, combined with knowledge from the environmental bioremediation community, can show how bioprocessing can be incorporated into the petrochemical industry.

Flexible feedstock sources beyond those derived from fossil fuels include large amounts of available biogenic and waste carbon inputs, such as municipal solid waste (MSW). The goal of integrating biobased pyrolysis oil made from lignocellulosic biomass or MSW into a petroleum refinery has been pursued for several decades (Chen et al., 2014; Talmadge et al., 2014), despite challenges with inorganic foulants and catalyst poisons common in biogenic and waste carbon. The chemical engineering community has a key opportunity to understand how biogenic and waste-based substrates affect current manufacturing infrastructure. And much more room is available to explore the concepts of "refinery integration" of stabilized, biogenic and/or waste-based, carbon-rich streams into the existing, mostly amortized petrochemical complex.

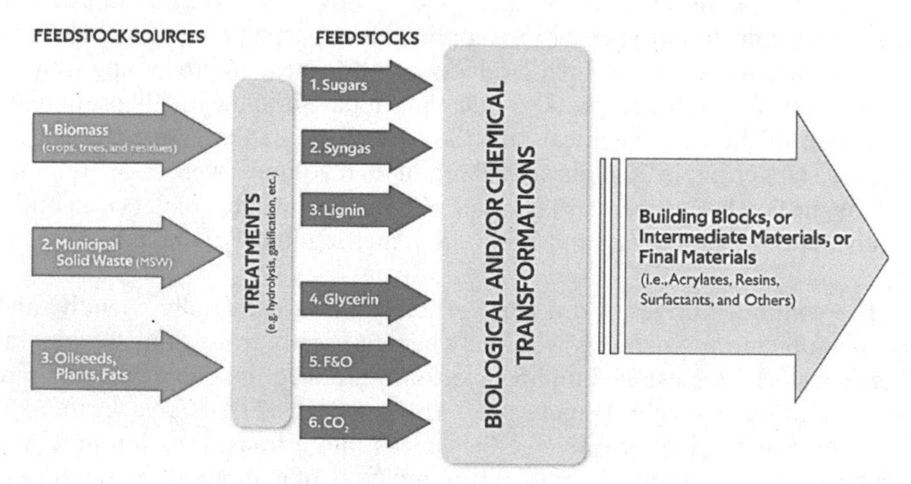

FIGURE 6-2 Possible treatment and transformation pathways for more flexible feedstock sources and feedstocks. NOTE: F&O = fats & oils.

More broadly, existing waste streams of biogenic and waste-based carbon could serve as useful feed streams for leapfrog technologies. Chemical engineers are already playing critical roles in the harvesting, densification, conversion, and scale-up of innovative processes for leveraging these biogenic and waste-based feedstocks. As discussed in Chapter 3, by 2030 there will be an estimated 1 billion dry tons of lignocellulosic biomass annually in the United States alone that can potentially be sustainably harvested (DOE, 2016). This potentially large feedstock, along with large amounts of other available wet waste (e.g., food waste, manure, sludge, fats, oils, and greases total are ~700 million tons per year; Milbrandt et al., 2018), is distributed across the United States. Not only does its use offer the potential to produce meaningful amounts of transportation biofuels to offset substantial GHG emissions (Chapter 3), but it also could serve as a key feedstock for chemical manufacturing. Products could include both direct-replacement biochemicals (Nikolau et al., 2008) and biochemicals that do not resemble their fossil-based counterparts but offer a performance advantage (Cywar et al., 2021; DOE, 2018a). MSW also offers a substantial and important feedstock, which again is highly heterogeneous.

The manufacturing of direct-replacement chemicals offers substantial opportunities for chemical engineers to develop scaled-out, distributed manufacturing systems and innovative, large-scale processes that can compete with the conversion of fossil resources. Conversely, performance-advantaged bioproducts could also serve as economic incentives to invest in new capital infrastructure at scale to displace the fossil carbon-based feed streams that dominate chemicals and materials production at scale today. These target bioproducts are an opportunity for chemical engineers to develop fully integrated, novel processes for transforming typically highly oxygenated feedstocks (e.g., sugars, aromatics derived from lignin, algae biomass) into novel molecules and new materials for which the properties often are not known a priori. Indeed, performance-advantaged bioproducts can offer potential benefits along the entire value chain (from feedstock to manufacturing, use, and end of life) relative to incumbent chemicals and materials, thus providing product design, economic, and environmental benefits (Cywar et al., 2021; DOE, 2018a). Systems-level approaches will be necessary to understand the ultimate potential of a given process concept or early demonstrations to reach scalability for manufacturing processes (Cywar et al., 2021).

From a process perspective, the conversion of heterogeneous feedstocks of essentially any type into valorized end products can be broadly categorized into the "fractionate-first" approach that the petrochemical complex has successfully adopted or a "one-pot" approach that attempts to convert all feedstocks simultaneously. The latter includes such conversion approaches as hydrothermal liquefaction, pyrolysis, and gasification. While TEA and LCA, along with the demonstrable technical feasibility of a given process, will ultimately and quantitatively inform how various processes are adopted, scaled, and enabled, many opportunities exist to define new flowsheets using emerging tools in electrochemistry, photochemistry, synthetic biology, integrated separations and catalysis, and many other tools that are familiar to chemical engineers. These approaches, and combinations thereof, present opportunities to significantly change the manufacturing landscape.

From a chemistry perspective, new feedstocks—especially those related to lignocellulosic and algal biomass, wet organic waste, and CO_2 and other waste gases—are often highly oxidized relative to conventional fossil feedstocks. Many common existing transformations add heteroatoms (e.g., nitrogen and oxygen) to hydrocarbon feedstocks to manufacture products. The use of new feedstocks that are more oxygen rich (and potentially contain other heteroatoms) offers the opportunity to develop new and novel transformation processes. These transformations could also take place in a condensed rather than a gas phase, the latter of which is typical of most conventional hydrocarbon processing. New separation technologies will be critical to realizing these transformations, as will catalysts that can enable the necessary reductive chemistry while remaining stable in aqueous environments.

Beyond innovative process and chemistry developments, there are fundamental research problems for chemical engineers to solve in the feedstock flexibility arena. A prominent example that hinders systems-level work is the lack of robust thermodynamic data in common process-modeling packages for the molecules found in biobased or waste-based feedstocks. Such data are plentiful for the hydrocarbon-rich feedstocks of relevance to petrochemical refining. However, the ability to model important thermodynamic properties of other molecules, which are often richer in oxygen and other heteroatoms, poses a substantial challenge, as does their incorporation into process simulators.

PROCESS INTENSIFICATION AND DISTRIBUTED MANUFACTURING

Many of chemical engineering's historical successes involve the efficient production of chemicals on a large and therefore economical scale. A world-scale ammonia plant, for example, generates 1,000 metric tons of ammonia each day. The design tools needed to scale up individual unit operations in these processes are well established and are a key focus of undergraduate chemical engineering education. Many opportunities exist to develop novel processes, which can potentially improve performance but have not always been used in commodity-scale plants.

Process intensification (PI) is designed to create improved chemical processes by moving beyond the idea that a single piece of equipment performs a single unit operation. In a traditional chemical plant, for example, a reactor and a distillation column might be used separately for a reaction and a separation step in an overall process. Reactive distillation (e.g., Taylor and Krishna, 2000) and membrane reactors (e.g., Iulianelli et al., 2016) are two alternative PI strategies that combine these two steps into a single process. PI also encompasses efforts to use nontraditional driving forces to accomplish unit operations—for example, microwave heating in reactions (e.g., Goyal et al., 2019) or the use of structured contactors in adsorption-based separations (e.g., Koros and Lively, 2012). Chemical engineers have numerous opportunities to use these strategies to develop innovative new processes, as well as to remove bottlenecks from existing large-scale processes. (Examples relevant to electronic-materials manufacturing are highlighted in Box 6-2). While PI presents many opportunities for innovation, it is also necessary to acknowledge that there are often trade-offs between PI and process flexibility, which may be more important for the flexible feedstocks discussed elsewhere in this chapter.

BOX 6-2
Process Intensification in the Electronic Materials Industry

Process intensification (PI) has been a major technology focus since the 1990s (Creative Energy, 2007). PI reduces the size and cost of chemical production modules. The short time scale for the commercialization of electronic materials and the rapid ramp-up of the demand cycle require manufacturing solutions that can be scaled up quickly to meet the demand. Prepositioning of modular assets becomes economical when the process equipment is compact, self-contained, and standardized (Bielenberg and Bryner, 2018; Stankiewicz and Moulijn, 2000).

An important solvent used in the electronic industry—propylene glycol methyl ether acetate (PGMEA)—has been the subject of numerous applications of PI to reduce waste, improve quality, reduce impurities, and lower costs, including catalytic distillation using a fixed-bed catalyst integrated into the distillation packing, which reacts propylene glycol monomethyl ether with methyl acetate (Hsieh et al., 2006; see Figure 6-2-1).

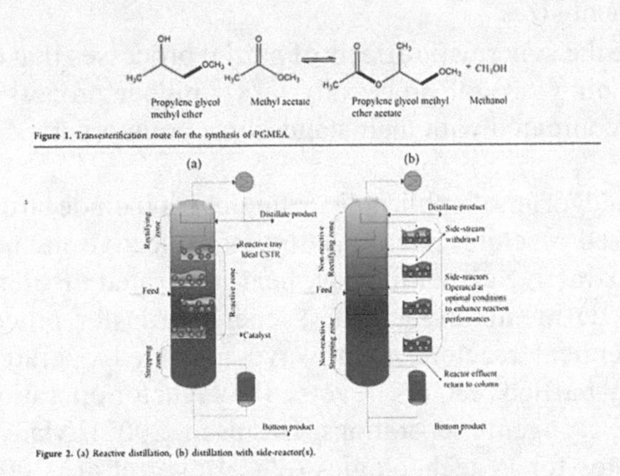

FIGURE 6-2-1 Reactive distillation supports a novel intensified process for the manufacture of a key solvent used in the electronics industry. SOURCE: Hussain et al. (2019).

Another example of PI in manufacturing of electronic materials is the use of external acceleration to eliminate the limitation of the gravity driving force in phase separation. Extraction and distillation can be compressed in small spaces using high-gravitational rotating packed beds. Nanoparticles are useful in many electronic-material processing steps, and monodisperse $CaCO_3$ nanoparticles can be made in high-gravity rotating packed-bed reactors. Breakthroughs are often achieved by applying existing technologies from different chemical and materials production industries to a new area (Cortes Garcia et al., 2017; Kang et al., 2018; Rao, 2015).

Nitrogen trifluoride is used as a cleaning and key etchant gas. The first large-scale industrial process used a two-step synthesis in which hydrofluoric acid was first fed to a traditional electrolysis cell to produce fluorine gas. The purified fluorine was then fed simultaneously with ammonia into a continuous stirred tank reactor, which produces nitrogen trifluoride and a solid waste stream of ammonium fluoride. In an intensified process, ammonia and hydrofluoric acid are added directly to a modified fluorine electrochemical cell to produce nitrogen trifluoride without a solid waste stream (Coronell et al., 1997; Hart et al., 2015; Krouse et al., 2016). This combination of reaction steps and the attendant simplification of process equipment reduce waste and lower capital cost.

Four principles for PI design have guided thinking about how chemical processes are developed (Harmsen, 2007; Tian et al., 2018):

- Maximize the effectiveness of intra- and intermolecular events. Improving process kinetics is a major principle for obtaining higher process performance, as it is usually the underlying limiting factor for low conversion and selectivity.
- Give each molecule the same processing experience, which results in products with uniform properties. Uniform product distributions facilitate waste reduction, which in turn reduces the efforts required for product separation.
- Optimize the driving forces at every scale, and maximize the specific surface area to which these forces apply. Doing so results in more efficient processes that use lower amounts of enabling materials, which then leads to reductions in equipment sizes.
- Maximize the synergistic effects of partial processes that enable multitasking. By combining several processing tasks, higher process efficiencies can be achieved compared with their stand-alone counterparts.

The general categories of applicable technologies include structured devices (e.g., structured catalyst-based reactors, microreactors, nonselective membrane reactors), hybrid processes (e.g., extractive crystallization, heat-integrated distillation, reactive distillation, selective/catalytic membrane reactors), energy transfer processes (e.g., rotating packed beds, sonochemical reactors, microwave-enhanced operations), dynamic processes (e.g., oscillatory baffled reactors, reverse flow reactor operation), and others (e.g., supercritical reactions, cryogenic separations; Harmsen, 2007). Many catalytic processes are also good candidates for PI technologies (e.g., Boger et al., 2003; Broekhuis et al., 2001; Cybulski and Moulijn, 1994; Kapteijn and Moulijn, 2020; Machado and Broekhuis, 2003; Machado et al., 2005; Nordquist et al., 2002; Welp et al., 2006, 2009). For example, slurry catalytic processes in the specialty chemical and pharmaceutical industry with gas and one or two liquid phases can be intensified by replacing the slurry catalyst with a fixed monolith catalytic reactor. These reactors can be installed in a pump-around loop to existing classic stirred-tank reactors to allow modularization and the ability to operate a single stirred tank with multiple beds using a monolith-loop reactor arrangement. Finally, microreactors are a more recent trend in chemical reactor synthesis. Their small channels allow for extremely accurate temperature control and high mixing intensity for single- and multiphase reaction processes. Scale is achieved by increasing the numbers of microreactor systems. The continuous flow reaction and separation networks allow for adequate production to meet most demand.

Traditional large processes increase product volume by scaling up. An alternative strategy achieves product volume by scaling out the deployment of many compact processing units in parallel. This is the key aim of modular manufacturing. Scaling out has benefits over scaling up in such processes as

- water treatment, both in municipal settings and for produced water from oil or gas wells;
- upgrading of natural gas from remote wells where pipeline infrastructure for gas transport is problematic;
- processing of bioproducts where transportation costs play a significant role in net GHG emissions;
- pyrolysis of waste polyolefins where the size of the pyrolysis reactors is limited by heat transfer considerations; and
- industrial sectors, such as pharmaceuticals or electronic materials, where highly valuable products are often produced in small quantities.

Although examples in pharmaceutical processes typically use highly controlled feedstocks, the other examples are cases in which modular processes need to function despite significant variations in the availability and location of the process feedstock. Additionally, scaling out is sometimes necessary because of technical realities, while in other cases it offers an economic advantage. Pyrolysis of waste polyolefins is operated in scaled-out plants because of the heat transfer considerations; however, the price of the pyrolysis oil increases significantly in this context. On the other hand, and in the same example, collecting plastic waste over a large area and transporting it to a pyrolysis plant add cost that the benefit of a scaled-up plant might not be able to counterbalance. In this latter case, a distributed network of pyrolysis plants makes economic sense.

PI and modular manufacturing are important areas in which the chemical engineering research community can provide intellectual leadership. A challenge for the academic community is that the successes (or failures) of work in either area are inherently determined at a process scale. Both demonstrating that a process is possible at "lab scale" and combining such work with process modeling and/or LCA are needed to support large investments. Also critical is the development of new materials and processes that can be deployed readily at the requisite scale and cost for use in the target processes. To give just one example, individual membrane modules used in current water treatment applications have surface areas measured in hundreds or thousands of square meters. Any putative new membrane that cannot be produced readily and economically at this scale will simply have no meaningful impact in water treatment, regardless of how superlative its performance may be (Koros and Zhang, 2017). This observation highlights the importance of pursuing research focused on the manufacturability of modular components, not just on the development of new high-performance materials. This is also an area in which close working relationships between academic researchers and industrial practitioners can be fruitful.

Rapid advances in additive manufacturing (e.g., 3-D printing; see Box 6-3) have also opened up a wide range of possibilities for generating new devices used in chemical processes. These possibilities are perhaps especially rich in the realm of PI and modular manufacturing. Because much work in additive manufacturing is taking place in the mechanical engineering and materials science communities, chemical engineers have numerous opportunities to combine expertise from these adjacent fields with application-specific needs in chemical processing.

BOX 6-3
Additive Manufacturing

Additive manufacturing (AM), typically accomplished by a form of patterned layer deposition, is increasingly used to create a range of products, from dentures to mobile homes. These methods (e.g., vat polymerization, material extrusion, material or binder jetting, powder bed fusion) provide versatility and customization beyond what has historically been achieved with conventional manufacturing. The ease, flexibility, and distributed access of AM enabled support for the health care community during the early months of the COVID-19 pandemic, particularly in the production of nasal swabs and other devices, such as face shields, that were not otherwise able to replace standard medical equipment (America Makes, AMT, and Deloitte, 2021).

Improvements in vat polymerization stereolithography in particular have enabled a continuous form of AM, offering an increase in speed and quality over layer-based approaches, albeit with the technology's own size constraints (e.g., de Beer et al., 2019; Tumbleston et al., 2015). This technology depends on the existence of an inhibition volume in which there is no polymerization between the product and the cross-linking light source to replenish the source polymer. Chemical engineers are particularly well suited to addressing the fluid transport challenges of this medium.

AM or 3-D printing has opened up new avenues for chemical engineers and materials scientists to create novel structures as research prototypes or products. A key enabling feature of 3-D printing has been the ability to customize the structure of the material in question and go from the design phase directly to manufacturing without having to go through the prototype phase. The last decade has witnessed particular growth in this field, with 3-D printing evolving from a "boutique" tool to one that can serve machine shop and manufacturing needs. Technological advances over the last decade have improved the speed and scale of 3-D–printed objects using various polymeric materials and metals. This capability is particularly important for printing custom objects from individualized specifications. The ability to transform digital structures into their physical form makes 3-D printing especially indispensable to 21st-century research and manufacturing.

3-D printing has also been actively pursued for printing organs. Organs, by nature, are characterized by 3-D architectures needed for their respective biological functions (e.g., chambers and valves for the heart, nephron tubes for kidneys). Current approaches for organ engineering rely on the use of appropriate templates that provide initial structural support for cell growth, but then rely on the natural ability of cells to self-organize into the right tissue architecture. 3-D printing provides an additional layer of architectural control. Combinations of materials and cells have been 3-D printed to produce organoids that capture some of the essential biological and structural features of their natural counterparts.

Opportunities for chemical engineers in AM broadly include supporting advances for faster printing with higher resolution, as well as use of multiple or more advanced and sustainable materials with attention to end of life. These efforts are best undertaken in partnership with mechanical and software engineers, collaboratively improving the technologies that scale these processes for widespread use.

Lastly, many modular processes (e.g., the treatment of stranded natural gas) are likely to occur in remote locations where monitoring or access by highly trained operators is limited. This fact, together with the issue of feedstock flexibility discussed above, highlights the need to develop modular processes that are inherently robust to process upsets and that take full advantage of advanced instrumentation and control strategies. Modular processes could leverage electrochemical power generated at small scale and on-site and/or biological manufacturing, which can often require lower heat and power inputs.

THE CIRCULAR ECONOMY AND DESIGN FOR END OF LIFE

The Industrial Revolution, beginning with the invention of the steam engine in the 17th century, enabled the use of raw materials and energy—which seemed to be infinitely available at the time—to make and eventually to mass produce products. The resulting economy was thus primarily linear (i.e., take → make → dispose), with resources being extracted or harvested, energy being used to make products, and products being disposed of at their end of life (Collias et al., 2021). The world's linear economy annually generated about 110 million metric tons of MSW in 1900, more than 1 billion metric tons in 2000 (including about 80 percent of consumer products, excluding packaging, disposed of after a single use), and about 2 billion metric tons in 2016, and this number is forecast to grow to 3.4 billion metric tons by 2050 (Hoornweg et al., 2013; Kaza et al., 2018). The consequences of the growth of the current economy are unsustainable.

The Circular Economy

A sustainable future requires a transition to a circular economy. In that model, resources are managed differently than in the linear economy, the way products are made and used changes, and new consideration is given to the fate of products and materials at their end of life. The circular economy uses waste streams as sources of secondary resources, and it incorporates the principles of green chemistry and engineering (Box 6-1; Anastas and Warner, 1998; Anastas and Zimmerman, 2003; Collias et al., 2021).

In the circular economy model, materials and products are made efficiently and reused, thus preventing waste. If new raw materials are needed, they are produced sustainably. This model represents a paradigm shift that is consistent with emerging consumer preferences for recycling and reducing nonrecyclable waste (Nielsen, 2015), as well as new government restrictions on pollution and waste. A circular economy can be facilitated by emerging digital technologies and product designs that track materials and products and extend their lives. Decoupled from the consumption of finite resources, it can drive economic growth with business, societal, and environmental benefits. In the long term, the circular economy will achieve cost savings in materials that result in cost savings in products and packaging. For consumers, the benefits of the circular economy will come from new developments and trends in such areas as urbanization, technologies (e.g., anaerobic digestion), information technology capabilities, online retail sales, business models, and packaging technologies (EMF, 2014).

The circular economy model encompasses two cycles: biological and technical (Figure 6-3). In the biological cycle, food and biological materials (e.g., wood) feed back into the system through recycling processes, such as composting and anaerobic digestion, that regenerate living systems, such as the soil. In the technical cycle, such strategies as recycling, reuse, repair, and remanufacturing allow the recovery and restoration of products, components, and materials (EMF, 2013a,b, 2014).

The circular economy is based on three strategies (EMF, 2017a,b) that are consistent with the principles of green chemistry and engineering:

- Strategy 1: *Design to avoid pollution and waste*—Reduce or eliminate GHG emissions and hazardous substances and the resulting pollution of air, land, and water. Also limit, if not eliminate, waste of materials during the manufacturing of products and packaging, as well as during the discarding of products and packages at their end of life.
- Strategy 2: *Extend useful life*—Design for durability, reuse, remanufacturing, and recycling to keep materials, products, and packaging circulating in the economy as long as possible, thus preserving energy and materials.
- Strategy 3: *Regenerate natural systems*—There is synergy between a circular economy and a biobased economy. In the biobased economy, biobased materials are made from biobased resources and/or with the use of biobased energy, and these biobased materials cycle between the economy and natural systems (to become biobased resources). Ideally, the circular economy uses renewable and avoids the use of nonrenewable resources.

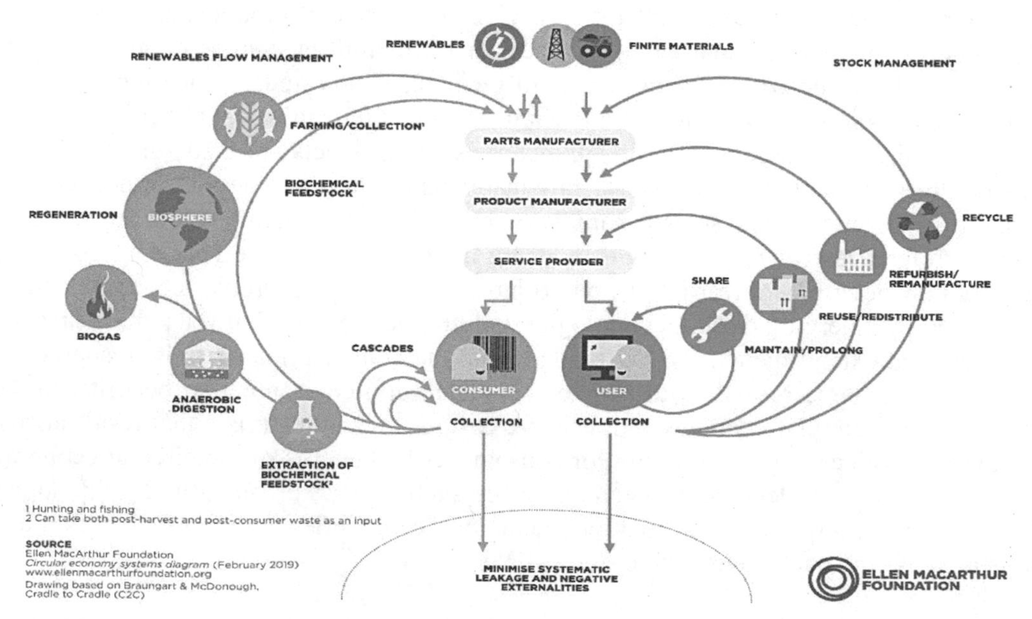

FIGURE 6-3 A circular economy system diagram showing the flow of materials, nutrients, components, and products (biobased materials are shown in green). SOURCE: EMF (2019).

Each of these strategies is described further below, with a focus on plastics as an example of an area in which chemical engineers will play a key role. The essential principles of the circular economy, however, apply beyond the manufacture of synthetic polymers.

Chemical engineers are uniquely positioned to solve problems associated with the three strategies of the circular economy. Their contributions could include redesigning processes and products to reduce or eliminate pollution, developing new ways to reduce and utilize waste, designing products to be used longer, and designing processes and products using sustainable feedstocks. Besides providing technical expertise and leadership in various areas of the circular economy, chemical engineers have opportunities to address challenges in the following four more specific technology areas:

- purification of materials with a large volume and a high rate of collection, such as paper, cardboard, polyethylene terephthalate (PET), glass, and steel;
- recycling of polymers with a large volume and a low rate of collection;
- utilization of by-products of manufacturing processes, such as used concrete, CO_2, and food waste; and
- development of materials with high value that currently have a small volume and a low rate of collection, such as 3-D printing materials and biobased materials.

Design to Avoid Pollution and Waste

Pollution and waste are associated with the production of all materials, products, and packaging. To show how concepts associated with the circular economy can be applied to reducing energy consumption and GHG emissions, it is useful to consider the opportunities for plastics.[1] More than 90 percent of the feedstock for the plastics industry is petroleum or natural gas. About 6 percent of the world's production of petroleum and natural gas liquids (NGLs) is used to produce plastics, with about half used as feedstock and the other half as fuel in the production processes (EMF, 2016). Another significant percentage of natural gas production is used as feedstock and fuel for plastics manufacturing.

Since the early 1950s, plastics have dramatically changed people's way of life, but at the same time, they pose an environmental challenge because of the means used for their disposal in various parts of the world. Since 1950, 8,300 million metric tons of synthetic polymers has been produced, and 4,900 million metric tons has been discarded (Figure 6-4). Resistance to degradation—one of the most important properties of many

[1] Here, for the purposes of simplicity, the terms "plastics" and "polymers" are used interchangeably. IUPAC (the International Union of Pure and Applied Chemistry) defines plastics as "polymeric materials that may contain other substances to improve performance and/or reduce costs." Plastics are manmade and include both thermoplastics and thermosets, whereas polymers can be manmade or occur naturally.

plastics—is the main cause of their persistence in the environment. Thus, a circular economy for plastics provides the best opportunity to continue enjoying the benefits of plastics in everyday life while reducing the environmental impact of their mismanaged disposal. This circular economy of plastics addresses both aspects of the first strategy listed above (i.e., both pollution and waste reduction).

The primary goal of the circular economy of plastics is that plastics never become waste, and instead reenter the economy. There are two main strategies for achieving this goal: create an effective after-use plastics economy, and drastically reduce the leakage of plastics into natural systems and other negative externalities (EMF, 2016). Steps to reduce the pollution and waste of plastics include selecting from and executing various options that are depicted in typical waste hierarchies (Billiet and Trenor, 2020). The most preferred option is rethinking and redesigning the product and package. If that option is not possible, other options, in order of preference, are as follows:

- Reduce the amount of plastic used in the product and package.
- Reuse the product and package.
- Recycle the product and package.
- Use the plastic in the product and/or package as fuel.
- Dispose of the plastic in the product and/or package in a managed landfill.

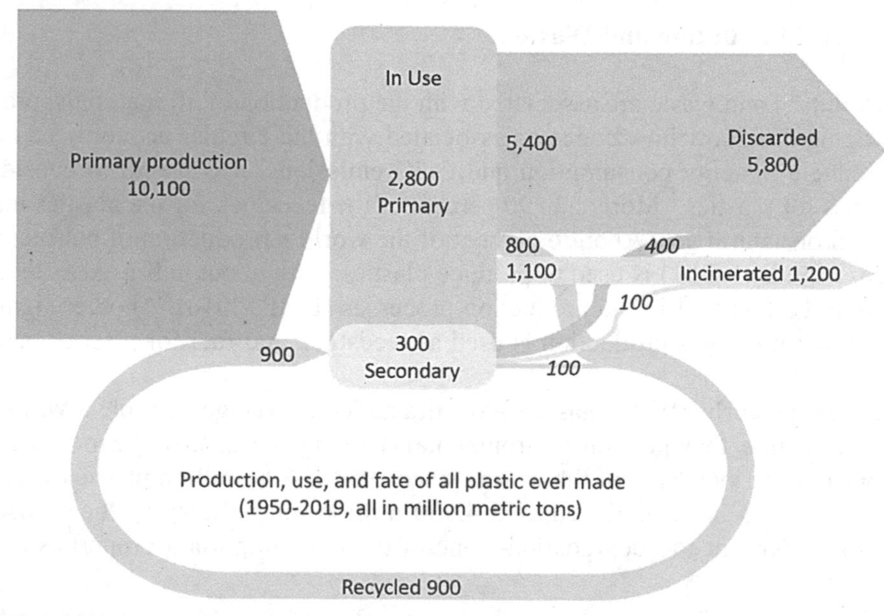

FIGURE 6-4 Global production, use, and fate of polymer resins, synthetic fibers, and additives, 1950–2019, in millions of metric tons. SOURCE: Geyer (2021).

The least desirable outcome is disposal of the plastic in a mismanaged way. Replacing fossil-derived plastics with biobased plastics (totally or partly) and using renewable energy in the production of fossil-derived plastics are some alternative options for reducing the carbon footprint of plastics. Chemical engineers have an opportunity to apply quantitative, systems-level thinking to this problem through the application of TEA and LCA to determine which options optimize emissions reductions while considering other trade-offs, such as water consumption, cost, and environmental justice. Chemical engineers also have an important role to play in increasing and improving the recycling of plastics. Opportunities exist in all recycling technologies, such as mechanical recycling; dissolution recycling; and advanced recycling methods, such as depolymerization, pyrolysis, and gasification (Figure 6-5).

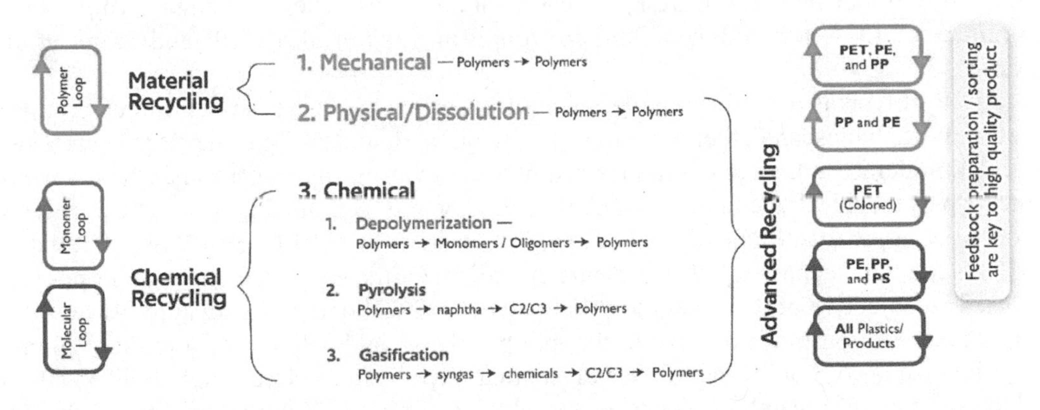

FIGURE 6-5 Processes (mechanical, dissolution, and chemical or advanced recycling processes) for recycling of plastics, recycling loops, and plastic feedstocks in each recycling process. NOTE: EoL = end-of-life; PE = polyethylene; PET = polyethylene terephthalate; PP = polypropylene; PS = polystyrene; WtE = waste-to-energy. SOURCE: Collias et al. (2021).

There are numerous opportunities to improve the quality of plastics produced and increase the range of plastics that can be mechanically recycled. Technologies that clean waste plastics and remove surface and bulk contamination are also the focus of chemical engineering work. Mechanical recycling of mixed polymers leads to the formation of polymer blends, and the associated technologies for compatibilization are important. Plastics present an additional challenge in that, unlike metals or glass in recycling, polymers change their structure. Advanced recycling will help recycle the polymers that mechanical recycling cannot, and could provide infinite recycling loops. An example of an LCA of mechanical recycling processes is described by Franklin Associates (2018).

Scaling up of dissolution recycling processes that produce food-grade plastics from waste plastics presents another opportunity for chemical engineers. Examples of such processes include Newcycling® by APK AG, PureCycle Technologies (Layman,

2019a,b), and CreaSolv® by CreaCycle GmbH (Maeurer et al., 2012). The main unit operations that chemical engineers can further develop, optimize, scale up, and commercialize are polymer dissolution, extraction, processing, layer separation, filtration, contaminant migration, and solvent diffusion and recycling (Pappa et al., 2001; Walker et al., 2020; Zhao et al., 2018).

Depolymerization of condensation polymers is applicable to polyesters (e.g., PET and polylactic acid [PLA]) and polyamides (e.g., nylon), among others. Hydrolysis, methanolysis, glycolysis, ammonolysis, aminolysis, and hydrogenation are typical chemical processes that, depending on the chemical used for the PET chain scission, degrade the starting polymers to either their respective monomers or oligomers. The product of depolymerization is purified to remove contamination and colorants, and then used to form new polymers. Catalyst or enzyme development, process optimization and scale-up, purification, and polymer processing are key areas with an intense chemical engineering focus. Examples of LCA for PET recycling are found in Shen et al. (2010) and Singh et al. (2021).

Pyrolysis is used for polyolefins, multilayer packaging, fiber-reinforced composites, polyurethanes, and other polymers that are difficult to depolymerize. Pyrolysis takes place at moderate to high temperatures and produces various hydrocarbons. Chemical engineers play a central role in the design of the reactor and its modeling. Catalytic pyrolysis (Cocchi et al., 2020; Miandad et al., 2016; Ratnasari et al., 2017; Zero Waste Scotland, 2013) is another technology that presents chemical engineering challenges and opportunities. The objective of this technology is to achieve C–C bond breaking at lower temperatures and reaction times relative to thermal pyrolysis, and to produce a higher-volume and higher-quality liquid fraction. Catalysts that have been explored include FCC (fluid catalytic cracking) catalyst, spent FCC catalyst, ZSM-5 (Zeolite Socony Mobil-5), HZSM-5, Cu-Al_2O_3, zeolites, Fe_2O_3, MCM-41 (Mobil Composition of Matter No. 41), coal fly ash, and mixtures of the above (Miandad et al., 2016; Ratnasari et al., 2017). In addition to thermal and catalytic pyrolysis technology vectors, plasma pyrolysis, microwave-assisted pyrolysis, and hydrocracking are other pathways for converting feedstock to pyrolysis oil for further conversion. A chemical engineering challenge in pyrolysis is to achieve process scale-up rather than process scale-out (i.e., scale via parallel units/modular design), which is common today, including with the advent of electrochemically driven reactors. This scale-up challenge results from the difficulty of achieving adequate heat transfer to the plastic (which acts as an insulator) as the volume of the reactor increases and its surface area does not increase proportionally. As a result, the typical annual capacity of pyrolysis reactors/plants is 5–10 thousand tons, and larger-capacity plants are constructed with many reactors in parallel rather than the more economical option of using larger reactors.

Gasification converts plastics to a gaseous mixture of CO and H_2 (syngas), CO_2, CH_4, and other light hydrocarbons via partial oxidation in the presence of steam and oxygen or air at less than the stoichiometric ratio (Higman and van der Burgt, 2008; Rezaiyan and Cheremisinoff, 2005).

Incineration can be useful for end-of-life plastics and other waste. In waste-to-energy (WtE) processes, for example, MSW is combusted, and the heat produced is used to make steam for generating electricity or to heat buildings. Besides producing electricity, WtE contributes to reducing the amount of material that would otherwise go to landfills and the resulting landfill emissions. In 2018 in the United States, about 12 percent of the 292 million tons of MSW was burned in WtE plants, generating 13 billion kWh of electricity in 2019 (EIA, 2020c). The percentage of MSW that feeds WtE installations ranges from 12 percent in the United States to 74 percent in Japan. Many large landfills generate electricity from the methane gas that is produced from the decomposition of biomass. Incineration and WtE processes can also potentially release hazardous chemicals and particles. These chemicals can affect air quality, affecting the neurological, respiratory, and reproductive systems of the human body and damaging the environment. Removing these chemicals from the incineration process is an important chemical engineering challenge.

Other potential solutions to the plastics disposal problem beyond the recycling technologies discussed above include the following:

- Use of biodegradable plastics and various enzymes and biodegrading organisms for plastics—LCAs are necessary to determine whether in some environments, the negative effects of the uncontrolled release of the biodegradation products into the atmosphere outweigh the benefits of using biodegradable plastics.
- Closed-loop recycling of polymers synthesized with in-chain functional groups that act as break points—For example, Häuβler and colleagues (2021) prepared a redesigned version of polyethylene with renewable polycarbonate and polyester in-chain functional groups that can be recycled chemically by solvolysis with a recovery rate exceeding 96 percent.
- Composting and anaerobic digestion—Compositing is a managed process of controlled decomposition of organic material by aerobic microorganisms producing compost (also called humus), CO_2, water, and heat. Anaerobic digestion is used primarily to process wet waste material with microorganisms that break down organic material in the absence of oxygen and produce biogas.

Extension of Useful Life

In addition to reducing environmental impact, there is also an economic incentive for keeping materials, products, and packaging in use as long as possible. The economic value of plastic packaging that is lost after a single use is estimated to be between \$80–120 billion, and the environmental cost of plastic packaging is estimated to be \$40 billion—more than the industry's total profits (McKinsey Sustainability, 2016). In designing products for longevity, the entire life cycle of the product and the value chain need to be managed. For example, if a physical unit is needed, design thinking might suggest making it more durable or making it easy to maintain by using designs that allow critical components to be replaced when worn out.

Apart from designing for easy disassembly, use of single-material packaging is at the top of the list of desirable options for a product at the end of life. Multimaterial packaging does not allow for mechanical and other types of recycling because the various materials cannot easily be separated. Such multimaterial packaging might include blends of different plastics or multilayer films, with each layer made of a different polymer. Separation technologies for multimaterial films (e.g., biomimicry-based adhesives between the materials or layers) and the development of monomaterial solutions that achieve the same performance as multimaterial solutions are two key areas in which chemical engineers will continue to contribute.

Regeneration of Natural Systems

A biobased economy includes both biobased materials and biobased fuels (fuel uses are discussed in Chapter 3). Sugars (e.g., glucose) are the typical feedstocks for a biobased economy, with biomass being the preferred primary source. Biomass is typically classified as generation 1 (e.g., sugarcane, corn, sugar beet, potato) or generation 2 (e.g., crop residues, grass, straw, wood, MSW). Because generation 1 biomass as a sugar source competes with food uses, generation 2 biomass is a better long-term source of sugars.

Generation 2 biomass contains three natural polymers: cellulose (35–50 percent of the biomass), hemicellulose (25–30 percent), and lignin (15–30 percent). Cellulose is a linear polymer composed of (1,4)-d-glucopyranose units linked by β-1,4-glycosidic bonds. Hemicellulose is a branched heteropolymer composed of hexoses, pentoses, and uronic acids linked by different bonds. Lignin is a high-molecular-weight, amorphous polymer composed of aromatic rings of phenyl propane. Various processes are required to separate biomass into its three main components, and then convert cellulose and hemicellulose to sugars, and lignin to other smaller chemical units. The challenges and opportunities for chemical engineering in the use of biomass are significant. Technically and economically successful hydrolysis of biomass and the production of low-cost sugars are key to enabling the biobased economy. Chemical and enzymatic hydrolytic processes need to be advanced in this space. Also, valorization of lignin is key to making overall biomass utilization successful.

Use of biomass for fuels, products, and power has been recognized by the U.S. Department of Energy (DOE) as a critical component of reducing dependence on volatile supplies and prices of imported oil (Biddy, 2016). The importance of chemicals in the economy is disproportionate to the amount of oil used to produce them. For example, only about 3–4 percent of petroleum is used to make chemicals, whereas the pretax value of the petrochemical products (e.g., plastics, detergents, paints, adhesives, cosmetics, pesticides) was about $375 billion in 2005—about the same as the pretax value of transportation fuels, which use about 71 percent of petroleum (Marshall, 2007). The production cost of biobased chemicals and its comparison with that of the corresponding petroleum-derived chemicals depend on the prices of biomass, petroleum, and natural gas, as well as production capacity, technology maturity, and other factors. As the current price of natural gas is relatively low (compared with the price of petroleum on an energy basis), the current production cost of the incumbent chemicals is relatively low relative to that of the biobased chemicals.

Over the past two decades, biorefineries have been proposed as a way to achieve favorable economics compared with typical petroleum refineries and downstream processing plants. The products of biorefineries can be fuels, chemicals, or both. The International Energy Agency (IEA, 2020b; IEA, ICCA, and DECHEMA, 2013) presented a vision of biorefineries based on eight feedstock platforms:

- Syngas—Biomass gasification yields syngas, which is converted into products using fermentation or other chemical processes.
- Biogas—Anaerobic digestion of high-moisture-content biomass (such as manure and waste streams from food processing plants) yields primarily methane, which can be used as a chemical feedstock.
- C6 and C5 sugars—C6 sugar streams and, to a lesser extent, C5/C6 mixed sugar streams can be used in fermentation processes to produce various chemicals.
- Plant-based oil—The basis of the oleochemical industry, with a by-product of glycerin, plant-based oil has been explored as feedstock for the production of chemicals.
- Algae oil—Algae has a higher productivity than plants because the entire biomass can be used to produce high-value products, such as high-protein food or feed, pigments, antioxidants, vitamins, and sterols.
- Organic solutions—When fresh wet biomass (such as grass or clover) is dewatered, an organic solution (nutrient-rich press juice) and a press cake (fiber-rich) are produced, and both can be used to make other chemicals.
- Lignin—Lignin's native structure suggests that it can be used to produce aromatic molecules; however, only a limited number of products from its derivatives, lignosulfonates and kraft lignin, have been produced so far.
- Pyrolysis oil—Biomass pyrolysis produces pyrolysis oil, which can then be upgraded to different chemicals.

Biobased plastics are also part of the circular economy of plastics. Currently, there is a biobased plastic-alternative material (either biobased, biodegradable, or both) for many fossil-derived conventional plastics, at least at laboratory or pilot scale (Siracusa and Blanco, 2020). There are three main groups of biobased plastics (Figure 6-6):

- biobased or partially biobased and nonbiodegradable plastics, such as biobased polyethylene (PE; e.g., Braskem's I'm green™), polypropylene, and PET and biobased polymers (e.g., polytrimethylene terephthalate [PTT]) (top left quadrant of Figure 6-6);
- biobased and biodegradable plastics, such as PLA, polyhydroxy alkanoate (PHA), polybutylene succinate (PBS), and starch-based blends (top right quadrant of Figure 6-6); and

- biodegradable plastics from fossil resources, such as polybutyrate adipate terephthalate (PBAT), polyvinyl alcohol (PVOH), and polycaprolactone (PCL) (bottom right quadrant of Figure 6-6).

Compared with the conventional fossil-derived plastics (bottom left quadrant of Figure 6-6), biobased plastics can potentially have a lower carbon footprint and additional end-of-life options, such as composting. The global production of biobased plastics was 2.1 million metric tons in 2020, representing less than 1 percent of total plastics production. Biodegradable plastics made up about 58 percent of the biobased plastics volume, with the remainder comprising biobased and nonbiodegradable plastics (European Bioplastics, 2021). LCA can be a valuable tool for evaluating the trade-offs between conventional fossil-derived and biobased plastics.

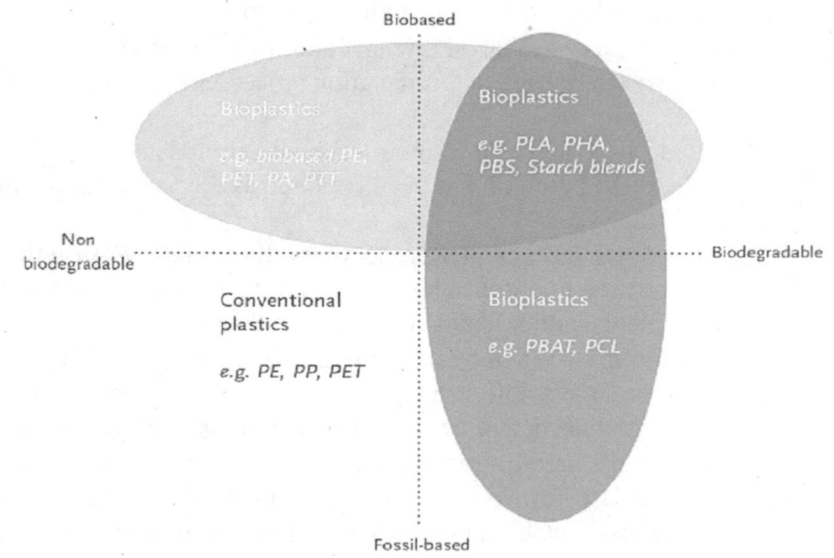

FIGURE 6-6 Qualitative plot of polymers based on their biodegradability (horizontal scale) and whether they are bio- or fossil-based (vertical axis). NOTE: PA = polyamide; PBAT = polybutyrate adipate terephthalate; PBS = polybutylene succinate; PCL = polycaprolactone; PE = polyethylene; PET = polyethylene terephthalate; PHA = polyhydroxy alkanoate; PLA = polylactic acid; PP = polypropylene; PTT = polytrimethylene terephthalate. SOURCE: European Bioplastics (2021).

The economics of the processes used to make biobased plastics depend heavily on the stoichiometry of the feedstock conversion to the product, and particularly on the oxygen content of the feedstock. For example, the theoretical mass yield of PE from sugar is about 30 percent, and the rest (70 percent) of the sugar mass is lost as water and CO_2 (see the oxygen chart in Figure 3-10 in Chapter 3). This process starts with sugar fermentation to ethanol, dehydration of ethanol to ethylene, and polymerization of ethylene to PE. This bio-PE cannot compete in price with fossil-derived PE because of the stoichiometry of the conversion reaction from carbohydrate to hydrocarbon and the economy of scale present in the production of fossil-derived PE. A similar argument holds for the production of biofuels from sugar or biomass. Note that the capacity of a typical petroleum refinery is 10–130 million metric tons per year (measured as the amount of petroleum that can be distilled in the atmospheric distillation units), whereas the capacity of an average ethanol fermentation plant is about 0.25 million metric tons, and that for ethanol dehydration to ethylene is smaller.

The biobased chemicals that are economically advantageous to produce from sugars are those that have an oxygen content similar to that of sugar or biomass, such as monoethylene glycol (MEG) and furan-2,5-dicarboxylic acid (FDCA), with mass theoretical yields of about 100 percent; acrylic acid (AA), with mass theoretical yield of 80 percent; and polyethylene furanoate (PEF), purified terephthalic acid (PTA), and PET, with theoretical mass yields exceeding 50 percent. Use of such feedstocks as lignin, CO, CO_2, and vegetable oils will be advantageous to produce chemicals with similar oxygen contents. Various LCA studies on biobased plastics have been reviewed (Walker and Rothman, 2020).

It is important for chemical engineers to be involved in the development of the biobased economy. More specifically, challenges and opportunities for chemical engineers in this domain include

- scalable and economical processes for producing biobased plastics (e.g., PEF and PBS) and biobased monomers (e.g., biobased terephthalic acid; Collias et al., 2014);
- new biodegradable plastics for the marine environment;
- scalable and economical technologies for deconstructing lignin to molecules that can be used as feedstocks for various chemicals, such as
  - bacterial, enzymatic, fungal, photocatalytic, hydrogenolysis, pyrolysis, oxidation via ionic liquids, and hydrolysis deconstruction technologies (Cao et al., 2018a; Glasser, 2019; Kellett and Collias, 2016; Ragauskas et al., 2014; Xu et al., 2019; Zakzeski et al., 2010); and
  - engineering of lignin to make it more amenable to low-energy chemical deconstruction, such as by using monolignol ferulate transferase to introduce chemically labile ester linkages into the lignin in poplar trees (Wilkerson et al., 2014), and the proposed work on use of CRISPR (clustered regularly interspaced short palindromic repeats)–based genome engineering to edit multiple genes simultaneously to optimize biomass processing (Chanoca et al., 2019);
- technologies that use CO_2 as a feedstock for various chemicals;
- reduction technologies for carbohydrate feedstocks to produce less oxygenated chemicals and hydrocarbons (e.g., dehydration of lactic acid to acrylic acid [Collias et al., 2018; Godlewski et al., 2014]; see Bozell and Petersen [2010] for the top 10 chemicals of the bioeconomy and the National Renewable Energy Laboratory [Biddy, 2016] for the 12 chemicals with prospects for near-term deployment);

- technologies that avoid energy-intensive processes, such as steam cracking; and
- technologies for converting waste plastics to higher-value chemicals or structures (upcycling), such as PE waste upcycling to long-chain alkylaromatics using tandem catalytic conversion by Pt supported on γ-alumina (Zhang et al., 2020b), lubricant and waxes (Celik et al., 2019; Rorrer et al., 2021), diacids (e.g. succinic acid, adipic acid) via oxidation, hydrogen (Uekert et al., 2020), and lithium-ion battery anodes (Villagomez-Salas et al., 2018).

CHALLENGES AND OPPORTUNITIES

A sustainable future will require a shift to a circular economy in which the end of life of products is accounted for, incorporating green chemistry and engineering. The chemical engineering profession emerged in large part to confront the urgent challenges faced more than a century ago in the then-burgeoning petroleum-refining industry. Petroleum is a highly heterogeneous organic resource. The continued drive toward more efficient, environmentally friendly, and cost-effective manufacturing processes will benefit from a much wider range of available feedstocks for use as building blocks to produce chemicals and materials. The challenge of feedstock flexibility offers chemical engineers an opportunity to develop advances in reductive chemistry and processes that will allow the use of oxygenated feedstocks, such as lignocellulosic biomass. Chemical engineers also have substantial opportunities develop scaled-out, distributed manufacturing systems and innovative, large-scale processes that can compete with the conversion of fossil resources. Fundamental research opportunities include the collection of robust thermodynamic data to improve the modeling of feedstock molecules that include oxygen and other heteroatoms.

Current challenges in process design include the need for improvements in distributed manufacturing and process intensification—areas in which the chemical engineering research community can provide intellectual leadership. Collaborations between academic researchers and industrial practitioners will be important for demonstration at process scale. In the transition from a linear to a circular economy, specific opportunities for chemical engineers include redesigning processes and products to reduce or eliminate pollution (e.g., Shi et al., 2021), developing new ways to reduce and utilize waste, designing products to be used longer, and designing processes and products using sustainable feedstocks.

Recommendation 6-1: Federal research funding should be directed to both basic and applied research to advance distributed manufacturing and process intensification, as well as the innovative technologies, including improved product designs and recycling processes, necessary to transition to a circular economy.

Recommendation 6-2: Researchers in academic and government laboratories and industry practitioners should form interdisciplinary, cross-sector collaborations fo-

cused on pilot- and demonstration-scale projects in advanced manufacturing, including scaled-down and scaled-out processes; process intensification; and the transition from fossil-based organic feedstocks and virgin-extracted inorganic feedstocks to new, more sustainable feedstocks for chemical and materials manufacturing.

7
Novel and Improved Materials for the 21st Century

- Chemical engineers have a critical role to play in the development of new materials and materials processes from the molecular to the macroscopic scale.
- Chemical engineers' integration of theory, modeling, simulation, experiment, and machine learning is accelerating the discovery, design, and innovation of new materials and new materials processes.

Chemical engineers are deeply involved in the design, synthesis, processing, manufacturing, and ultimately disposal of materials of all kinds, and the connections between the fields of materials science and chemical engineering are numerous and diffuse. A taxonomy might suggest that materials scientists and engineers are concerned mainly with materials structure, properties, and performance, while a chemical engineer would likely focus less on performance and more on materials processing, but there are as many exceptions to that generalization as there are examples. In industry, the distinction is usually unmeasurable. Full coverage of the myriad dimensions of chemical engineering's role in the materials economy would require its own report, but this chapter highlights several important aspects and key opportunities for the future. For example, materials research in academic chemical engineering departments will include work in polymer science, rheology, catalysis, biomaterials, nanomaterials, electronic materials, self-assembly, and soft matter; several of these subjects have been discussed in earlier, application-focused chapters of this report.

Chemical engineers have been responsible for many advances in materials design and development. An example is the reverse osmosis membrane for water desalination made of cellulose acetate and developed by chemical engineers in the 1950s (Cohen and Glater, 2010). Likewise, many of the polymeric materials used in regenerative engineering and drug delivery have emerged over decades from the laboratories of chemical engineers. Other contributions include new catalysts and zeolites (Ahmed, 2007), Gore-Tex®, and the automobile catalytic convertor.

Chemical engineers have played a key role in materials processing and system design that enable plants to make useful amounts of materials safely. They also are developing solutions for the consequences of decades of plastic generation, which has left the world awash in plastic waste. And they bring rigorous life-cycle assessment (LCA) to the processes that produce materials, as examined in greater detail in Chapter 6.

The economic and environmental burden of electronic waste rivals that of plastics—indeed, the fastest-growing segment of the global solid waste stream is electronics and electrical waste (Kaya, 2016)—and poses a challenge even more complicated to address. Consumer products such as mobile phones and personal computers (PCs) contain more than 1,000 different chemical products, and more than 260 million PCs were sold in 2019 alone (Gartner, 2020). Gold, silver, copper, and palladium are the most valuable metals found in the waste stream, but they are challenging to recover. To gain perspective on the magnitude of the problem, consider that an economically viable gold mine yields 5 g of gold per ton of ore, while 1 ton of discarded cell phones can yield as much as 150 g of gold, 100 kg of copper, and 3 kg of silver (Nimpuno and Scruggs, 2011).

This chapter explores four areas of materials research, design, and production in which chemical engineers are particularly active: polymer science and engineering, complex fluids and soft matter, biomaterials, and electronic materials. These four areas are but a subset of materials work done by chemical engineers, and a deeper survey would yield its own report. Other areas that the committee did not explore include advanced (non–petroleum-based) performance fluids, such as lubricants or high-temperature heat transfer fluids; mixture formulation in general (beyond complex fluids); materials for separations beyond water purification (discussed in Chapter 4); and construction materials such as concrete and asphalt,[1] an area that could lead to new collaborations with civil engineers. In addition, there may be a role for chemical engineering in quantum materials and quantum information technologies.[2] Catalytic materials and applications in the energy transition are discussed in Chapter 3.

POLYMER SCIENCE AND ENGINEERING

In the 100 years since the publication of Hermann Staudinger's landmark paper *Über Polymerisation* (1920)—which marked the beginning of an ability to design plastics with infinitely tunable properties, moldability/processability, and remarkably low expense—polymers have infiltrated every aspect of people's lives. The purification of water, the preservation of food, the clothing people wear, the diapers on babies, the vehicles used for transport, the components of computers, and the delivery of medicines to the body all contribute to reliance on a global polymer industry whose products have grown in volume to more than 350 million metric tons per year since 1950 (Figure 7-1). Chemical engineers have been central to the development of this industry from its inception because of their integrated understanding of chemical synthesis and catalysis, thermodynamics, transport/rheology, and process/systems design. Indeed, the understanding of polymer rheology and design of processing equipment to handle highly viscous polymer melts represents a triumph of chemical engineering. Over this 100-year timeframe, the field of polymer science, including the contributions of chemical engineering, has become increasingly molecular in focus, leading to the ability to create new materials that integrate new polymer chemistries, topologies, and assemblies.

---

[1] The National Academies' Committee on Repurposing Plastics Waste in Infrastructure will explore the effectiveness and utility of plastic waste in asphalt mixes and other materials used in infrastructure, which may present opportunities for chemical engineers to collaborate with other disciplines.

[2] The National Academies' Committee on Identifying Opportunities at the Interface of Chemistry and Quantum Information Science is currently examining the research needed to make progress in this area.

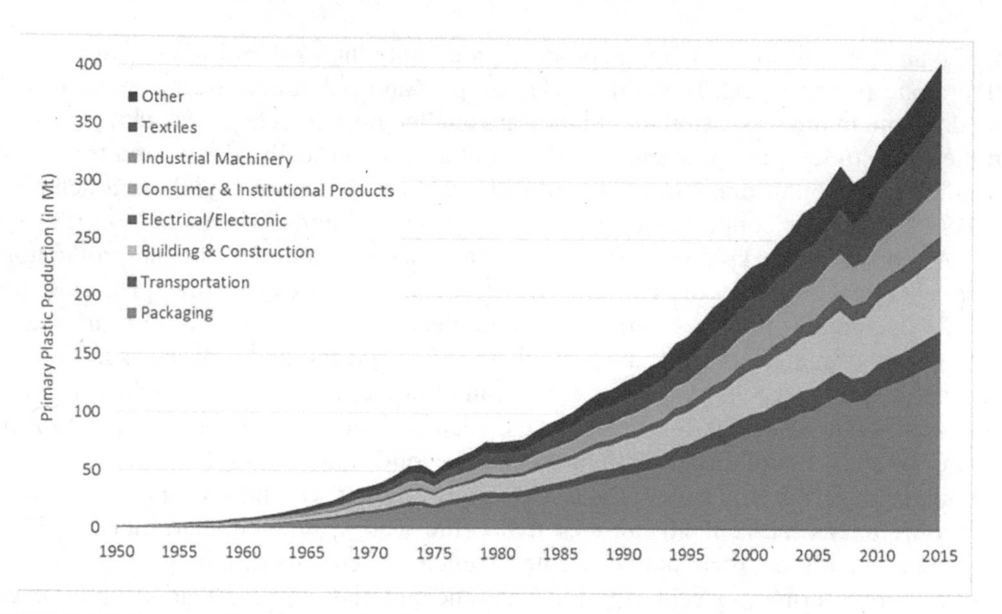

FIGURE 7-1 Increase in primary plastic production between 1950 and 2015. The figure shows end uses including textiles, industrial machinery, consumer and institutional products, electrical/electronic, building and construction, transportation, and packaging. SOURCE: Geyer et al. (2017).

Chemical engineers' backgrounds in polymer chemistry, thermodynamics, and kinetics have been especially well suited to designing and understanding the interplay between thermodynamic and kinetic driving forces acting through the atomic- (monomer), nanoscopic- (sequence, chain shape), and mesoscopic-length scales that lead to self-assembly—ultimately determining the macroscopic functional properties of a polymeric material. Block copolymers are one example of such hierarchically assembled materials. The ability to control the properties of the individual blocks to induce hierarchical order and control polymer macroscopic properties has enabled such applications as thermoplastic elastomers, semiconductors, nanolithographic masks and patterns, and ion-conducting solid-state electrolytes. Directed self-assembly of block copolymers led to the ability to make nanometer-scale patterns with high fidelity across macroscopic areas to rival stereolithography. New opportunities arise as one considers new assemblies in which the building blocks are more complex, are potentially dynamic, and change under external stimuli. In this way, both the molecular-scale structure and the macroscopic properties could adapt with time and situation to lead to truly active and responsive materials. Chemical engineers are well positioned to address the molecular engineering of hierarchical assembled polymers, which will rely on the design of chain conformation, mesoscopic structure, and macroscopic function in an integrated manner. For example, noncovalent interactions (e.g., metal-ligand, electrostatic, hydrophobic, hydrogen-bonding interactions) lead to dynamic, transient bonding with multiple handles to control multiple length and time scales simultaneously.

As laid out in the National Science Foundation workshop report *Frontiers of Polymer Science and Engineering*, making the next leap forward in the design of functional materials will require understanding, predicting, and utilizing molecular building blocks, sequence, conformation, and chirality to direct mesoscopic structure on desired time and energy scales to control macroscopic polymer properties (Bates et al., 2017). Further, while the ability to predict structure across length scales of six orders of magnitude has improved dramatically, it is now necessary to learn how to predict properties and behavior in order to guide material design. Within the next 20 years, the capability to design polymers from the bottom up for target applications will likely advance to a point at which the desired properties and behavior of a polymer can be specified and, using a combination of multiscale simulation and artificial intelligence (AI), the building blocks and the processing strategy can be designed to make it.

The ability to rationally embed information and function in synthetic macromolecules by controlling monomer sequence opens the door to polymers with the sophistication of biological molecules (Figure 7-2) and potentially with the ability to meet "the grand challenge of precision control over polymer structure and function" (Bates et al., 2017). Recently, strategies for controlling monomer sequence have evolved to enable synthesis of sequence-controlled synthetic polymer chains at gram scale (Lutz et al., 2013), allowing for molecular weight, architecture, and chain end control through highly efficient coupling reactions and step-wise automated synthesis and purification. Two fundamental questions that need to be answered if the potential of these advances is to be realized: What monomeric sequence should be incorporated? and What level of control is necessary to yield new structures and properties? (Bates et al., 2017). Now that almost anything can be made, what should be made? Even a single 80-mer chain composed of two different repeat units permits 6 x 10^{23} distinct sequences. In this Avogadro-scale design space, predictive tools that allow selection of specific targets of assembly become essential, as a design driven by intuition is no longer useful.

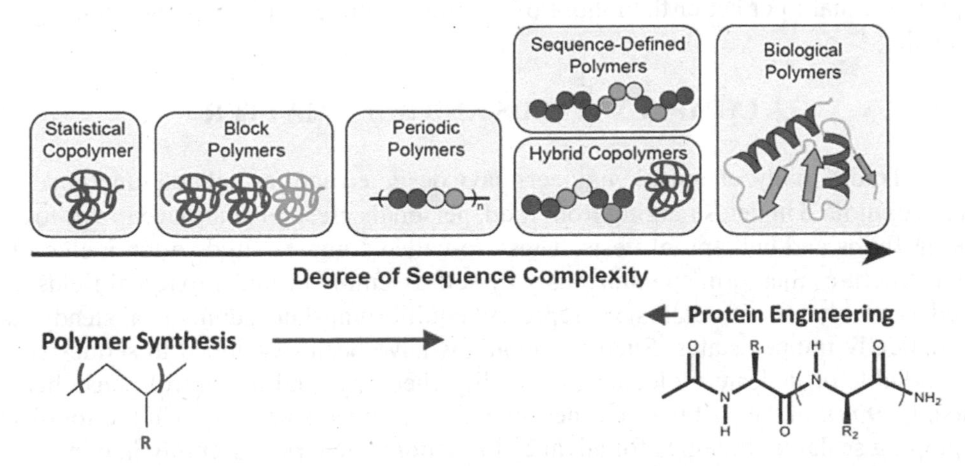

FIGURE 7-2 Increasing complexity in the synthesis of polymers ranging from copolymers to biological polymers. SOURCE: Modified from Rosales et al. (2013).

While the future of designed, molecularly engineered polymers-on-demand is exciting, responsible care of the environment demands a paradigm shift to include life-cycle considerations in the design imperatives. The combination of ultralow cost, scalability, and utility has led to the widespread use of plastics in every aspect of modern life. As a result, society has accumulated a huge debt of polymers that will never degrade. Global production of plastics has risen exponentially in the 70 years since their first mass production, from less than 2 Mt in 1950 to 359 Mt in 2018 (PlasticsEurope, 2019), and production is projected to double again in the decade ahead (Jambeck et al., 2015). Plastics are found in industries as diverse as food, medical devices, electronics, transportation, and construction. Packaging is one of the most common applications for plastics, and most packages are used once and then discarded after a life cycle of much less than 1 year (Teuten et al., 2009).

Mismanagement of plastic waste has contributed to persistent, visible environmental pollution (Rahimi and García, 2017), the formation and widespread dispersal of microplastic fragments, and leaching of contaminants that impact the health of ecosystems (UNEP, 2021). Only about 9 percent of plastic waste is recycled, and almost all is converted via mechanical recycling to lower-value materials (downcycling; Geyer et al., 2017). The combination of widespread use of plastics and the failure to deal with their end-of-life phase has brought the plastic-waste problem to the forefront of the world's attention. The extreme aspect ratio and entanglement of polymer molecules give plastics the unique properties that led to their widespread use over the last century, but these same features—combined with the breadth of designer chemistries (and designer properties) of commodity plastics—present significant challenges to the recycling or upcycling of plastics to monomers or high-value products (see Chapter 6).

The development of feasible depolymerization processes requires fundamental understanding of catalysis, polymer chemistry, and the rapidly changing melt rheology during processing, and all of this knowledge is in the domain of chemical engineering. Similarly, chemical engineers are well poised to develop alternative, scalable plastics with properties equal to or better than those of current commodity plastics and with a greener life cycle.

COMPLEX FLUIDS AND SOFT MATTER

Traditionally, chemical engineers have designed and formulated functional fluids that are exploited in fields ranging from food, personal care, and pharmaceuticals to active braking fluids and bulletproof vests. These so-called complex fluids often include functional structures that form spontaneously by self-assembly or under external fields by directed assembly. The structures can represent equilibrium states, dynamical steady states, or kinetically trapped states. Such formulations have been developed to sequester sparingly soluble or delicate molecules, to modify rheology, and to control phase behavior (Larson, 1999). In the past two decades, chemical engineers have been at the forefront of developing scalable strategies for advanced functional material assembly in complex fluids and soft matter. The advent of nanotechnology lent urgency to this area, as the promise of advanced functionality relied on the development of efficient and scalable schemes for incorporating microscopic and colloidal building blocks into larger structures (Box 7-1). Although this area of research is the modern version of colloid science and chemistry, its academic center in the United States now rests firmly in chemical engineering departments.

BOX 7-1
Smart Materials

Passive daytime radiative cooling (PDRC) surfaces (see Figure 7-1-1) have been proposed as coatings for buildings to reduce the indoor temperature to below ambient, thereby reducing dependence on air conditioning. PDRC surfaces rely on reflection from structures embedded in the coating, and radiation in the long-wave infrared window of the spectrum to send radiated energy out of the Earth's atmosphere (Catalanotti et al., 1975). These surfaces typically have complex designs incorporating elements that can include alternating layers of materials (Raman et al., 2014; Rephaeli et al., 2013), emitter particles, and carefully placed metal mirrors (Chae et al., 2021; Zhai et al., 2017). However recent advances have revealed a new path to their design: an effective PDRC surface was generated from a porous polymer film (polymethyl methacrylate, poly[vinylidenefluoride-co-hexafluoropropene]) that exploited broadband scattering from the bubbles for reflection of sunlight and the emission of the polymer itself for radiation (Mandal et al., 2018; Wang et al., 2021c). The design of such scalable functional structures is an area in which chemical engineers can contribute in new ways to functional materials for energy management.

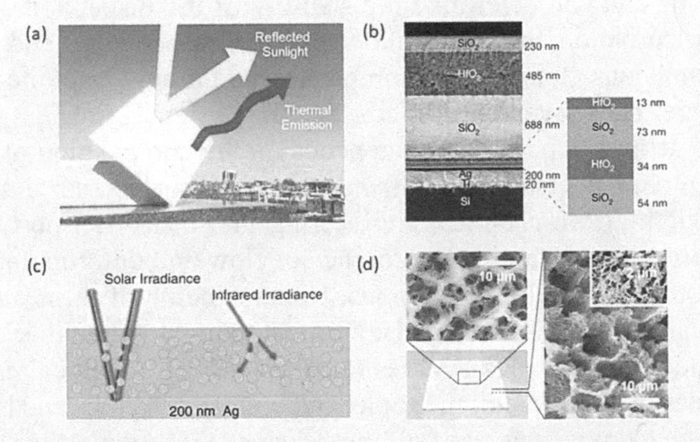

Figure X: Passive Daytime Radiative Cooling Surfaces (a) High reflection and emission enable a net radiative loss from the surfaces to generate cooling below ambient temperatures. (b) Scanning electron microscope image of a photonic radiative cooler. It consists of seven layers of HfO_2 and SiO_2, on top of 200 nm of Ag, a 20-nm-thick Ti adhesion layer, and a Si wafer substrate. (c) Schematic of hybrid PDRC materials with silica particles randomly distributed in polymethylpentene matrix backed with a thin silver film. The silver film diffusively reflects most of the incident solar irradiance, whereas the hybrid material absorbs all incident infrared irradiance and is highly infrared emissive.
(d) Micrographs showing top and cross-section views of porous P(VdF-HFP)HP PDRC film that is made by a scalable process potentially useful at architectural scales. Inset shows the nanoporous features.

Figure a and d: Mandal, Jyotirmoy, et al. "Hierarchically porous polymer coatings for highly efficient passive daytime radiative cooling." Science 362.6412 (2018): 315-319.
Figure b: Raman, Aaswath P., et al. "Passive radiative cooling below ambient air temperature under direct sunlight." Nature 515.7528 (2014): 540-544.
Figure c: Zhai, Yao, et al. "Scalable-manufactured randomized glass-polymer hybrid metamaterial for daytime radiative cooling." Science 355.6329 (2017): 1062-1066.

FIGURE 7-1-1 Passive daytime radiative cooling surfaces. (a) High reflection and emission enable a net radiative loss from the surface to generate cooling below ambient temperatures. (b) A scanning electron microscope image of a photonic radiative cooler material shows its seven layers. (c) A schematic of hybrid material that absorbs all incident infrared irradiance. (d) Micrographs show the top and cross-section of a porous film. SOURCES: Mandal et al. (2018); Raman et al. (2014); Zhai et al. (2017).

Surfactants

Surface active agents (surfactants) are found in nearly every household product and pharmaceutical formulation and are used industrially in processes ranging from emulsion polymerization to enhanced oil recovery. Historically, surfactants have been synthesized from petroleum feedstocks, although considerable progress has been made in the synthesis of surfactants from plant-based feedstocks. For example, nonionic surfactants made with sugar-based glucoside hydrophilic moieties instead of petroleum-sourced ethylene oxide groups have the advantage of coming from a renewable resource. In the same spirit, plant-based oils can be used to synthesize the hydrophobic portions of surfactant molecules. Catalytic conversion of biomass can create surfactant molecules with properties (e.g., tolerance in hard water) superior to those of conventional materials (Park et al., 2016).

Surfactants can be used at low concentrations to modify and control the surface tension and other properties of interfaces. Chemical engineers have made leading contributions to understanding of the statics and dynamics of far-from-equilibrium soft materials, including multiphase fluid systems that contain dispersed droplets or bubbles, surfactants, and other adsorbed materials. Examples include droplets in emulsions, bubbles in foams, and dispersions of one polymer in another (a polymer composite). These systems are typically dominated by complex dynamics at the interfaces. For example, surface-active molecules or materials self-assemble by adsorption on the surfaces of droplets and bubbles. These surfactant monolayers generate rich stress conditions that alter the effective stress of composite systems and can determine the stability of the dispersions to a dynamic perturbation. This perturbation can occur under processing or aging, and can determine the behavior of emulsions, foams, and composites commonly exploited in fields as diverse as personal care, pharmaceuticals, foods, and the design of tires.

The response of the system to a perturbation depends on the composition of the interface, which evolves over time. Understanding of the dynamics of surfactants and adsorbing species is built on insights from chemical engineering into mass transport and thermodynamics. This transport often occurs in the presence of flow and deformation of the interface and can depend strongly on flow conditions. Understanding of these stress conditions comes from chemical engineering's long history of studying the fluid mechanics of multiphase flows with nonideal complex stresses (Scriven, 1960). Chemical engineers have addressed these issues by understanding droplet and bubble dynamics and have developed paradigms with which to understand the nonlinear dependence of interface mobility and surfactant concentration (Stebe et al., 1991) and to reveal the modes of droplet extension, deformation, and breakup (Stone and Leal, 1989). In extensional flows, drops break up to form tiny droplets with diameters far smaller than that of the parent drop (Taylor, 1934). Chemical engineers have revealed how the presence and distribution of surfactant dictate this occurrence (Anna et al., 2003; Eggleton et al., 2001; Pozrikidis, 1997). These are classic examples in which far-from-equilibrium effects occurring in self-assembled structures determine the dynamics, stability, and processing conditions of soft-matter systems.

The self- and directed assembly of materials on interfaces remains an exciting field. Interfaces are inherently open systems in contact with two bulk phases, and are sites for accumulation of (denatured) proteins, macromolecules, lipids, and particles, all of which can self-assemble, interact, form structures, and influence dispersion behavior. Chemical engineers continue to design interfacial probes to understand interface dynamics (Crocker et al., 2000), including those that exploit Brownian motion in the interface (Squires and Mason, 2009) and those driven externally (Reynaert et al., 2008). Chemical engineers design formulations for delivery to interfaces, with broad impacts that include the design of lung surfactant replacements for premature infants (Alonso et al., 2004) and of surfactant formulations for oil spills (Owoseni et al., 2014). Many of the concepts developed in these inquiries are now used to study materials assembly at complex fluid interfaces (Kaz et al., 2012) for functional materials and sensing applications (Sivakumar et al., 2009).

At higher concentrations in aqueous solutions, surfactant molecules self-assemble into a variety of micellar aggregates, and at even higher concentrations form various liquid crystal phases. The hydrophobic core of a micelle can be used to solubilize a hydrophobic molecule, and micellar solutions are the base of many formulations for water-insoluble drugs. Under some conditions, micelles can grow into long, flexible cylinders that entangle and give rise to highly viscous and viscoelastic solutions that are useful, for example, as fracking fluids for oil fields (Samuel et al., 2000).

The combination of a hydrophilic material (water) and a hydrophobic material (oil) with a surfactant in most cases creates an emulsion that is thermodynamically unstable. Appropriate mixing and shear, together with use of the right kind and concentration of surfactant, can produce emulsions that are long-lived or easily reemulsified by the consumer. Such emulsions, in which the dispersed drops are micron sized or larger, are common in consumer products and the food industry. Their structure may be water droplets in oil (W/O) or oil in water (O/W). It is possible as well to produce multiple emulsions, such as a water drop in an oil drop within another water drop (W/O/W). Additional processing can drive the droplet size smaller and create nanoemulsions, in which droplets are as small as 10 nm (Helgeson, 2016). Multiple nanoemulsions can also be made, and nanoemulsions are very attractive as drug delivery vehicles (Zhang et al., 2018a). Finally, under some circumstances, water, oil, and a surfactant can form a thermodynamically stable solution called a microemulsion, which contains microdomains of oil and water. It is an unfortunate nomenclature, but the characteristic domain size in a microemulsion is smaller than that of a nanoemulsion.

The formulation and optimization of complex fluids draws on core chemical engineering concepts from thermodynamics, transport, and kinetics while recognizing the key role of molecular forces. Chemical engineers are likely to continue to be at the forefront of this work.

Nanoparticles

Among the most exciting developments in soft matter in the last 20 years is the design, synthesis, and assembly of nanoparticles—colloidal particles ranging in size from one to a few thousand nanometers. Just as whole new classes of polymer architectures can now be designed and synthesized through molecular engineering principles and made into complex functional materials, nanoparticles represent a new class of building block with tremendous potential for functional materials. The size of nanoparticles means they are controlled by the same laws of statistical thermodynamics that control polymers and complex fluids. Beyond the micron-sized polystyrene (PS), polymethylmethacrylate (PMMA), or silica colloidal particles long studied by chemical engineers, today's nanoparticles can be made from a variety of materials (e.g., gold, silver, CdTe/S/Se, PbS/Se). Nanoparticles such as these are grown in solution, often as faceted nanocrystals with polyhedral shape. The size and shape of nanoparticles can be controlled using various techniques, yielding hundreds of different possible shapes with nearly monodisperse size distributions. Lithographic techniques now make possible essentially any particle shape. By combining shape with anisotropic interparticle interactions, practically any nanoparticle building block is possible. Spheroidal particles of PS or PMMA can be coated anisotropically with gold or platinum, or made of multiple materials, to form patchy particles called Janus particles. A Janus particle—where the two halves of the particle have different interactions with the solution—is the particle equivalent of a block copolymer. Particle analogs of triblock copolymers are also possible, as are particle analogs of star and other copolymers. These patchy particles are conferred valency by the surface pattern resulting from the synthesis and subsequent coating or functionalization (Glotzer and Solomon, 2007). This valency, combined with particle rigidity, gives rise to unexpected self-assembled structures, many of which are isostructural to atomic and molecular crystals but on larger length scale (Figure 7-3).

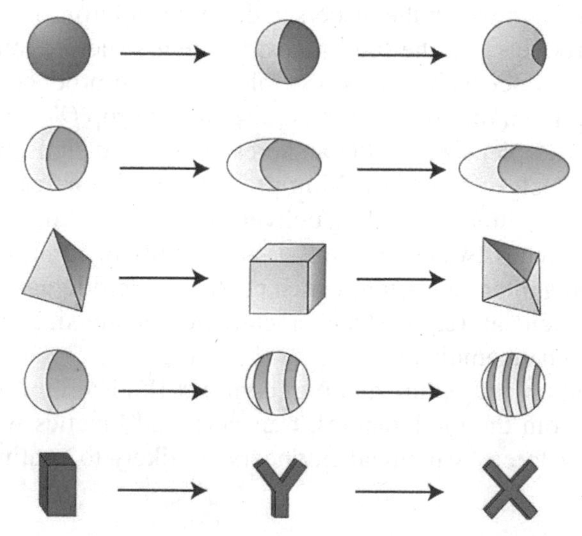

FIGURE 7-3 Examples of patchy particles made possible by combining nanoparticle shape with interaction anisotropy through surface patterning obtained via, e.g., grafted ligands, complementary DNA, or gold coating. Each row represents a homologous series of patchy particles resulting from changing only one alchemical variable (from top to bottom: Janus balance, aspect ratio, faceting, pattern discreteness, branching). SOURCE: Adapted from Glotzer and Solomon (2007).

In the near future, it will be possible to make nanoparticles of any size and shape, out of almost any combination of atomic elements or in a core/shell structure, and functionalized with any type of surface ligands—including alkanes, dendrimers, DNA, and proteins—into any arbitrary surface pattern. These patchy particles are next-generation building blocks for self-assembly into complex and functional structures. For example, DNA-functionalized noble metal nanoparticles are programmable atom elements that provide a powerful path to a wide range of self-assembled colloidal crystal structures (Kim et al., 2006; Laramy et al., 2019; Tian et al., 2020b). Mixtures of particles functionalized with complementary DNA linkers can form colloidal cocrystals, in principle, of arbitrary complexity limited only by the shapes of the particles and human imagination. Even one type of particle with self-complementary linkers can, because of its anisotropic shape, self-assemble into complex structures, such as a colloidal clathrate crystal with more than 100 particles in its unit cell (Lin et al., 2017). The interplay between enthalpy and entropy and between thermodynamics and kinetics is central to assembly engineering of new materials from nanoparticles and falls squarely within the expertise of chemical engineering. The same question posed above for block copolymers—Now that we can make anything, what should we make?—applies here as well. Theory and computer simulation are essential for finding the "sweet spots" in the vast design spaces available for patchy particles. Data science, and in particular deep learning with neural networks, has an important role to play in nanoparticle design for self-assembly. Alchemical potentials (van Anders et al., 2015) and other methods (Sherman et al., 2020) will enable inverse design of nanoparticles for targeted applications.

Beyond the self-assembly of complex structures, the next decade will see increased effort to understand, design, and engineer the thermodynamics and kinetics of assembly processes. Engineering assembly processes, or "assembly engineering," will make possible functional materials, reconfigurable materials, materials that rely on metastable structures, and soft metamaterials. Assembly engineering will also make possible materials inversely designed to have unique combinations of properties not typically found together, enabling wholly new classes of materials for a wide range of applications.

Assembly engineering may also involve directed assembly—for example, using external electromagnetic fields that dynamically rearrange suspended colloids to modify suspension rheology (as used in active brakes) and optics (as used in electronic paper e-ink). Sometimes, the field is one that comes from the complex fluid itself. The most obvious of such fields is surface tension; capillarity lies at the heart of some of the most widely adopted schemes for organizing colloidal building blocks. Nanoparticles are often spread and compressed on air–water interfaces of Langmuir troughs to form self-assembled structures trapped at the air–water interface. Nano- and microparticles are commonly deposited on solid surfaces using capillary assembly methods that rely on the collection and deposition of the building blocks near the three-phase contact line. So great are surface tension and the change in energy when particles adsorb that adsorbed particles can become trapped on the interface. Pickering emulsions are droplets stabilized by kinetically trapped monolayers of particles at their interfaces. Bijels are their bicontinuous counterparts, stabilized by jammed particle layers. The structures formed using surface tension can be exploited directly, can be polymerized, or can serve as templates for other advanced materials. For example, polymerized bijels have been explored as hollow fiber membranes, and metallized bagels have been explored as high-surface-area electrodes.

Janus and other patchy particles can be made into "active" particles that are self-propelled. One approach is to coat half the particle with a material (e.g., palladium) so that the particle is propelled via catalytic activity when the material encounters a "fuel" (e.g., hydrogen peroxide). Another approach is to drive nanoparticles with external magnetic or electric fields (Han et al., 2018). Collections of active particles can produce complex emergent behavior far from equilibrium, including swarming and motility-induced phase separation, even in the absence of interparticle attractive interactions (Marchetti et al., 2013; Takatori and Brady, 2015). The field of active matter is a burgeoning one that, as it relies on nonequilibrium thermodynamics, falls naturally within the portfolio of chemical engineering research. By combining assembly engineering with active nanoparticles, chemical engineers are well positioned to create novel materials and material machines with robotic function at the nanoscale.

BIOMATERIALS

The design of biomaterials has seen strong growth in chemical engineering as a large number of advances have led to clinical translation that leverages degradable and biologically derived polymer systems. Materials design has had an influence in areas ranging from controlled localized release systems to small interfering RNA (siRNA) delivery, and a significant impact on the advancement of medicine and improvement in patient health. Key areas for advances in the biomaterials field include regenerative engineering, wound healing, systemic and localized delivery of nucleic acids, and the delivery to and detection and imaging of regions of the body that present unique barriers or opportunities.

The biomaterials field can also provide solutions in areas beyond human health. Applications with unique promise include areas that support plant and animal science, as well as the design of plant- or organic-based materials systems that could enable biologically derived or biomimetic polymers. This section focuses primarily on health-related applications, with a brief discussion of other areas in which biomaterials will be important. In addition the recent National Academies report *Frontiers of Materials Research: A Decadal Survey* (NASEM, 2019d) includes an extensive discussion of research opportunities in biomaterials.

Materials for Regenerative Engineering

The need for methods for generation or repair of tissues and organs is compelling. The ability to create materials systems that can be tuned to mimic cellular environments, including the extracellular matrix of tissues and organs, and the biological cues they present opens up the pathway to deriving organs and repairing tissues. Wound healing; the regeneration of bone and cartilage; the replacement of key organs, such as skin or liver; and the repair of tissues, such as cardiac tissue, are just some of the applications enabled by the generation of materials matrices and components that can be combined to support appropriate living cell types. Key to the function of biomaterials for these applications is the ability to provide settings with the appropriate physical, chemical, and mechanical cues to enable single- or multicellular systems to arrange, order, and structure into organized tissues. The array of chemistries available for generating tissue scaffolds is vast; however, there are overarching requirements for low material toxicity, minimal immunogenicity, and the ability to match mechanical properties of the original organs or tissues. In many cases, moreover, the biomaterial needs to be capable of undergoing controlled degradation as native tissue is evolved to replace the synthetic material scaffold. The complexity of the biological tissues that require replication introduces additional constraints on the materials systems that can be used for these applications.

Polymeric materials, including a number of polymer systems that form hydrogels, have played a critical role in the design and development of biomaterials. These advances have been made possible by the range of synthetic backbones available; the ability to "design in" degradability based on hydrolytic susceptibility, pH, or the presence of protease-susceptible bonds; and the ease of functionalization of polymeric constructs—all of which make them excellent scaffolds for presenting a range of different proteins, peptides, and other molecular cues on their surfaces. Along with polymers, a range of mineral composite materials—including hydroxyapatite; porous silica; and other engineered materials, such as carbon nanomaterials and inorganic or hybrid materials systems—have provided biophysical cues, such as mechanical stiffness, and sources for biomineralization of hard tissues, such as bone.

Key advances include the development of dynamic synthetic hydrogels whose chemistry can be reversibly activated using light or physical or chemical stimuli to induce changes from one state to another. The more traditional chemically cross-linked hydrogel network is static, which prevents these networks from undergoing remodeling or supporting different tissue development or growth stages. However, more recent capabilities have allowed the incorporation of photoresponsive groups that enable 3-D patterning of surface functional groups. Such groups can act as cellular cues, develop reversible cross-linking to modify mechanical stiffness, or enable materials that can reversibly exhibit different mechanical states (Rosales and Anseth, 2016). The use of dynamic chemical bonds and secondary interactions could enable such changes in biological state as hypoxia, cell state, or the presence of certain signaling molecules to trigger changes in a way that enables cycling (Figure 7-4).

More recent efforts have also been directed toward engineering materials that can incorporate and support multicellular systems—essentially allowing for the generation of 2-D and, more particularly, 3-D cellular cocultures supported by materials that enable cell–cell interactions like those in biological systems. To a large extent, work has focused on direct replacement or replication of tissues or organs; however, there has also been an interest in engineered multicellular living systems (M-CELS; Kamm et al., 2018), which include in vitro cellular cultures that can exhibit additional, unique, or modified functions, such as on-demand production of a biochemical upon sensing of a trigger molecule, or actuation of cellular strips upon a given electrical or chemical stimulus. These concepts require the design of materials systems that can provide not only physical support but also a matrix that will enable cell–cell configurations that best enable function.

Demands for multicellular systems, whether for synthetic organs or living machines, include the need to achieve larger and more complex materials systems that can support angiogenesis and the development of a vascular system that can provide oxygen and nutrients to cells while in culture. For these reasons, some of the key challenges in this area include supporting endothelial and smooth muscle cells while promoting the sprouting and growth of stable vessels; degradable materials systems more readily accommodate a growing vascular system. Organization of cells of different types is also required for sufficient function; biomaterials that can be patterned, aligned, or otherwise manipulated to organize cells selectively are of great interest for these applications.

Additive manufacturing, or 3-D printing, is an important tool for biomaterials synthesis. Biological inks, or bio-inks, which involve the integration of cells into biologically compatible materials using water-based solutions or suspensions, can be used with additive manufacturing methods to create complex patterned structures containing specific cells in precise arrangements, as with organs and functional tissues (Murphy and Atala, 2014). Because different tissues exhibit quite different mechanical and chemical properties, a range of different kinds of bio-inks are needed to address these various needs. Furthermore, needs for additional materials include ways to generate cellular suspensions in droplet form while protecting cells from extreme shear forces during printing. The materials used need to have sufficient cohesion to maintain shape and enable adhesion to the scaffold. Ultimately, the softer scaffolds typically used may also suffer from mechanical failure during assembly because of the weight of the printed scaffold, and this balance between mechanical support and cell compatibility sometimes leads to trade-offs (Decante et al., 2021). It is anticipated that the development of new and more readily manipulated materials systems for bio-inks that address these issues will be a focus of chemical and materials engineers.

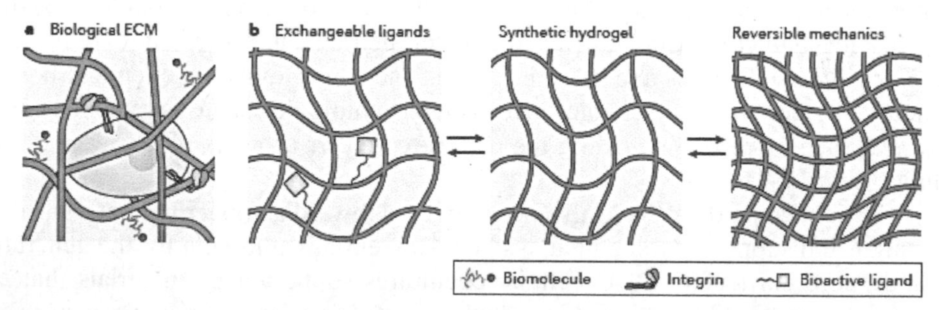

FIGURE 7-4 Biological extracellular matrix and synthetic strategies involving reversible chemistries. (a) The native extracellular matrix (ECM) is a heterogeneous fibrillar network that provides biochemical and physical cues to cells. (b) Synthetic hydrogels are traditionally static, polymeric networks (middle panel). Dynamic hydrogels can capture aspects of the native ECM by temporally controlling ligand presentation (left panel) or reversibly cycling through changes in mechanics (right panel). SOURCE: Rosales and Anseth (2016).

Additional needed developments include the ability to incorporate vasculature into these systems—a key issue requiring more than precision cellular placement. While cellular systems can be printed in minutes, it can take days for precursor or stem cells to develop appropriate stable vasculature. It is therefore difficult to incorporate blood vessels directly into such tissues, and patterning of endothelial cells is insufficient for creating functional systems. Finally, a greater ability to mimic the native extracellular matrix environment is necessary to achieve systems that can develop and evolve to generate integrated organs. Although it is possible to imbed growth factors and other proteins into existing matrices, it would be preferable to pattern them by printing at higher resolutions (Ng et al., 2019).

It is important to note the importance of the above techniques not only for tissue regeneration but also for organ-on-chip applications, which replicate organ function using microfluidic devices. Such applications are discussed in Chapter 5.

Materials for Drug Delivery

A second large and significant area of growth for biomaterials is the generation of materials systems that can deliver drugs to the body with control over the timing of release; the release location; and the targeting of specific tissues, cells, or organs (see also Chapter 5). The past few decades have led to a detailed understanding of the chemistries of biodegradable materials capable of achieving desired time-release profiles and of the thermodynamics that control self-assembled materials, such as block copolymers and liposomes. A range of accomplishments in both more traditional and new materials systems enable drugs to be encapsulated effectively and to retain their efficacy while decreasing undesirable side effects and cellular or organ-level toxicity.

A large array of functional polymers, lipids, silica, and inorganic hybrid systems have been developed that enable different modes of preprogrammed delivery or delivery on demand in response to specific stimuli. Recent developments have led to therapies that involve the administration of biologic molecules, ranging from proteins and peptides to nucleic acids, such as siRNA and messenger RNA (mRNA). Although many advances have been achieved in the field of controlled drug delivery, several key challenges and opportunities remain following the discovery of new therapeutic approaches and the introduction of new experimental and computational tools to aid in the understanding of these materials.

Systemic Delivery and Nanomedicine

The development of nanoparticle systems—which include polymeric and liposomal nanoparticles, as well as inorganic and hybrid nanomaterials—offers another opportunity for chemical engineers to make an impact in biomaterials. Applications of nanoparticles are promising, particularly for cancer treatment, for which nanoparticles are believed to have an advantage in sequestering toxic drugs from the rest of the body and targeting tumors. Unfortunately, although safety and efficacy have improved and cancer nanomedicines are currently in clinical trials, the delivery of nanoparticles in patients is complex and presents several challenges.

The surface chemistry, shape, size, and mechanical properties of nanomaterials affect their circulation half-life within the blood stream, their ability to access tumors by crossing the endothelial barriers that make up the vascular wall, and their downstream release characteristics as they progress from circulation to accumulation in the tissue of the target tumor. Nanomaterials can be designed to present a range of different surface ligands, including synthetic and native biomolecules, peptides, and proteins, such as antibodies that exhibit strong affinities for surface-bound proteins or other biomarkers on the surfaces of target cells, such as tumor cells. With more detailed investigations of the impact of physical and chemical features and nanoparticle composition, the field is recognizing that these characteristics can yield significant differences in nanoparticle transport and distribution in the body. This increased understanding will guide how nanocarriers are deployed against a targeted disease such as cancer.

Nascent opportunities in the cancer field include nanoparticles that can contain combination therapies addressing two or more vulnerabilities found to be synergistic in each tumor type. Of particular importance is the ability not only to encapsulate dual therapies in a singular particle system, which can present unique materials challenges, but also to control when these therapeutics are released, as in some cases, synergy depends on the timing and staging of release. An advantage of nanoparticle systems is the ability to tune these release profiles and ensure that target cells receive a sufficient dose of each drug.

Nanomedicine beyond Cancer

Although cancer is by far the disease setting most studied in nanomedicine, nanocarriers have also been used in several other biomedical applications. Nanoparticle systems show potential for the targeted delivery of antibiotic or antiviral therapeutics to infected regions of the body (Yeh et al., 2020), particularly in the case of inflamed and infected regions such as lung or cardiac tissue. Additionally, nanoparticles can be designed to "home" to immune cells in circulation or in the lymph nodes based on nanoparticle ligands designed to bind to cellular surface markers. In this case, nanoparticles are fundamentally interesting for the delivery of vaccines. Several organs present unique challenges for delivery that have been addressed using nanomedicine. For example, nanoparticles with dense, highly hydrated polyethylene glycol (PEG) brushes have been developed to penetrate the unique barrier of corneal tissue for eye treatments, leading to commercial products for addressing macular degeneration. As another example, because nanoparticles are naturally filtered through the liver, they also prove to be useful systems for delivery of drugs to the liver. As nanomaterials are found to be highly versatile and broad in material properties for adapting to challenging organs, their use is anticipated in several additional areas, including avascular cartilage tissues and the ultimately challenging blood–brain barrier, an area that will be extremely impactful for the delivery of treatments for neurological disorders.

High-Throughput and Data Science Approaches to Discovery

A more systematic approach to nanoparticle design can ensure that the field advances in a manner that contributes meaningfully to patient care. High-throughput methods have been devised for generating large numbers of nanoparticles with differing characteristics (e.g., Dahlman et al., 2017). When such approaches are coupled with the ability to perform large-scale in vitro screens, it is possible to discern nanocarrier candidates that exhibit increased efficacy for delivery. The potential to generate significant databases that include interactions of nanocarriers with a range of tumor-associated cells, including patient-derived cells, can also provide the basis for machine learning approaches for understanding and guiding the design of nanocarriers that target tumor or tumor-associated cells, such as stromal or immune cells. These studies need to be coupled with in vivo studies to understand the ultimate transport and trafficking properties and targeting efficacy; such studies may also be designed in a high-throughput fashion using DNA barcoding or chemical barcoding methods to label nanoparticles, in conjunction with state-of-the-art imaging and detection methods coupled with appropriate cell-sorting algorithms. It is also possible to consider these methods as a means of determining possible biomarkers for patients whose tumors readily take up nanoparticles or whose tumor vasculature may exhibit the enhanced permeation and retention effect, thus making it possible to identify candidates who will benefit most from nanoparticle therapies. These approaches represent a renaissance in understanding of and much more rapid evolution of nanoparticle design. Future work will extend recent accomplishments to a much broader set of nanocarrier compositions and structures, allowing a greater amount of nanomaterials discovery toward tailored nanoparticle function.

Nucleic Acid Delivery

Perhaps the most important and impactful recent advance in drug delivery is the ability to deliver and transfect nucleic acids, including mRNA, siRNA, and DNA. The ability to directly encode cells to produce specific proteins or silence genes is powerful because it enables the direct programming of cells, including the use of CRISPR (clustered regularly interspaced short palindromic repeats) components. Although viral vectors can be very effective in transfecting genetic material, they also pose health and safety risks, as well as issues of scale for manufacturing, reproducibility, and tunability. Challenges to the delivery of nucleic acids are multifold; they include the need to compact a very large and highly negatively charged macromolecular species to a size compatible with cell uptake and the need to create a protective coating upon encapsulation of the nucleic acid to protect against exposure to proteases in the bloodstream. It is also important for systemic delivery approaches that the encapsulating materials system have a neutral or negative charge because of the cytotoxicity and protein opsonization resulting from positively charged nanomaterials. Ultimately, key susceptibilities for delivery of these biologic systems have been degradation of the RNA/DNA molecules themselves due to the presence of RNAse or DNAse enzymes in the body. Recent advances in functionalization of the nucleic acids can render siRNA cargoes fairly inert to breakdown under physiological conditions; however, although mRNA can be protected with functional groups, it remains more susceptible to breakdown and requires packaging that protects against degradation. Additional issues requiring attention are the targeting of nanocarriers to desired organ or cell types and the ability to direct trafficking of the nanoparticle to the cytosol or nucleus, where the cargo can be available for direct transfection or translation.

Cationic lipid nanoparticles (LNPs) have been found to exhibit the above properties while generating stable complexes that protect nontarget cells and tissues from exposure to the nucleic acid cargo during distribution throughout the body. The positive charge of the lipids interacts with negatively charged RNA backbones to generate uniform nanoparticles, usually with a net positive charge. If the cationic lipids are combined with pegylated lipid molecules, the PEG can act to dilute and shield positive charge, making the resulting structures increasingly stable while avoiding interactions with serum proteins and circulating cells, such as monocytes, that lead to clearance and destabilization. Key to the successful delivery process is the need for the cargo to exit the endosome as it buffers to lower pH. Cationic LNPs are the basis for effective endosomal escape, enabling release of RNA or DNA to the cytosol; it is thought that the lipids aid in compromising the endosomal membrane and lead to its disruption and release. The first LNP therapeutics were introduced by Alnylam as an siRNA therapeutic and involved a cationic lipid complexed with a chemically stabilized RNA interference (RNAi) molecule. Significant advances in the composition and structure of the lipids used for LNPs enabled the development of the Moderna and Pfizer mRNA vaccines for COVID-19 and form the basis of a powerful platform for RNA medicine. Novel and improved formulations and production of LNPs will be important for future vaccine manufacturing.

There are some limitations to the use of cationic LNPs. Lipids tend to traffic to specific regions of the body, including the liver, because of the apolipoprotein E (ApoE) receptor present on liver cells, and it can be challenging to design lipids with appropriate biodistributions for targeting other organs following intravenous injection. Lipids also provide fewer chemical handles for introducing smart or responsive chemistries, and they do not allow for staged or more complex release systems. Polymeric systems excel in providing a large design space while enabling the presence of high-charge-density segments by using block or segmented block copolymers or polyelectrolytes. Although some polymers have been found to be highly effective transfection agents, a trade-off between efficacy and toxicity has resulted from the impact of their high positive charge. One of the important biomaterials challenges in the upcoming decades will be the discovery and design of synthetic vectors that can rival the transfection efficiencies of viral delivery while remaining highly safe.

ELECTRONIC MATERIALS

Chemical engineers have played a central role in the discovery, design, and production of the materials (e.g., polymers, semiconductors, glasses) needed to create electronic devices. The growth in demand for these devices, and thus these materials, shows no signs of slowing. The development of new materials for these applications will require sophisticated research and development, as discussed here, and will likely require elements of AI and data analytics for the materials development (see Chapter 8).

Semiconductor Manufacturing

The common process steps in manufacturing semiconductor devices are insulator deposition, lithography, etching, wet cleaning, metallization, and chemical–mechanical planarization, although not necessarily in that order (Table 7-1). Common challenges for these steps include ever-increasing purity requirements in response to increased upstream/downstream process sensitivities, environmental considerations in response to global requirements and compatibility, means of maintaining consistent manufacturing and packaging processes, and approaches for managing and improving global distribution. There can also be challenges specific to individual molecule or formulation types, such as the stability of the formulation and its sensitivity to air and the safe handling of flammable or toxic materials.

The critical processes for materials manufacturing include synthesis, purification (including distillation and filtration), formulation blending, and packaging and materials compatibility. Each of these processes may be considered during the discovery phase of a new material but are fully accounted for during the transition to mass production.

Several trends are expected to continue. First, the number of new materials being introduced to the semiconductor and adjacent electronic industries will continue to increase, and more rapid molecule or formulation identification will be required to meet the industry cadence. In addition, some historically important processes may become obsolete because of environmental restrictions and a lack of supply due to supply chain interruptions (especially in the case of rare earth elements) or general shortages (as in the case of helium).

Second, materials will need to be increasingly pure, with the purity levels of metals reaching the low parts-per-trillion range, and greater synthetic process control will be needed to reduce already low-level organic and inorganic impurities. The enhanced purity requirements will drive improvements to in-line analytical process monitoring and real-time process control, as well as other advances in manufacturing. Chemical engineers can play a role in each of these areas. Finally, the markets for most major materials used for semiconductor manufacturing are expected to grow (Tremblay, 2018), driven by increasing use of materials in manufacturing of both logic and memory devices, as well as overall industry growth.

Challenges in Electronic Materials Discovery

Research and development of electronic materials occur in sequence but have different inputs, outputs, timelines, and risks. For suppliers of materials to the electronics industry, the input to the research process is an established material need from device makers, and the output is enabling knowledge, ultimately in the form of a product concept to satisfy that need. The product concept is the input for the development process, which determines how this concept can be realized as a salable product manufactured economically and reliably at scale.

TABLE 7-1 Primary Semiconductor Manufacturing Processes, Common Materials Used in Each Process, and Chemical Engineering Processes Required

Manufacturing Process Step	Type of Processing	General Material Classes	Current Material Challenges	Related Chemical Engineering Processes
Deposition	• Plasma-enhanced • Chemical vapor • Atomic layer • Spin-on • Electroplating • Physical vapor	• Organosilane • Silicon-containing polymers • Organometallics • Metal-containing formulations	• Safe handling • Environmental and purity	• Synthesis • Purification • Packaging • Chemical distribution
Etching and dopant gases	• Plasma-assisted etching	• Inert and reactive gases • Halogenated gases • Mixed specialty gas blends	• Safe handling • Environmental and purity • Packaging technology	• Synthesis • Purification • Packaging
Lithography	• Spin coating	• Formulated-polymer blends • Solvents • Metal-containing polymeric blends	• Environmental and purity	• Polymer synthesis • Distillation • Purification
Wet cleaning	• Spin coating • Immersion bath	• Aqueous, semiaqueous, and solvent-based formulations • Acids, bases • Solvent	• Environmental and purity	• Chemical mixing • Purification • Filtration • Packaging
Chemical mechanical planarization	• Spin coating	• Particle-containing aqueous formulations	• Environmental and purity	• Chemical mixing • Purification • Filtration • Packaging

SOURCE: Internal knowledge from EMD Electronics, 2021.

The unmet current and future needs for advanced materials for the electronics industry are being addressed through close collaboration with foundries and independent device manufacturers (IDMs), precompetitive consortia, and industry roadmaps. Industry roadmaps, now in their third generation (Gargini et al., 2020), take the longest view in identifying key trends, industry challenges, potential solutions, and required innovations over the next 15 years. These roadmaps have been invaluable in accelerating industry advances by advancing the best technologies to keep pace with Moore's law. Material suppliers with appropriate resources can confidently make long-term strategic research and business decisions based on these roadmaps.

Since the formation of SEMATECH as an industry–government partnership to maintain U.S. leadership in the semiconductor industry (Irwin and Klenow, 1996) and its evolution into an international precompetitive consortium (Carayannis and Alexander, 2004), precompetitive strategic partnerships have been instrumental in the electronics industry to advance the technologies needed to realize industry roadmaps (Logar et al., 2014). Medium-term needs for materials suppliers can be understood through participation in these consortia. Near-term manufacturing needs can be understood only through close collaboration with the foundries and IDMs. Collaborative innovation within the semiconductor industry confers an advantage despite the challenges of maintaining relationships and controlling proprietary information (Kapoor, 2012). Materials suppliers can receive timely feedback on new product performance and integration issues, which enables faster implementation of new processes for device manufacturers (O'Neill and Zheng, 2019).

Once a material need has been specified, the research process of discovery begins. That process has evolved over the past several decades but follows the basic cyclical scientific method of relevant field study, hypothesis development, testing, and assessment. In the past, this cyclical method was a slow, trial-and-error process. Leveraging the arrival of more powerful computers and improved theoretical calculations in the second half of the last century, the current so-called third age of quantum chemistry now enables calculations at least as accurate as the results of physical experimentation (Langhoff, 1995; Richards, 1979). Thus, there is now a rational approach to the design of materials, one that entails creating new molecules with desired properties using predictions from models based on atomic or molecular structures. The accuracy and utility of these models are directly tied to the rapid scaling of computational power, which itself was enabled by the discovery of new processes and materials used by the semiconductor industry. These improvements led to better and faster computational results, such that screening experiments can be conducted reliably in silico (Hafner et al., 2006; Hautier et al., 2012), limiting the synthesis targets for physical experimentation to compounds or mixtures with the best prospects of achieving the material needs of the industry. Additionally, laboratory experimentation has been enhanced by the commonplace computer control of experimental apparatus and analytical instrumentation (Ford, 1982). Thus, advanced computing has increased efficiencies in the testing phase with high-throughput screening and combinatorial material synthesis. These approaches revolutionized pharmaceutical research but have now been demonstrated for discovery of new materials in semiconductors, especially when the new material can be synthesized and evaluated as a thin film (Takeuchi et al., 2005).

Strategies for Development of Electronic Materials

Chemical engineering merges chemistry fundamentals and material physics with the hardware and equipment required to produce the desired commercial chemical products. The electronic materials manufacturing industry elevates this challenge to a new level not experienced with the manufacture of many traditional chemical products. Comparison with the evolution of the pharmaceutical chemical industry offers a useful guide for the advancement of the electronic materials industry.

Purity and product uniformity have improved throughout the progress of the electronic materials industry. Early in the history of integrated circuit (IC) manufacturing, materials were used as they were available in the standard reagent and commodity chemicals market. Later, IC manufacturers, using statistical tools to characterize chip-manufacturing performance, drove the quality and specifications for materials far beyond what was generally available. Thus, the development of purification technologies is a primary driver for new products across the industry, and also drives the need for continuous quality verification using process analytical technology and supporting statistical tools that must be upgraded to handle the massive amounts of data generated with automation. The future will look more and more toward AI to enhance problem solving and material and process design (see Chapter 8).

Materials used in the electronics industry are constantly changing in response to the new requirements of manufacturers. Many of these materials are unique in structure, can be very hazardous, and require customized reactors and processes to produce. These reactor and separation unit operations require a great deal of creative technology design. The development of robust processes can be accelerated when capital costs are kept low and process equipment can be kept small. Process-intensification methods, modular continuous reactors, microreactors, and purification platforms can speed up product commercialization, with assets prepositioned globally (Tian et al., 2018).

Many of the raw materials used to manufacture electronic chemical products are available only as commodities with limited purity, far below the necessary requirements. Efforts are increasingly focused on the development of purification technologies using traditional and novel purification equipment, methods, and schemes. Additionally, the stringent requirements for product purity require that suppliers of electronic materials evaluate and characterize the subtle and increasingly important impact of equipment materials of construction on process chemistry. The industry organization SEMI provides standards for purity, analytical methods, and processing for many raw materials.[1]

Metallic impurities are notoriously problematic in all electronic materials, limiting the materials of construction suitable for manufacturing and separation equipment. Even extremely low levels of surface corrosion can lead to unusable products. Polymeric materials, on the other hand, can leach low levels of unbound polymers or plasticizers.

---

[1] See https://www.semi.org/en/products-services/standards.

Glass reactors and process equipment can be vulnerable to extremes in pH, which can cause leaching of silica, boron, and metals. Surface science is a critical and ever more important discipline, and chemical engineers will play a role in the development of new products and processes for this industry.

Ion-exchange resin processing is commonly used to remove metal impurities in aqueous and organic raw materials. The diverse chemical functional groups available on resins provide a wide range of metal-removal selectivity. The choice of resins and combinations of resins can be made specific to the metals (and their oxidation state) to be removed. Preparation and pretreatment of the resin bed are important to reduce polymer contamination. Materials often need to be diluted before processing to achieve the necessary removal purity and then reconcentrated (Alexandratos, 2009; Silva et al., 2018).

Other purification methods include many variations on distillation, filtration, and extraction. They are usually applied for small-scale continuous processes, and include packed-column distillation, molecular distillation, wiped-film and spinning-band distillation, sublimation, liquid extraction using Karr columns and rotating-disc columns, crystallization, adsorption for both liquid- and gas-phase materials, ion exchange, crossflow and flow-through filtration on novel modified membranes and plastics, and simulated moving-bed chromatography. Hydrogen peroxide, for example, is a critical material used in chemical mechanical polishing, and a new approach to reverse osmosis is proposed as a novel method for purification (Abejón et al., 2010; Gao et al., 2017).

Cleaning procedures as they pertain to such equipment as piping, pumping, and storage vessels are frequently underappreciated. Especially in the pharmaceutical, fine chemical, and electronic chemicals industry, cleaning of equipment between changes in product, raw material lots, or batches is often a complex process. Cleaning processes require detailed metrics, and in many cases may be as complex and time-consuming as the chemical synthesis or purification process. Chemical surface preparation of unit operation equipment and packaging containers will become a project in surface science and a subset of the overall process development.

CHALLENGES AND OPPORTUNITIES

Chemical engineers can contribute to the materials development domain across a range of material types and applications. In particular, they have a unique role to play in the continued development of polymer science and engineering because of their understanding of chemical synthesis and catalysis, thermodynamics, transport and rheology, and process and systems design. Chemical engineering is the logical home for research and development of complex fluids and soft matter. The combination of molecular-level understanding and thermodynamic and transport concepts yields important insights and enables advances. The science and application of nanoparticles by chemical engineers in both industry and biomedicine are rapidly accelerating, offering the opportunity for application breakthroughs. Chemical engineers play an essential role in advancing the development of biomaterials for both regenerative engineering and organ-on-a-chip technology, and chemical engineering principles are at the heart of understanding and improving targeted drug delivery both spatially and temporally. As the United States has lost dominance in the area of semiconductor processing, chemical engineering expertise around reactor design, separations, and process intensification has become critical to the success and growth of the electronic materials industry.

Recommendation 7-1: Federal and industry research investments in materials should be directed to

- **polymer science and engineering, with a focus on life-cycle considerations, multiscale simulation, artificial intelligence, and structure/property/processing approaches;**
- **basic research to build new knowledge in complex fluids and soft matter;**
- **nanoparticle synthesis and assembly, with the goal of creating new materials by self- or directed assembly, as well as improvements in the safety and efficacy of nanoparticle therapies; and**
- **discovery and design of new reaction schemes and purification processes, with a steady focus on process intensification, especially for applications in electronic materials.**

8
Tools to Enable the Future of Chemical Engineering

- Current and future chemical engineers will need to navigate the interface between the natural world and the data that describe it, as well as use the tools that turn data into useful information, knowledge, and understanding.
- Some emerging and future tools will be developed in other fields but will have a significant impact on the work of chemical engineers; others will be developed directly by chemical engineers and have an impact in science and engineering more broadly.
- Some tools and capabilities will be evolutionary, with gradual and predictable development and applications, while others will be revolutionary and will change chemical engineering research and practice in ways that may be difficult to predict or anticipate today.

Earlier chapters concentrate on specific aspects of the discipline of chemical engineering (and by extension, of its students, academicians, researchers, and practitioners); this chapter is concerned with the tools and capabilities that will be required in the field's endeavors in the future. Consequently, the content of this chapter cuts across the topics covered in earlier chapters and is motivated by and organized around a defining question: What existing and anticipated tools and capabilities will drive chemical engineering and chemical engineers of the future?

Some of these tools and capabilities currently exist in some form but will evolve in the future, some are emerging as this report is being written, and yet others will need to be created and will coevolve with chemical engineering over the next few decades. Some of these tools will emerge from current and future trends in technologies, with little or no involvement of chemical engineers in their development even though they will have significant impacts on chemical engineering. In other cases, chemical engineers will be involved directly in the development of tools and technologies that will advance science and engineering more broadly. Some of these tools and capabilities will be evolutionary in the sense that their development and application will be gradual and more predictable; others will be revolutionary in that either their development, their application, or both will alter the landscape of chemical engineering research and practice in ways that may be difficult to predict or anticipate today.

Whether engaging in cutting-edge medicine or manufacturing; revolutionizing the design, optimization, and operation of efficient energy systems that respect environmental constraints; discovering and designing new materials; or innovating in any of the other endeavors discussed in earlier chapters, the chemical engineer of today and tomorrow will, conceptually, need to navigate the interface between the natural world and the data that describe it. What devices (physical or otherwise), tools, or capabilities will make it possible to turn data into useful information, knowledge, and understanding in the future? While the discussion here could address a virtually endless list of tools and capabilities—many of which, when used in combination, will drive innovation—the focus is on four categories: data science and computational tools, modeling and simulation, novel instruments, and sensors.

DATA SCIENCE AND COMPUTATIONAL TOOLS

Chemical engineering has been a data-intensive field from the very beginning. The chemical industry was built on systematic measurement and cataloging of experimental measurements, including thermodynamic properties, phase diagrams, rate constants, flow rates, and heat- and mass-transfer coefficients. Early steam tables from the 1930s exemplify the importance of data-intensive approaches in chemical engineering. These data led to the development of correlations, equations of state, and process simulators, which have allowed industry to design, optimize, and innovate. Until the late 1990s, chemical engineering datasets were stored primarily in print—notebooks, publications, or books. After a glossary of definitions, the first technical chapter of Perry's classic chemical engineering handbook is a compilation of physical property data (Green and Southard, 2019). The National Institute of Standards and Technology (formerly the National Bureau of Standards) was entrusted by the federal government with the collection of such data, which were subsequently published in the form of much-sought-after compilations and books.

Chemical engineering is undergoing a transformation driven largely by the way data can be acquired, curated, shared, and utilized. Many labor-intensive measurements traditionally performed by humans—from calorimetry, to composition measurements using chromatographs equipped with autosamplers, to more sophisticated systems capable of carrying out and quantifying chemical reactions—are now routinely automated, making it easier than ever to acquire data in the laboratory (Sanderson, 2019). Open-source databases are now commonplace in such areas as biology (Karsch-Mizrachi et al., 2012); thermophysical properties; materials properties (CHiMaD, 2021; Jain et al., 2013; NIMS, 2021; PRISMS Center, 2021); and climate, water management, and agriculture (University of Hertfordshire, 2021). The venerable steam tables, once available only in print, can now be accessed on the internet (NIST, 2021). These transformations are propelling advances on multiple fronts, but they are just the beginning. For example, the chemical industry only recently began the process of digitizing data previously stored over the course of many decades in laboratory notebooks. However, the infrastructure needed to ingest and organize such data most efficiently is not yet available, limiting the potential use of these vast stores of existing data. The way that infrastructure is developed over the next decade will determine how rapidly and how effectively companies and researchers can benefit from gaining access to vast amounts of information from multiple sources.

Moving forward, industry will develop systems that include not only internal, proprietary data but also data mined from the literature and from publicly accessible and commercial databases. Natural language processing tools are improving at a rapid pace, now capable of extracting sought-after information, such as experimental data needed to validate computer models, from the published literature (Olivetti et al., 2020). How one uses data and integrates data from multiple sources to generate valuable information is expected to become a major differentiator in highly competitive industries. Recent examples that illustrate this point include the optimization of water and energy use in paper mills (Ahmetović et al., 2021); the optimization of raw materials production from natural sources, such as switchgrass (Galán et al., 2021); and the production of renewable fuels from multiple sources (König et al., 2020). If they are to contribute to the development of such tools, chemical engineers will need to develop skills to exploit new possibilities as they emerge (Rzhetsky et al., 2015).

The growing ubiquity of data, along with increasingly powerful computers, networks, and algorithms, is creating opportunities for integration of data from disparate sources on scales never before imagined. It is now possible to monitor every aspect of a chemical plant through a myriad of sensors capable of collecting data continuously, and it will become increasingly common to couple such "lower-level" process data to other types of "higher-level" data pertaining to global supply chains, raw materials availability and characteristics, failure data, responsible water management (Caballero et al., 2020), or climate conditions throughout the world. This capability will enable chemical engineers to optimize enterprise-wide performance at levels not envisaged a decade ago (Hubbs et al., 2020). The COVID-19 pandemic has revealed the fragility of worldwide supply chains, arising from their inherent interconnectedness and the "just-in-time" operating philosophy of many businesses. Chemical engineers will have enormous opportunities to reimagine and redesign these systems to exploit the new ubiquity of and instantaneous access to data to achieve unprecedented robustness. Modern approaches to the design of all processes and products will revolve around the instant availability of relevant data. Growth in the volume of worldwide data is creating both urgency and opportunity for chemical engineers to develop the tools needed to assess data quality and in turn transform all the data into information and knowledge.

Other opportunities for chemical engineers will arise from combining technical and sociological data to achieve advances that could not have been anticipated just a few years ago. Examples include quantifying the dynamics of failure in scientific endeavors or technology translation (Yin et al., 2019) and reorganizing in real-time teams that are optimal for a given collaborative task (Wang et al., 2020). Combining sociological data with, for example, real-time geographic data on pollution, energy distribution, or water availability will allow chemical engineers to integrate sociological implications into engineering considerations to ensure that technology solutions are equitable and inclusive and do not disadvantage any population groups.

Indeed, the possibilities for integrating data and information from disparate sources and domains of inquiry are limitless. A massive amount of data will be available on climate, pollution, water transport, and the environment; on global and local supplies of raw materials; on how processes operate or perform when materials are altered; on such outcomes as quality and consumer demand for the corresponding products; and on human health (e.g., heart rate, blood pressure, oxygen level), both aggregated and individualized.

What new solutions, technologies, and industries will be possible through clever integration of disparate data? The following are some concrete examples.

Process automation. Highly automated processes will need to be developed based on vast sensing capabilities that previously did not exist, allowing real-time process management and dynamically optimized performance. This domain will grow rapidly as part of Industry 4.0.[1] Given its long history with data-driven approaches, the chemical industry will be quick to benefit from such innovations, while other industries, such as food and biopharmaceuticals, may be slower to benefit because of regulatory issues and relative conservatism in the adoption of new technologies.

Human health. Chemical engineers will aid in advancing the biomedical sciences to facilitate the practice of personalized medicine by addressing the human body as a system, informed by the massive amounts of data generated by medical devices and fitness trackers. Wearable devices invented by teams of chemical engineers and medical professionals will synthesize a multitude of real-time data streams to enable the instantaneous transmission of health information to physicians and phones. In the future, such information may be used to prevent everything from headaches to heart attacks. With their expertise in microfluidics, nanotechnology, and process control, combined with their deep understanding of biological processes within a systems framework, chemical engineers will drive invention and innovation in this space.

Additive manufacturing. Additive manufacturing enables on-demand production of materials, metamaterials, and parts. New data science tools will enable ever-more-rapid prototyping and scale-up, real-time assessment of performance in situ, and judicious adjustments based on continuous feedback from finite-element models running in parallel and in real time. New tools may one day make it possible to adapt manufacturing processes autonomously and in real time and to use immediately available raw materials based on local supply chain data.

Materials and molecular design. The rapid design of molecules or materials for targeted applications will be enabled by the development of widely available "living" databases that are updated automatically with experimental data on newly developed materials, and/or with computational data providing predictions about such materials and incorporating updates as they become available. Similarly, understanding of structure–property relationships ranging from the molecular to the manufacturing scale will enable "inverse design and synthesis," whereby one specifies the end-use characteristics and performance objectives for materials, and the computational tool determines the molecular components required to achieve them. Such tools will enable precision material design from the nanometer to meter scale.

Quantitative structure–property relationships (QSPR) and quantitative structure–activity relationships (QSAR). Statistics and probability—the classic tools of data science—also have a long history in chemical engineering, as exemplified by the use

[1] Industry 4.0, or the fourth industrial revolution, denotes the automation of traditional manufacturing and industrial processes.

of statistical methods for QSPR and QSAR—mathematical relationships linking, for example, a particular molecule to its structure or pharmacological activity, respectively (Cherkasov et al., 2014).

Automated/self-driving laboratories. As discussed further below, a future in which artificial intelligence (AI) is used to power automated laboratories is already taking form. To illustrate this concept, consider a polymer material designed to achieve certain end-use performance objectives. Given the raw materials and laboratory process needed to manufacture this material, a robot produces it by running the predesigned process, subsequently characterizing the polymer, and making any necessary feedback corrections automatically until the desired material performance is achieved. Of course, realizing this scenario in practice will require first developing all the sensors and robotics, along with the chemistries and characterization, necessary to enable high-throughput operation. To carry out syntheses, today's automated laboratories require extensive human intervention for setup and programming. Methods for automated extraction of chemical synthesis actions from experimental procedures documented in papers and laboratory notebooks have been described that can enable automatic execution of arbitrary reactions using robotic systems (Vaucher et al., 2020). IBM Research has described a system[2] in which a dataset of chemical procedures derived from cloud-based literature systems is integrated with cloud-based control of a robotic synthesis machine. In a live demonstration, this system identified a route to synthesis of 3-bromobenzylamine and carried out the process using air-sensitive reagents (Extance, 2020).

The preceding examples represent only a fraction of the ways in which the generation of and access to data will change the way chemical engineers work. Tomorrow's chemical engineers will need to be capable of using the massive amount of data of all sorts whose ready availability will continue to increase. It is easy to imagine a not-too-distant future characterized by data-on-demand, where data on any subject, at any level of granularity, will be readily and instantly accessible. Such a future suggests profound and exciting opportunities for chemical engineers, who are trained in process integration and systems-level thinking—skills that will be required to synthesize disparate data streams into information and knowledge.

Artificial Intelligence

AI is one of the modern tools of data science that is rapidly transforming all fields of science and engineering, including chemical engineering. Over the next few decades, AI is expected to transform all industry sectors in profound ways; indeed, the ways in which AI is used will differentiate companies with respect to gaining a competitive edge. Whereas prior revolutions in chemical engineering were driven by reductionist models, there is ample evidence that the next revolution surely will be data-driven and powered by AI.

AI is a general term used to describe "machines" (fashioned primarily by models and algorithms) designed to mimic human intelligence. With their unique combination of

---

[2] See https://rxn.res.ibm.com.

knowledge and skills—in particular in mathematics, statistics, computers, analytical thinking, systems integration, and process control—chemical engineering researchers have been among the first to integrate AI into their toolset, and now into the discipline's curriculum. Engineering colleges are challenged to keep pace with the increasing demand for courses in AI, and in particular in machine learning (ML), the subfield of AI concerned with the development and use of computer systems to learn adaptively from patterns contained in data, using algorithms and statistical models, without human supervision. At the same time, chemical engineering departments are under pressure to teach data science and ML in the context of their discipline, in part because of changing expectations and needs of employers for those who hold chemical engineering degrees. Each year, an increasing number of chemical engineering graduates take nontraditional jobs as data scientists at such companies as Facebook, Airbnb, Microsoft, Google, Uber, Amazon, and more. In a growing number of instances, these companies are hiring chemical engineers for their data science teams preferentially over computer scientists and even data scientists because they recognize that a large part of "doing" data science lies in asking the right questions using existing AI tools rather than, for example, inventing new AI machines.

While chemical engineering has been data-driven since its inception, what is new to the discipline is the explosive growth in the use of AI—and in particular ML—in making predictions from data. Perhaps the most rapidly growing use of AI by chemical engineers is in the area of deep learning—a subset of ML referring to the use of deep neural networks (DNNs), a class of ML methods whereby multilayered neural networks (NNs) are trained to perform tasks or discover hidden correlations in datasets far too large to be handled any other way. Simply put, NNs use structures motivated by networks of biological neurons to learn, through iteration (training), the relationship between inputs and outputs. Because the outputs are labeled and defined explicitly, the learning is said to be "supervised," as opposed to "unsupervised learning," exemplified by clustering, where the output is not predefined explicitly, and the algorithm's task is to discover groupings within data. In NNs, the network architecture is essential to its success. Examples of NN architectures used in chemical engineering research today include convolutional NNs (CNNs, used, for example, to learn images); generative adversarial networks (GANs); recurrent NNs (RNNs); and long short-term memory networks, where each descriptor is indicative of the unique characteristic of the NN structure in question. DNNs can have millions of parameters (node weights) to train and therefore require massive amounts of data to make useful predictions—a challenge for many research problems. Nevertheless, NNs and DNNs have seen success in a number of application areas within and adjacent to chemical engineering. For example, Shell employed four-layer gated RNNs to predict valve failures a month prior to failure (Gupta, 2018).

While ML concepts have existed for decades, only in recent years have computer chip speeds and architectures become fast enough for ML to become sufficiently practical to realize its promise. In addition to enabling faster computations, advances in computers have made it easier to collect, store, and manage large datasets. Furthermore, significant advances in the theory of statistical learning have led to new ML algorithms. The explosive adoption of ML by nearly every technology sector and research area has been fueled by the development and sharing of powerful, open-source, easily accessible libraries that simplify many of the most cumbersome tasks. This accessibility and ease of use of libraries is in large part responsible for the rapid adoption of ML by chemical engineers.[3] Such libraries encode expertise outside the typical knowledge domain of chemical engineers, providing naturally effective ways of expressing and working with complex computations, and thereby facilitating the implementation of data science and ML pipelines.

The relationship between science and engineering disciplines and AI software development has been mutually beneficial: the aforementioned libraries have seen massive investments and improvements because of the value they have provided to data science as a discipline, while AI's successes with scientific and engineering problems, among others, are driving continued increases in the use of ML software. Science and engineering generally can expect to continue to reap the benefits of industry investments in AI. A 2020 report from Grand View Research suggests that the AI industry will see a compounded annual growth rate of 40.2 percent between 2021 and 2028 (Grand View Research, 2021).

Today, chemical engineering researchers at universities are already using deep learning alongside experimentation and computer simulation. Examples of the many problems for which chemical engineers are making exciting breakthroughs using ML include mapping equilibrium phase diagrams, predicting system failure well in advance, evaluating metabolic networks, designing new molecules and materials, developing immunotherapies, understanding cellular processes and disease, and controlling nonequilibrium processes (Dobbelaere et al., 2021; Sanchez-Lengelingand Aspuru-Guzik, 2018; Venkatasubramanian, 2019). The swift adoption of AI by chemical engineering researchers is reflected in the steep rise in the number and frequency of American Institute of Chemical Engineers (AIChE) fall meeting presentation titles and abstracts containing AI-related terms. Between 2006 and 2020, the total number of presentations involving work using data science in some form increased from 0 to nearly 400, while the fraction of such presentations at a given meeting increased from 0 percent to nearly 7 percent (Figure 8-1). (In each dataset, an abstract is counted only once if the term of interest appears.) The appearance of no other words or two-word phrases increased as rapidly over this period. For comparison, the two most commonly found technical words in fall meeting AIChE abstracts are "bio" (in ~30 percent of all abstracts) and "nano" (in ~20 percent of all abstracts), frequencies that have not grown substantially since 2006.

AI has many potential future roles in chemical engineering. The following sections describe two areas in which the impact of AI in chemical engineering is expected to be profound: manufacturing and materials discovery and design.

[3] Examples of such libraries written for the Python program language include NumPy (performs numerical calculations), pandas (enables data cleaning and analysis), Scikit-learn (implements ML models), SciPy (provides numerical routines and solvers common in science and engineering), TensorFlow (provides a scalable platform for performing ML calculations), Matplotlib (creates data visualizations), and Keras (simplifies the definition and training of neural networks for deep learning).

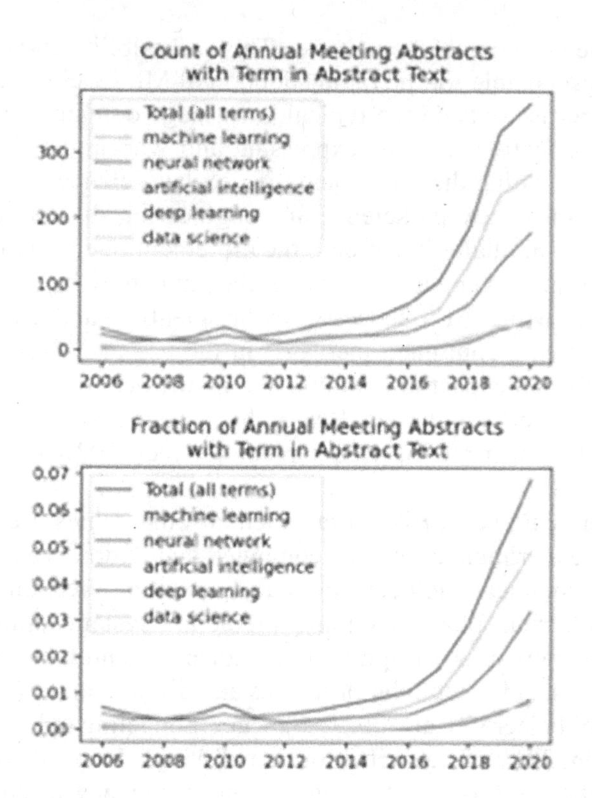

FIGURE 8-1 Count (top panel) and fraction (bottom panel) of American Institute of Chemical Engineers annual meeting abstracts that include terms related to data science, 2006–2020.

Artificial Intelligence and Data Science Applications in Manufacturing

Enabled by the internet of things (IoT), the availability of and role played by data in large-scale manufacturing have changed dramatically in recent years. It is now reasonable to expect that ubiquitous real-time sensors will be a core part of any future manufacturing operation, and that in general, relevant detailed information will be available at all scales, from the individual plant to the overall supply chain. The result will be a rich set of opportunities not only in the productive use of data but also in the methods for generating high-quality data and for maintaining the security of these processes. In each of these three areas, manufacturing industries are innovating rapidly without waiting for academic research to lead.

Productive Use of Data with Artificial Intelligence

An immediate implication of the above trends is the existence of a "data flood" in many manufacturing settings. Numerous AI tools have been developed to use such data effectively. A major challenge for chemical engineers, however, is that most data science methods originated in fields in which the implications of "off-spec" predictions are less dramatic than would be the case at a large-scale chemical plant. A related challenge is that many of these methods are not transparent, with components that have no recognizable connection to the physical characteristics of the process in question. Their "black box" predictions are therefore not readily interpretable, a challenge extending beyond manufacturing. These challenges highlight important research and development (R&D) directions for chemical engineers in combining data science methods judiciously with the best aspects of traditional, physics-based models. An underappreciated aspect of data science with particular importance in manufacturing is data quality, curation, and provenance. Important R&D challenges lie in automated assessment of data quality—for example, in the detection of sensor faults in large networks of sensors.

Innovative Modes of Obtaining High-Quality Data

Some aspects of introducing the IoT into manufacturing are as conceptually simple as measuring a process variable, such as temperature, at hundreds of points in a plant instead of at a handful of points. However, there are important areas in which R&D could enable collection of critical data that are currently challenging or impossible to obtain. In a traditional facility, sensors are placed in fixed, predetermined locations. Efforts are already underway in industry to collect data from a variety of locations using mobile sources, such as robots or autonomous vehicles. Chemical engineers have opportunities to collaborate with adjacent disciplines to explore the full implications of these concepts for manufacturing operations. They have opportunities as well in the development of imaging methods that provide low-cost, real-time insight into conditions inside inaccessible spaces (e.g., in reactor vessels) or at nanometer scales, perhaps by leveraging insights from associated areas, such as medical imaging.

Cybersecurity Implications in Manufacturing

Even as the IoT opens up unprecedented avenues for control and management of manufacturing facilities, these same avenues create enormous cybersecurity challenges. Well-publicized incidents of cyber-enabled attacks on physical resources in chemical manufacturing have already taken place. Significant needs for R&D include adapting general principles from the field of cybersecurity to the specific challenges of large-scale manufacturing. Just as data science methods can be used for fault detection, similar methods could be used to detect possible breaches of cybersecurity. It is notable that, as is the case with more familiar aspects of process safety, cybersecurity is a need shared by all parties in the chemical industry, suggesting that public–private consortia focused on precompetitive R&D would be well suited to addressing this challenge.

Artificial Intelligence Applications in Materials Discovery and Design

The nascent fourth paradigm of science has evolved rapidly over the past decade, enabling the use of big data and ML for the design and realization of new, better materials

(Tolle et al., 2011). In the semiconductor industry, for example, ML is enabling the identification of property patterns and anomalies for new materials. Chemical engineers are using ML to predict "sweet spots" for synthesis of new semiconductors, alloys, polymer materials, and nanomaterials through training on previously obtained experimental and simulation data. Force fields—necessary for molecular-dynamics simulations of materials—are well known to fail to predict some material properties (e.g., dynamic properties) accurately while reproducing others (e.g., thermodynamic properties) successfully. Chemical engineers are using ML to generate ab initio and phenomenological force fields that can accurately describe multiple material properties simultaneously, thereby enabling more complex simulation studies of phase transformations in materials, for example. They are designing new catalysts using a combination of electronic structure calculations and ML. Nonequilibrium materials processes (such as self-assembly and crystallization) are challenging to study for many reasons. Combining ML with molecular-dynamics simulations of nanoparticle self-assembly into complex structures, chemical engineering researchers are discovering microscopic details of assembly pathways that can be used to engineer products with fewer defects. In the future, both reaction engineering and assembly engineering will benefit from deep learning and other AI tools.

Scientific inquiry typically begins with a hypothesis generated by a researcher. Using AI successfully for hypothesis generation requires a high-quality training dataset. Depending on the maturity of the field, the exercise may produce a large set of well-characterized existing materials, or the result may be no more than a potential solution space with gaps. Fortunately, these gaps can be partially filled through modeling and simulation, using, for example, ab initio computational methods. Examples of areas in which high-throughput computational screening of materials have been demonstrated include optoelectronic semiconductors (Luo et al., 2020), thermoelectric materials (Sarikurt et al., 2020), and metal organic frameworks (Daglar and Keskin, 2020). In many cases, such AI-based hypothesis generation can be computationally expensive. In the future, the emerging field of quantum computing could greatly reduce the cost of these calculations or enable solutions to problems that would otherwise be intractable using classic computing architectures (Cao et al., 2019). However, technical and programming challenges need to be overcome before quantum computing can become a tool in the chemical engineering toolbox (Almudever et al., 2017).

Once a dataset of relevant materials and their properties (measured or calculated) is assembled as a training set for ML, generative AI models can be used to fill out the design space and suggest optimized synthesis targets, even if they were not part of the original training set. While AI-based discriminative learning (that is, the ability to discriminate among different kinds of data instances) is now commonplace (Ng, 2016), recent improvements in generative modeling (the ability to generate new data instances) are now allowing the generation of hypothetical new materials (Elton et al., 2019; Maziarka et al., 2020). Improved generative algorithms can restrict the output of hypothetical materials to those that are physically possible as manufacturable product (Dan et al., 2020). In the near term, human experts will be required to select potential, viable synthesis candidates from the set of predicted materials.

In the future, the testing phase of the research process also may be enhanced by AI. Standardized machine-readable representations of chemical species (simplified molecular-input line-entry system [SMILES]) and reactions (SMILES arbitrary target specification [SMARTS]) have been developed. These representations have been recognized as analogues to a context-free language (Sidorova and Anisimova, 2014). Atoms and molecules (SMILES) are the letters and words, while reactions (SMARTS) are sentences. Thus, state-of-the-art computer processing for language may be used to improve models for predicting chemical reactions and properties (Whiteside, 2019). Earlier approaches to predicting chemical reactions relied on expert-crafted reaction rules and heuristics to describe potential retrosynthetic disconnections (Socorro et al., 2005), with mixed results. More recent approaches are taking advantage of modern AI algorithms and the large reaction corpora, such as the U.S. Patents and Reaxys databases, to construct purely data-driven prediction engines (Coley et al., 2018). Beside retrosynthetic approaches, AI algorithms for solving the forward problem of predicting products given a set of reactants and reagents have been described (Schwaller et al., 2019).

Computing

From mainframes, to desktops, to cloud computing, to graphics processing units (GPUs), chemical engineers have been among the first to exploit the most powerful computer architectures available. The first uses of computers in chemical engineering can be traced back to the early 1950s (Wilkes, 2002). Since the early 1990s, chip speed has doubled roughly every 18 months. As a result, computing ability—in terms of both time to solution and complexity of problems that can be handled—has reliably increased by roughly three orders of magnitude every decade. What took months to compute in 2000 takes only minutes today and will likely take only a small fraction of a second in 2030. At the same time, the cost of computing has plummeted, democratizing and increasing access to supercomputing resources. The implications of this increase in computing capability and concomitant decrease in cost are profound. It now costs next to nothing to store the output of large simulations for subsequent interrogation offline, or it may now be faster to rerun a simulation when new calculations are desired or to perform real-time calculations as the simulation is running. This flexibility afforded by the broad availability and speed of both computing and data storage means that it is becoming faster and easier to explore vast, multidimensional design spaces; predict phase behavior in complex multicomponent or reactive systems more accurately; or apply ML to large, heterogeneous datasets. It also means that for many chemical engineering problems, simulation and experimental length and time scales coincide. This makes it possible for experiments to inform model development directly and for computation to be a true equal partner with experiment in R&D.

Several relatively recent game changers have emerged in the computing tools available to chemical engineers. An example is the GPU. Driven largely by the multi-billion-dollar computer gaming industry, GPUs are now ubiquitous throughout scientific computing and are today a key component of modern supercomputers. Central processing units (CPUs)—the original main processors in personal computers—have a handful of compute "cores," exhibit low latency, and excel at serial tasks, but can handle only a few operations at a time. By grouping tens of thousands of moderately fast CPUs or hundreds of powerful CPUs together in a single supercomputer, computations can be carried out in parallel. Since the first parallel computers appeared in the early 1990s, chemical engineers have exploited their capabilities to great effect.

CPU-based parallel computing was the cutting-edge tool in scientific computing until about 2010, when GPUs emerged in personal and business computers. Since the invention of the GPU by NVIDIA in 1999, every desktop or laptop computer has contained a GPU to process graphics displayed on the screen more quickly than was possible with the CPU, which now handles all remaining tasks. In contrast to CPUs, GPUs contain many cores, are high-throughput, excel at parallel processing, and can handle many thousands of operations simultaneously. Game developers and professional graphics designers exploited and drove R&D to increase on-chip parallelism to render ever-more-realistic images in real time. NVIDIA's public release of the parallel computing language CUDA in 2007 made the power of GPUs accessible to the science and engineering communities, which then could take advantage of GPUs by inserting a few simple commands into their existing codes.

Today, nearly all major scientific software has been or is being ported to or rewritten for GPUs and GPU/CPU hybrid supercomputer architectures, which are expected to be the primary architectures for at least the next decade. In molecular simulation, for example, open-source software such as LAMMPS,[4] HOOMD-Blue,[5] GROMACS,[6] and NAMD (Kass et al., 2011) runs on GPUs, enabling operations up to several orders of magnitude faster than those using CPU-based codes. Together with standards provided, for example, by Dockers,[7] computational science and engineering applications can be developed and tested on low-cost laptop and desktop computers; scaled up to run on local clusters, on federally funded computer facilities (e.g., XCEDE),[8] or on cloud-based service providers; and then scaled up again to run on the fastest, most powerful supercomputers (including the current fastest open-science supercomputer, Summit,[9] at Oak Ridge National Laboratory). The growing use of software "containers"[10] greatly facilitates porting simulation code from one architecture to another, making it easier for chemical engineering researchers to use state-of-the-art software without having to know the nuts and bolts of the computer they are using.

Another commercial driver of GPU technology is AI, with GPUs enabling the use of deep learning. The ubiquity of GPUs for deep learning has been accelerated by Tensor Cores, which not only speed up the linear algebra and large matrix operations at the heart of NNs but also carry out, as a single operation, multiple operations common in gaming. By enabling faster graphics and faster, more powerful AI algorithms, the video games

---

[4] See https://www.lammps.org.
[5] See https://hoomd-blue.readthedocs.io/en/latest/#.
[6] See http://www.gromacs.org.
[7] See www.docker.com.
[8] See www.xsede.org.
[9] See https://www.olcf.ornl.gov/summit.
[10] See https://www.docker.com/resources/what-container.

industry, the financial sector, and nearly every major industrial sector are today indirectly driving breakthroughs in computing for science and engineering. GPUs are also game changers for processing of imaging data, fueling everything from improved computed tomography (CT) scans to self-driving cars.

The continued applicability of Moore's law—achieved by still-decreasing nanometer-scale chip features, continued densification and thus multiplication of transistors on single chips, continued densification of chips on computer boards, faster networking and switch speeds for faster communications and input/output, new and faster computing algorithms, declining costs, and ever-increasing accessibility—will continue to enable chemical engineers to tackle bigger, more difficult, and more complex problems. Yet while today's computing power continues to increase, advances in AI are now pushing the limits of what is currently available, motivating the search for alternative options with even greater computing capability (Olson et al., 2016). The holy grail of computing is quantum computing, which until recently has been somewhat of a laboratory experiment but is now becoming more mainstream as an option available through some cloud providers. In traditional computing, a transistor holds the information as a 1 or 0 depending on whether there is a current applied (1) or not (0). In the case of quantum computing, because the qubits are in a superposition of states, they can hold a continuum of values between 0 and 1. Thus, in contrast to traditional computing's binary system, quantum computers can theoretically hold much more information within each qubit. Given the reliance on managing quantum states to control the computation, the quantum computer's surrounding environment needs to be tightly controlled against vibration and temperature variation. This is one of the fundamental challenges in quantum computing today, responsible for its high error rates and limiting its adoption. Accordingly, scientists across the globe are working to reduce error rates and foster confidence in quantum computer results. Today's approaches to this error-reduction problem have been based on the use of several materials, including superconducting materials and ionic fluids, but the expectation is that mainstream implementation may require a material not yet developed. It is widely expected that the requirements for strict environmental controls and high-purity materials for fabricating today's computer chips will become even more stringent as the computer industry moves toward exascale and quantum computers. These challenges offer additional opportunities for chemical engineers.

MODELING AND SIMULATION

Mathematical modeling—the development and use of mathematical equations as a convenient surrogate for studying, understanding, and even designing a physical entity—has been an indispensable tool for the modern chemical engineer. In general, mathematical models range from simple equations based on material and energy balances of large-scale industrial equipment or empirical correlations connecting one observed phenomenon to another to complex multiscale models that span the entire length scale, capturing atomistic molecular interactions and integrating up to macroscale phenomena.

Traditionally, modeling has almost always been connected to simulation in which a computer is used to solve the set of modeling equations representing a physical system so the behavior of that system can be explored under various conditions. However, simulation has evolved to be understood in the more general sense of a computer representation of a physical entity with or without the use of an explicit mathematical model. As a consequence, any computational representation of a physical system, whether via explicit mathematical equations or a computer simulation, is understood to be a "model" of the system in question.

Simulation also refers to systems in which interaction potentials between particles are developed to describe many interacting species, and emergent collective behavior is revealed. These approaches are leveraged to understand particles ranging from the quantum to colloidal scale, with detailed or approximate potentials. The results of these simulations often inspire reductionist modeling in terms of mesoscale mathematical models and can be used to understand or predict the parameters in these mesoscale models. Chemical engineers have developed and applied these approaches to understand biomolecular interactions essential to drug design and particle interactions and assembly at the heart of nanotechnology. This cross-talk between simulation to reveal emergent behavior and to inform model development can be further developed to embrace data-driven and equation-free approaches to simulating interactions across scales in complex, far-from-equilibrium systems.

Chemical engineers continue to innovate and drive the field of molecular dynamics (MD) simulation, which was invented in 1957 at Argonne National Laboratory and further developed over the next several decades, primarily in the United States and United Kingdom, by chemists and chemical engineers. Today, the most widely used MD codes—including LAMMPS, HOOMD-Blue, NAMD, and GROMACS, all led by U.S. researchers—are open source. Chemical engineers have developed some of the most popular molecular simulation algorithms, such as free-energy methods (Frenkel and Ladd, 1984), thermodynamic integration (Kofke, 1993), enhanced sampling methods, ensemble simulation techniques (Panagiotopoulos, 1987; Yan and de Pablo, 1999), and acceleration algorithms (e.g., bond-boost; Pal and Fichthorn, 1999). New generalized ensemble simulation techniques, such as digital alchemy (van Anders et al., 2015), use MD and Monte Carlo simulation for inverse design of materials building blocks for targeted structures and properties.

Rapid advances in computer technology and an unprecedented increase in computing power and data storage, combined with the ready availability of user-friendly modeling and simulation software packages powered by extremely efficient computational methodology, have facilitated modern modeling and simulation in chemical engineering. It is now possible, for example, to begin with simple rules for interactions among particles (electrons, protons, atoms) and simulate emergent behavior. On the other hand, such emergent phenomena may be so complex that the computer model itself is an "equation-free" model developed to capture the behavior on the basis of large volumes of experimental observations (using such data analysis tools as ML algorithms; Kevrekidis and Samaey, 2009). It is possible, therefore, that in the future, the distinction between first principles–based and data-based modeling/simulation will become sufficiently blurred as to become almost irrelevant. The following sections describe the role that the dual tools of modeling and simulation will play in chemical engineering in the future in the specific areas of education, research, and industrial applications.

Education Applications

Since computers were first introduced into chemical engineering, (Katz, 1966), simulation has been an integral part of chemical engineering education; this trend will continue and will likely accelerate in line with the advent of faster and more sophisticated computer hardware and software and transformative approaches (such as equation-free models). The three traditional central components of simulation are (1) the mathematical/computational model or interaction potentials upon which the simulation is based, (2) the algorithms used to carry out the simulation and/or solve the equations, and (3) the computer hardware and software used to carry out the necessary computations and display the results.

The chemical engineering classroom of the future will leverage modeling and simulation to expand student understanding. In the past, student intuition was developed in advanced undergraduate courses using simple canonical examples of chemical processes that permit analytical closed-form solutions. Advances in simulation are already making it possible to complement and expand these pedagogical approaches. For example, molecular-scale phenomena, phase transitions, and fluid flow can all be taught effectively using computer simulations. With virtually limitless computing power, the challenge is to convey this information in consumable form so that students understand the limitations of simple models; develop intuition; and become adept at leveraging this information to design, control, and exploit the systems of interest. Simulation allows students to learn through concrete examples how electronic interactions give rise to interatomic and intermolecular interactions; how these interactions produce structure, properties, and behavior; and how changes at the nanometer scale manifest at the mesoscopic and macroscopic scales.

The future will see improvements not only in the knowledge base upon which to build the models but also in the ability to compile, validate, and deploy such information rapidly and efficiently. Achieving the goal of conveying the information from simulations in "consumable form" will be facilitated by synergistic integration of model development and computation within unified simulation software, coupled with efficient graphical output capabilities and user-friendly application programming interfaces (APIs). Different parts of the chemical engineering curriculum will require different suites of modeling and simulation tools (e.g., MATLAB/SIMULINK for process control, ASPEN for process design, computational fluid dynamics [CFD] tools for fluid dynamics, MD and Monte Carlo simulation for thermodynamics). There is no question as to the increasing role of modeling and simulation tools in the education of chemical engineers of the future. The committee imagines a future in which modeling and simulation tools, in combination with AI and virtual reality technology, will afford students the experience of operating an entire chemical manufacturing plant virtually, in a manner that transcends even that which flight simulators provide to pilots in training today.

Research Applications

Chemical engineers have used simulation to contribute significantly to understanding of systems in which interaction potentials are developed to describe interacting atoms, molecules, polymers, nanoparticles, and larger particles, and emergent collective behavior is revealed. These approaches make it possible to predict the behavior of matter, whether a catalyst for a chemical reaction or a material for a specific application. Electronic structure calculations are used at the smallest level, which can inform classic force fields to access longer length and time scales. Chemical engineers have made leading contributions using such approaches in a wide variety of systems. Examples in biomolecular systems include protein folding and aggregation, hydration, and ligand–receptor interactions; self-assembly of lipids; and the role of hydration in ion complexation. In surfactant systems, examples include prediction of equilibrium mesostructures used to formulate personal care products or to template advanced materials. In nanotechnology, the same simulation methods are used to predict the self-assembly of "colloidal molecules" and patchy particles to form aggregates and assemblies for next-generation soft materials or their directed assembly in the presence of biasing fields.

Mesoscale models, which exploit coarse-grained potentials or even continuum free-energy functionals, are also used to access emergent behaviors (e.g., for protein aggregation in lipid bilayers, or colloidal assembly in complex fluids at length and time scales inaccessible to the most detailed simulations). The challenges for the future include better connecting simulation from one scale to another and developing interoperable simulation tools to inform model development for the systems scale. The coarse-grained or continuum-scale descriptions are valuable in that they include descriptors and parameters that may be related to experimental observations or may be exploited in systems-level mathematical models. Close coupling of these simulations with experiments can form a virtuous circle of simulation, prediction, and identification/determination of mesoscale parameters and experiments to improve understanding of complex systems.

Significant challenges remain in simulation of systems that are far from equilibrium. For example, it is currently challenging to move from the molecular scale, to the mesoscale, to the cellular or tissue level in biomolecular engineering; to relate surfactant mesostructure to rheology in systems under flow; to understand reacting systems with changes in mesostructure and physicochemistry over prolonged time scales; or to understand driven assembly of colloidal suspensions in complex soft matter. As simulation costs decline and data-driven approaches mature, cross-talk between simulation to reveal emergent behavior, mesoscale description, and mathematical models will advance. Such advances will facilitate, for example, environmental modeling, making it possible to start from detailed chemical reactions at the quantum level and integrate up to represent large-scale, macrolevel behavior, which in turn will enable high-fidelity climate prediction over periods of years.

Other rapidly developing tools that will be important for chemical engineering include CFD, computational quantum chemistry, and predictive reaction kinetics. Improved CFD models will contribute to understanding of turbulence. Direct numerical simulation (DNS) and large-eddy simulation (LES) for fluids and methods for two-phase flows, combined with faster computers, are yielding new insights. Computational quantum chemistry has become invaluable to chemical engineers for modeling force fields, electronic materials, catalysts, thermochemistry, and safety analyses. Chemical engineers are deeply involved in developing methods and codes that leverage these calculations with ML. Finally, tools of predictive reaction kinetics are advancing rapidly. Reactive force fields such as ReaxFF and QMDFF and improvements in ab initio molecular dynamics promise to enable model-based chemical experiments.

Chemical engineers can also meet the challenge of capturing and manipulating the information from simulation and experiment in forms that provide mechanistic insight to inform the design, control, and exploitation of the systems of interest. As research advances on this front, so will pedagogy, as these new approaches and capabilities will be integrated into the classroom.

One other tool that could revolutionize chemical engineering research is the development of interoperable models and software. Currently, there exist software for predicting properties, other software for materials design that use the properties, and yet different software for simulating the material and predicting its performance in end use. Integrating these disparate systems within a framework that facilitates interoperability would greatly speed up materials discovery and design.

With the growing power of AI and its adoption in chemical engineering, chemistry, physics, and materials science, molecular simulation is poised for a revolution. Standard approaches to developing interatomic force fields may become a thing of the past, replaced by the use of ML to develop force fields (Chmiela et al., 2018; Noé et al., 2020). Already there are a growing number of success stories. A recent study (Batzner et al., 2021) shows that with the use of equivariant neural networks, massive datasets are not required for deep learning, and that high accuracy can be achieved with far less computational effort than previously thought.

Industrial Applications

Chemical Manufacturing

Modeling and simulation have jointly played a significant role in the modern chemical and biomanufacturing industries, although applications have focused primarily on large-scale processes and equipment, mainly for operator training and for the design and implementation of advanced control systems. The increasing operational complexities and shrinking economic margins in the petrochemical industry, coupled with stringent environmental and quality demands on the manufacture of specialty chemicals and polymers, have spurred increased use of modeling and simulation. While the pharmaceutical industry currently lags behind the chemical industry, fundamental changes in regulatory requirements are motivating greater use of mathematical models and simulation in that industry, especially in the rapidly growing biomanufacturing sector. With the advent of such Food and Drug Administration initiatives as process analytical technologies and quality by design, an increasing number of leading pharmaceutical companies are incorporating process modeling and simulation systems into their manufacturing enterprises.

One concept penetrating pharmaceutical manufacturing is that of the "digital twin," described most simply as a virtual representation that serves as the real-time digital counterpart of a physical object or process. Park and colleagues (2021) provide a brief review of potential bioprocess applications.

With increasing capacity for collecting, curating, warehousing, and visualizing massive quantities of process data and with easy access to relevant ancillary data, the next challenge is to develop systems that will integrate process operating data seamlessly with ancillary data from the supply chain, policy, economic trends, global markets, and climate predictions to optimize production planning, scheduling, and operation. This capability will become especially indispensable for enterprises with multiple manufacturing facilities distributed across the globe. Such modeling is also important for scaling up, especially for large-scale plants that represent a large capital investment. The computer hardware technology necessary to facilitate the implementation of such systems currently exists and is advancing rapidly; the necessary software is sure to follow in the coming years.

Systems are currently being developed to allow the industrial practitioner to connect molecular-level models to large-scale process models and insert process models within the context of a facility. Consider as an illustrative example a dynamic model of a CO_2 capture unit within a power plant that is capable of predicting the plant's performance as the electricity output changes dynamically. Imagine that the systems-level model of the power plant is integrated with the market to enable the user to determine optimal plant configurations and product slates in a given market. Such a simulation system will, among other things, allow users to explore options for how best to design and optimize energy production/conversion systems. A beta version of such a system currently exists and offers a glimpse into what is possible in the future (Arent et al., 2021).

To be most useful, however, such systems need to provide different kinds of interfaces for different categories of users, making customized information available in real time as appropriate for each user: technical information for the research engineer and the practicing engineer, and high-level economic and business information and policy implications for decision makers. The committee envisages such systems also possessing the capability to incorporate past experiences (e.g., failures, safety incidents). Chemical engineers already feature prominently in, and in some cases have founded, companies that are commercial vendors of process modeling, simulation, control, and operations software, hardware, and services (e.g., the pioneering process design software ASPEN was developed and the company ASPENTECH founded by Larry Evans, a professor of chemical engineering at the Massachusetts Institute of Technology). Chemical engineers are likely to contribute significantly to the development, implementation, and commercialization of these systems of the future.

Life-Cycle and Technoeconomic Analyses

Life-cycle assessment (LCA) and technoeconomic analysis (TEA) are systems-level analysis tools used by chemical engineering primarily for new chemical technologies; they represent a completely different category of industrial applications of modeling and simulation. As mentioned throughout earlier chapters, the ability to use these tools will enable chemical engineers to ensure that their innovations have a net lower environmental impact relative to alternatives. LCA is fundamentally about the mass and energy balances of processes and is a standardized methodology for assessing the environmental impact of a product over its entire life cycle (i.e., production, use, and disposal) (Billiet and Trenor, 2020; Muralikrishna and Manickam, 2017; Nuss, 2015; Weisbrod et al., 2016; Zhang et al., 2018b); it can also be used to compare two or more products objectively. LCA has a fixed structure and follows the standards in International Organization for Standardization (ISO) 14040:2006 (ISO, 2006). It can be carried out as a

- cradle-to-grave analysis, which provides the full LCA from the raw material extraction step, through the use step, to the disposal step (grave);
- cradle-to-gate analysis, which provides a partial product LCA from the raw material extraction step to the factory gate (i.e., before the product is transported to the consumer); or
- cradle-to-cradle analysis, in which the end-of-life disposal step is a recycling process.

TEA, a methodology for analyzing the technical and economic performance of a process or product, combines engineering design, process modeling, and economic evaluation. It depends heavily on the technology readiness level (TRL) and includes goal, scope, and scenario definition; cost estimation; market analysis; profitability analysis; and results interpretation. Specifics on TRL titles, descriptions, tangible work results, and workplaces for the chemical industry can be found in Buchner et al. (2018). As the TRL increases, data availability and accuracy increase, and the TEA becomes more reliable.

Integrating LCA with TEA yields what is called environmental technoeconomic assessment (Thomassen et al., 2019), most useful when further combined with energy and mass balances, thermodynamics (Banholzer and Jones, 2013), government policies (e.g., potential carbon tax, packaging recycling content targets, greenhouse gas [GHG] ceilings), social acceptance assessment, and project control structure. This overall integrated assessment structure affects technology directions and needs to be incorporated into the thinking of chemical engineers as they develop new products and processes in the circular economy, discussed in Chapter 6 (see the example discussed in Zimmermann and Schomäcker, 2017). Appropriate modeling, simulation, and data assimilation tools tailored specifically to facilitate such large-scale systems analysis do not currently exist, creating an area of opportunity for chemical engineers as they facilitate the transition from the linear to the circular economy. AI will play an important enabling role in this regard (McKinsey Sustainability, 2019), allowing chemical engineers to harness information contained in large datasets to learn faster and more efficiently from highly complex systems. Specifically, AI can enable chemical engineers to

- design circular products and materials (via ML-assisted design processes for rapid prototyping and testing),
- operate circular business models (via AI's predictive capabilities from historical datasets), and

- optimize circular infrastructure (via improving the processes for sorting and disassembling products and recycling materials).

Medical Applications

Chemical engineers have contributed significantly to understanding of the molecular systems and mechanisms underlying many of the body's physiological functions. From signal transduction in cells (how signals are passed from the surrounding environment into the nucleus of a cell) to changes in gene expression in response to these signals, ultimately resulting in the regulation of various biological processes, chemical engineers have developed models that have facilitated the development of systems biology (Kinney et al., 2019). In an earlier generation, Bischoff and his coworkers pioneered the use of compartmental models to study the macroscale distribution of drugs in the human body (Bischoff, 2015). Mathematical models of metabolic pathways combined with novel measurement techniques have facilitated unprecedented high-fidelity analysis of human metabolism (Lachance et al., 2021). At the physiological level, the recent development of the artificial pancreas, an insulin delivery system for those with type 1 diabetes, is based on modeling, simulation, and control mechanisms developed primarily by chemical engineers in conjunction with clinicians (Dassau et al., 2017).

Understanding of human physiology is rapidly approaching a sufficiently quantitative level to permit the development of multiscale simulators of the entire body, from single cells, to tissues, to organs and organ systems, to the entire organism. Such systems can then be used for a limitless number of applications, including clinical trials. These so-called in silico trials will save a significant amount of money and time by allowing prediction of the clinical outcomes of drugs at various stages of development on the one hand, and on the other, facilitating the design and synthesis of complex biological drugs and predicting the behavior in vivo. Such simulation systems can also be used to implement personalized medicine, with the potential to revolutionize the practice of medicine and drug development (Ogunnaike, 2019) (see Chapter 5).

The building blocks necessary to actualize this class of simulators already exist in the form of single-cell models, multiscale biological system information, metabolic analyses, carbon labeling, flux analysis, single-cell genomics, high-throughput screening, and so on. With rapid advances in biological/medical knowledge, data-collection capabilities, and computer hardware and software technologies, the development of such medical simulation systems may occur sooner rather than later.

NOVEL INSTRUMENTS

Chemical engineers have had a transformative impact on instrument development, especially in the establishment of fundamentals underpinning measurement and characterization, in the development of hardware, and in early adoption. Analytical methods and visualization tools are two particular areas that have benefited from the participation of chemical engineers. These methods and tools have had an impact on chemical engineering research and practice in two distinct ways. First, in some cases (e.g., super-high-resolution microscopy and single-molecule detection), the tools have improved signal-to-noise ratios and improved accuracy and/or precision, thereby making higher-fidelity characterization possible. Similarly, advances in atomic-force microscopy and transmission electron microscopy have provided new information about synthetic as well as biological objects with unprecedented precision and accuracy. Second, improved speed in genomics, proteomics tools, and flow cytometry has allowed more rapid and more comprehensive explorations of the parameter spaces of interest to the researcher. The impact on biology and medicine has been particularly significant, since the high-throughput capabilities of these tools now make the routine collection of massive amounts of data a practical means of addressing biological complexity.

The field of tool development offers opportunities for chemical engineers to contribute to the development of next-generation instruments that will provide both fundamental and practical insights not possible today. For example, methods that track protein abundance dynamically inside a cell will provide unprecedented information about cellular function. In contrast, current protein analysis methods are rather simplistic, with capabilities limited to quantifying specific, predetermined proteins one at a time by antibody-based methods or mass spectrometry. Even if current proteomic methods were to be enhanced to become high-throughput, the technological basis would still be such that analysis could be performed only ex situ, requiring the removal of proteins from their native environment, with a two-fold undesirable consequence: some of the key information about in situ characteristics would be altered, and some of the essential features of protein–protein interactions would be lost. For proteins for which information on dynamic behavior is accessible, advances in visualization of such dynamic characteristics have yielded unique insights into cellular transport and cytoskeletal networks. Another example of tools that will have a significant impact on chemical engineering is dynamic in situ super-high-resolution microscopy, especially electron microscopy. Current transmission electron microscopy methods can provide only temporally averaged information.

The ability to track individual cells in real time in the body represents another opportunity to obtain potentially game-changing insight into the body's metabolism in both healthy and diseased states. Current methods for assessing cellular composition can provide only infrequent snapshots of macroscopic information at discrete points in time. Histology remains the primary clinical mode of assessing the cellular composition of tissues. Such assessment is rather limiting because it provides only a two-dimensional static view of cellular composition. The ability to visualize cellular motion in the body in real time will provide useful information about the immune response to infections, sepsis, and metastasis, all of which involve the complex, coordinated motion of multiple cell types.

Tools for building materials with molecular-scale precision will open up new opportunities in nearly all fields associated with chemical engineering. With such tools, unique structural features that are currently infeasible will be possible because the tools will offer the precision and control of 3-D printing and allow building of materials at the individual-molecule scale, making it possible to control functional attributes and identify and address manufacturing defects in real time.

"On-chip" systems offer a cost-effective means of generating information, synthesizing materials, and screening them for their important interactions. Examples of such systems include a chemical-plant-on-a-chip, human-on-a-chip, organ-on-a-chip, or colloidal-particle-generator-on-a-chip. (Opportunities for cell-, organ-, and organism-on-a-chip are discussed in Chapter 5.) Such chip-based methods offer many key advantages. From a synthesis point of view, they offer the advantage of small volumes and subsequently reduced costs. This attribute is helpful especially for research operations that aim to generate large libraries of materials in small quantities (for example, chemical libraries). Reduced scales are especially significant if the chemicals involved are toxic, since the on-chip systems use very low volumes of material. Such on-chip systems are also beneficial for colloidal synthesis. Traditional macroscopic methods of colloidal synthesis are often subject to a high degree of heterogeneity associated with interfacial phenomena. On-chip methods are very precise and have been used to synthesize colloids with more uniform dimensions. Such on-chip systems have also played an important role in the deployment of sensors that provide finer control over the flow and interactions of samples and reagents. Human-body-on-chips, which include systems comprising cellular chips representing one or more organs, are also poised to make a strong impact on drug development.

All on-chip systems share some common features and opportunities for impact from chemical engineering. First, microfluidics is a central component of all on-chip systems, and many of the fluids of practical interest (e.g., blood in the case of body-on-chips, or colloidal suspensions in the case of nanoparticle synthesis) are complex. Consequently, an understanding the flow of complex fluids in confined spaces is of prime importance. Second, all on-chip systems consist of a complex interconnection of a number of individual components, each complex in its own right. The design, analysis, and effective deployment of these systems offer opportunities for chemical engineers to contribute their expertise in fluid dynamics, reaction engineering, tissue and cellular engineering, and process systems engineering.

SENSORS

Sensors for Process Monitoring

Sensors play an important role in the design, development, and monitoring of processes and systems in all areas of chemical engineering. Development of "smart sensors," "ubiquitous sensors" (tool or enabling technology depending on specific use), "smart actuators," and "smart transmitters" hold the potential to transform a wide variety of manufacturing applications in the future. Such sensors can be classified broadly into those that measure physical parameters (e.g., temperature or volume), chemicals (e.g., reactants or cell culture media), or states (e.g., cell viability or particle sizes).

Physical Parameters

Sensors that measure such physical parameters as temperature, pressure, or flow rate have long been central to process monitoring in the chemical industry. With the advent of inexpensive, lightweight, and wearable types enabled by the development of flexible electronics, sensors that monitor physical parameters have found widespread applications. For the most part, sensors that monitor physical parameters are sufficiently mature and have already been incorporated into many practical applications. A key remaining opportunity moving forward lies in real-time analysis of data from such sensors for advanced process and product optimization. An emerging opportunity is in the adoption of these sensors and sensor networks for remote operation and remote laboratories. While remote operations have been used in the past largely for processes with safety concerns or with inherently limited access, the practical limitations imposed by COVID-19, which made remote operations inevitable, have brought this technology to the forefront not only for manufacturing processes but also for instruction in laboratory courses, representing an area of potential future impact for chemical engineers.

Chemicals

Sensors with which to measure the composition of chemicals play an important role in collecting molecular-scale information about a monitored system. For example, chemical analysis of reactants/products has long been used to assess reaction progress. Decades of advances in analytical tools have provided an array of methods (e.g., nuclear magnetic resonance [NMR], Raman spectroscopy, high-performance liquid chromatography, gel permeation chromatography, mass spectrometry [MS]) for assessing the molecular composition of complex chemical mixtures. Two key opportunities exist for the future. First, while the ability to measure (sense) chemicals in large laboratory equipment is practically unlimited, capabilities to do so using small sensors that can be readily inserted into the system for continuous monitoring are limited. Such sensors often require the development of new sensing mechanisms, robust hardware to implement those mechanisms, and novel technologies to enable continuous measurements in a seamless way. In addition, sensors for specific molecules (e.g., protein glycan groups) may require the development of molecularly and/or biologically based sensors interfaced with an appropriate device to generate the required measurement in the form of an electronic signal.

A second opportunity is the use of AI and ML to extract more meaningful information from complex data. Measurements offered by analytical tools such as NMR and MS are rich in information, and the information extracted from these measurements is heavily dependent on the question posed. The availability of new AI/ML algorithms can facilitate the extraction of new information from existing data, offering the opportunity to develop means by which new tools can extract information from preexisting data.

State

Another emerging opportunity lies in the capability to sense a state, which is in a relatively early stage of development. In this context, the "state" of a process or system refers in general to characteristics of the process or system that otherwise cannot be captured in a simple single parameter—for example, the viability of a phenotype of cells for application in protein or cell manufacturing, or particle-size distribution in a granulation or agglomeration process. Sensors capable of going beyond the current acquisition of physical parameter measurements have played and will continue to play a major role in the monitoring of chemical and biochemical processes. Future opportunities lie in the development of sensors for measuring the abundance of specific molecules and the establishment of a network of such sensors supported by advanced data analytics tools to enable optimum extraction of information contained in the massive quantities of acquired data. The development of self-learning, automated, self-monitored reactors is yet another opportunity that will depend critically on the availability of smart sensors.

Although significant progress has been made in real-time monitoring for quality control during production in dissolution, crystallization, drying, and other important unit operations, the implementation of similar monitoring during cell growth and protein production is lagging. For example, most antibodies and other therapeutic proteins require posttranslational modifications, such as glycosylation and disulfide bond formation. The product profile approved by the regulatory agency includes a range of allowed modifications typically not determined until postpurification. In addition, the extent of other, undesired modifications—such as oxidation, aggregation, charge variants, and truncation—needs to be determined and quantified because these modifications affect drug performance in end use. Two approaches for addressing these product characterization needs are (1) modifying existing analytical methods and using data science methods to determine the composition of complex mixtures (e.g., cell culture supernatant), and (2) developing altogether novel analytical optical and bioelectronic sensors. For example, infrared (IR)–based sensors are often used for measuring chemical composition, primarily because they can be used in situ for the most part. However, using Fourier transform infrared spectroscopy or near IR requires offline sensing and spectral deconvolution. Determining the level of contaminants, such as heavy metals, in a bioreactor feed is an important aspect of biomanufacturing. The development of sensors capable of providing these critical measurements requires not only development of the actual hardware of the sensor but also establishment of the fundamental parameters that define the state. Correct placement of sensors is another practical challenge that needs to be addressed. As new sensors are developed, proper placement within a network and the connectivity within the configuration will be central to successful implementation. The availability of transmitters to transmit the necessary data within the network with the requisite resolution will be critical. The development of such tools will require advances in hardware related to wireless transmission, as well as data analytics tools for efficient and effective information extraction.

Sensors for Health Monitoring

Design for Biological Applications

In the field of health monitoring, care for patients with diabetes has been impacted by sensor innovation through the development of continuous glucose monitoring technology. Monitoring of blood glucose levels, which, decades ago, depended on infrequent laboratory-based sampled measurements, is now carried out virtually continuously with wearable sensors (Lipani et al., 2018). This development has been made possible by significant advances in glucose sensing chemistries, a deeper understanding of the biocompatibility of sensors, and the development of electronics that enable a user-friendly patient interface. Such developments have transformed the sensing paradigm from "sample to the sensor," to "sensor to the sample," to "sensor on the patient." Sensing technology was once based on large analytical equipment, requiring that the patient sample (e.g., blood) be withdrawn and shipped to a central analytical laboratory (e.g., laboratory-based glucose measurements). As sensors became smaller and portable, they went where the sample was (e.g., home-based glucose tests), which eliminated the sample shipment process. As sensors get even smaller, they can go directly on the patient (e.g., wearable glucose sensors), thus eliminating the step of collecting the sample and putting it into the sensor. These continuous sensors utilize a needle-based sensor inserted subcutaneously to measure the glucose concentration in the extracellular space just under the skin. Efforts have also been made to develop completely noninvasive means of measuring glucose concentrations using spectroscopic methods or noninvasive collection of tissue fluid through the skin, including sweat-based sensors. Sweat-based sensors measure the composition of various analytes present in the sweat (including such small molecules as glucose and hormones, and such large molecules as proteins) and use these measurements to assess the patient's physiological state (Chung et al., 2019). Recent advances in sweat-based sensors are fueling optimism that the development of means of measuring additional analytes, including hormones, toxins, and allergic responses in the body, will likely have a similar transformative impact on health.

In addition to molecule-based diagnosis, wearable sensors now routinely perform a variety of other diagnostic measurements, such as temperature, blood oxygenation, and pulse (Gao et al., 2019). The ability to measure these key vitals through a wearable sensor has already transformed the collection of human physiological data, facilitating unprecedented acquisition of information about human responses in real-life situations in real time and continuously. Traditionally, these measurements were possible only using large, bulky sensors often connected to stationary electronic processors and displays.

Therapeutic drug monitoring is another area in which the impact of novel sensors can be transformative. Current drug-dosing regimens are rudimentary, based largely on average male body mass and drug trials that ignore the effects of race, ethnicity, gender, and many other patient-specific characteristics. The ability to measure drug concentrations in the body in real time can allow clinicians to achieve tighter control over pharmacokinetics, thus potentially reducing harmful variability in drug distribution and preventing adverse events due to large excursions from desired concentration levels. Successful implementation of such technology will require the development of drug-specific, nonfouling, long-term implantable sensors that provide real-time feedback to clinicians. Significant progress is already being made in this regard. For example, an aptamer-based sensor that can measure certain chemotherapeutic agents has been developed and tested at the preclinical level. The challenges associated with expanding the scope of such technologies for a broad range of drugs offer future opportunities for chemical engineers.

Data Handling and Processing

The ability to extract new, better information from existing data promises to have significant impacts on health and medicine. As one example, consider medical diagnosis, which has long relied on qualitative analysis of images derived from magnetic resonance imaging (MRI), CT scans, radiology, and histology. Tools that enable better image processing can extract better information and yield more accurate diagnoses compared with qualitative human analysis. Studies have already demonstrated that AI-based analysis of histology is likely to be more accurate and less likely to miss a rare event. At the same time, AI-based analyses and diagnoses can help eliminate human bias in such procedures (although if not implemented correctly with appropriate training data, AI can itself amplify biases).

Extrapolation of current trends into the future suggests an ever-increasing capability of wearable sensors to collect even more massive amounts of data about human behavior in healthy and diseased conditions. These data will certainly include such basic vitals as temperature, pulse, and oxygenation, which in themselves are sufficient for certain applications. However, the available data could contain real-time measurements of such quantities as blood glucose, hormones as an indication of stress, and alcohol as a measure of intoxication, among others. Some of the technical challenges to overcome in realizing such capabilities include development of sensors that can work efficiently with small amounts of analytes that are available in sweat and determine compositions with appropriate accuracy.

A challenge also exists in correlating these measurements accurately with blood concentrations, in particular with respect to the transport of analytes from the blood into the interstitial fluid. An opportunity exists to collect massive amounts of data and establish correlations across the population in a way that is simply not possible through the collection of small amounts of data. Can model-based analysis be used to predict catastrophic physiological events from these data? Can drug reactions be better understood or predicted through appropriate analysis of such massive amounts of data? Given the well-documented challenge posed for drug design and development by the heterogeneity of drug response across a population, the prospect of personalization of drug therapies presents an exciting opportunity for reducing adverse events without compromising therapeutic efficacy. However, actualizing this concept will require having adequately informative data available to support such personalization at the design stage and the reduction of adverse events at the follow-up stage. Continuous physiological monitoring offers the potential for producing precisely this sort of data.

CHALLENGES AND OPPORTUNITIES

The development of tools that synthesize available data in real time and frameworks or models that transform data into information and actionable knowledge could become one of the key contributions of chemical engineering to society over the next decades. It is easy to imagine a not-too-distant future characterized by data-on-demand—where data on anything, at any level of granularity, will be readily and instantly accessible. Such a future suggests profound and exciting opportunities for chemical engineers, who are trained in process integration and systems-level thinking—skills that will be required to synthesize disparate data streams into information and knowledge.

The systems thinking, analytical approaches, and creative problem-solving skills of today's chemical engineering graduates give them a distinct advantage in using AI in real-world contexts. The evolution of AI in the next decade will have enormous implications not only for the types of problems chemical engineers will be able to solve but also for how they will develop those solutions. Chemical engineers are poised to contribute significantly to the development of modeling and simulation tools that will influence education, research, and industry. They will continue developing and disseminating methods, algorithms, techniques, and open-source codes, making it easier for nonexperts to use computing tools for scientific research.

The increasing operational complexities and shrinking economic margins in the petrochemical industry, coupled with stringent environmental and quality demands on the manufacture of specialty chemicals and polymers, will continue to drive the increased use of modeling and simulation. While the pharmaceutical industry currently lags behind the chemical industry in its use of simulation tools, fundamental changes in regulatory requirements are motivating its greater use of mathematical models and simulation, especially in the rapidly growing biomanufacturing sector.

Recommendation 8-1: Federal and industry research investments should be directed to advancing the use of artificial intelligence, machine learning, and other data science tools; improving modeling and simulation and life-cycle assessment capabilities; and developing novel instruments and sensors. Such investments should focus on applications in basic chemical engineering research and materials development, as well as on accelerating the transition to a low-carbon energy system; improving the sustainability of food production, water management, and manufacturing; and increasing the accessibility of health care.

9
Training and Fostering the Next Generation of Chemical Engineers

- Chemical engineers are in high demand across most professions and job levels, and chemical engineering provides an excellent foundation for many career paths.
- The undergraduate chemical engineering curriculum has served the discipline well and has continued to evolve in response to scientific discoveries, technological advances, and societal needs.
- Educational attainment in chemical engineering for women and Black, Indigenous, and People of Color (BIPOC) individuals has remained essentially unchanged for more than a decade.

The chemical engineering curriculum today provides a robust foundation of tools and practices founded in an understanding of systems and molecular-level phenomena, including fundamental concepts of mass and energy balances, transport phenomena, thermodynamics, reaction engineering, control, and separations. Although the core subjects of the curriculum were first built around manufacturing processes, primarily petrochemical, they can be applied in most fields and professions. Indeed, previous chapters have highlighted how these concepts can be and have been applied across a wide variety of applications in energy, environmental sustainability, health and medicine; manufacturing; and materials development. As a result, chemical engineers are in high demand across most professions and job levels, and chemical engineering provides an excellent background for many career paths (NAE, 2018; see Figure 9-1).

In the past, chemical engineers tended to find industrial employment in manufacturing and process engineering. The connections between basic research and the workplace were usually made through industrial research and development (R&D) organizations that were aligned with internal business units with needs for both operational efficiency and new products or process developments. Chemical engineers would often begin a career in R&D or manufacturing and move, over time, into senior technical and business leadership positions. Faculty members would engage with industry as consultants and through university–industry collaborations. The net result of that model was a feedback loop from the market back to basic research at universities.

More recently, a strong shift in academic research topics in the field has occurred, driven primarily by changes in federal funding priorities, leading to a movement away from process research and toward basic and applied scientific research. This discovery-focused research includes areas, such as materials and life sciences, relatively new to chemical engineering. At the same time, many companies have globalized, shortened their time horizons, and reduced or eliminated longer-term research programs and laboratories.

The resultant growing gap between university research and market needs is often referred to as the "valley of death."

During the past decade, the world has undergone a major technology-led transition enabled by global networks, computing power, sensors, artificial intelligence, and machine learning. This transition will continue and likely accelerate, creating an ongoing need for new skills and capabilities. As technology continues to transform how work is performed, a growing need for collaboration skills—in communication, interdisciplinary teamwork, and project management—can be anticipated, as can educational needs yet to be identified. And all of these trends will require lifelong learning.

This chapter examines the current state of chemical engineering education, including the broader context of the existing academic education model (Box 9-1) and the value of the current undergraduate core curriculum. The committee proposes strategies for growing and diversifying the profession—both of which are essential to the field's survival and potential for impact—by making it more broadly accessible.[1] Following a discussion of the aspects of undergraduate and graduate education that will need to change to prepare the next generation of chemical engineers, the chapter turns to emerging trends that are shaping new models of learning and innovation for the future.

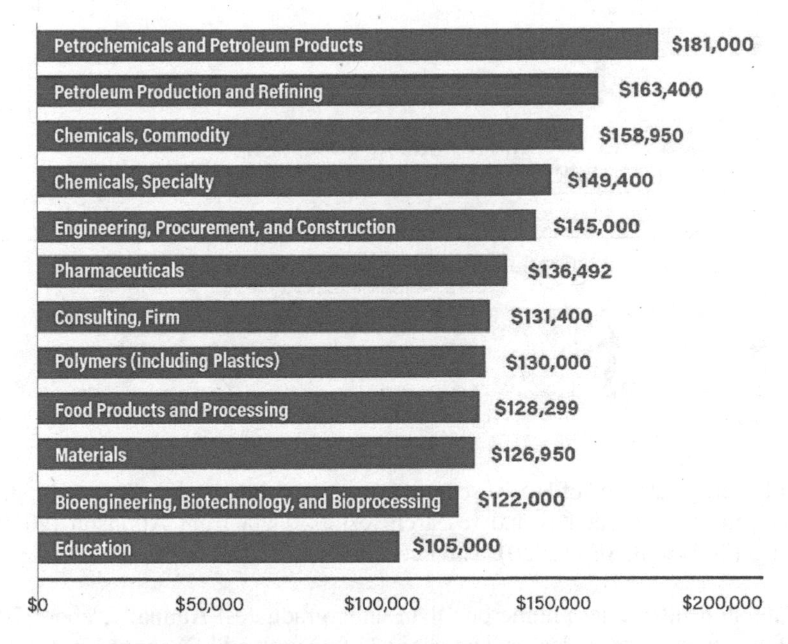

FIGURE 9-1 Career paths available to chemical engineers in a range of industries, shown here with median salary in industry categories with at least 30 respondents to an American Institute of Chemical Engineers salary survey. SOURCE: AIChE (2021).

[1] E.g., https://www.aiche.org/chenected/2021/02/ideal-path-equity-diversity-and-inclusion.

BOX 9-1
The Existing Academic Education Model

Academic institutions are tasked with three important societal responsibilities—education, research, and service—whose fulfillment has long brought, and will continue to bring, tremendous societal benefit. A simple "mass and energy balance" of U.S. universities is shown in the figure below.

Undergraduate education and graduate research and education are two distinct but connected functions of research universities. Undergraduate programs collectively receive more than $100 billion in tuition and state support for public institutions, and produce about 2 million college graduates each year (Atkinson, 2018; Kastner, 2018; NCES, 2021; NCSES, 2020; The Pew Charitable Trusts, 2019). More than 30 percent of the U.S. population over age 25 has a college degree (U.S. Census Bureau, 2019b), putting the United States among the top countries—but not at the top—in the number of college graduates per capita. The increasing costs of and inequities in access to education are major challenges that require attention to ensure that universities meet the societal needs of the future. In this regard, the widespread use of remote teaching during the COVID-19 pandemic may offer models for educational practice, while potentially improving access and decreasing cost.

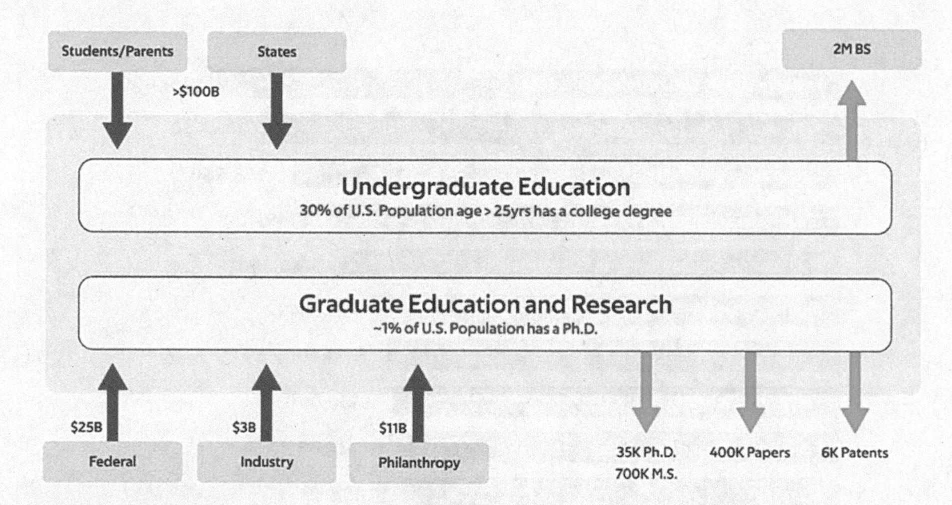

A "mass and energy balance" of U.S. research universities showing the flow of funding streams and various outputs of graduates and research results. Data from Atkinson (2018), Kastner (2018), NCES (2021), NCSES (2020), The Pew Charitable Trusts (2019).

Graduate students are less numerous than undergraduates. Annually, about 700,000 students graduate with a master's degree, and about 35,000 with a PhD; about 1 percent of the U.S. population holds a PhD (NCSES, 2018; U.S. Census Bureau, 2019a). Since the mid-1970s, graduate research in STEM (science, technology, engineering, and mathematics) fields has been funded primarily by federal grants, although the relative fiscal contributions of philanthropy and industry to academic research are growing at some institutions.

Chemical engineers with PhDs have driven research and development (R&D) in a variety of fields, including energy, water, food, health, and biotechnology. In the field of biotechnology, they have played a significant role in advances that include antibody design, biologics manufacturing, cell therapies; nanotechnology, nucleic acid therapeutics, and tissue engineering, among others. Collectively, the biotechnology/pharma industry accounts for a productivity of just over $1 trillion/year—an achievement not possible without the graduate education provided by universities. A fraction of graduating PhDs in chemical engineering pursue careers in academia and support the growth of the undergraduate education enterprise. These two career paths together support a clear and important connection between graduate education in the field and society at large.

Generally speaking, PhD programs in chemical engineering focus on teaching graduate students how to conceptualize and carry out research by using an individual research project as a training ground. The output at the end of the degree program encompasses both trained researchers and the products of their research (frequently in the form of published papers). A third key outcome of graduate research and education is invention, captured tangibly in patents. U.S. universities account for about 6,000 granted U.S. patents each year, representing about 4 percent of the total annual patent output in the United States (NSB, 2018; USPTO, 2021).

Recent years have seen a growing interest among both students and faculty in championing the translation of academic ideas into practice. There are many successful examples of commercialization of academic inventions driven by the entrepreneurial enthusiasm and expertise of their inventors. However, sole reliance on the business skills of individual inventors significantly limits the efficiency of translation and will certainly leave many important inventions behind. Thus it is important for universities to complement and collaborate, rather than compete, with industry. Academic–industry consortia provide a natural opportunity not only to support academic research but also to bring inventions into the marketplace and bring professional managers to spin-off companies. Universities can improve translational efforts by accepting and recognizing such endeavors and measuring their impact; by offering stronger and longer protection of intellectual property, thus allowing technologies to mature; and by investing in inventions emerging from their programs to develop and derisk them prior to licensing, encouraging inventors to address translational hurdles, and incorporating translation and entrepreneurship into the education they provide. Such efforts can provide additional avenues for researchers to pursue societal impact, enable universities to build stronger bridges with industry, and allow both universities and society to extract more value from a currently underutilized resource.

THE UNDERGRADUATE CORE CURRICULUM

Throughout its history, chemical engineering has been defined as a profession by its core undergraduate curriculum, a curriculum that has for more than a century prevented the "spalling of the profession" (p. 573, Scriven, 1991). At the same time, this core undergraduate curriculum has evolved with the incorporation of new topics reflecting emerging areas of impact and relevant practice, as well as ongoing dialogue about the very nature of the profession. The resilience of the discipline in the face of change reflects the nature of its core curriculum and how it brings together the underpinning sciences (chemistry, physics, mathematics, biochemistry, and biology) into an interdisciplinary problem-solving context. Together, the enduring nature of this canon and its history of adaptation and impact speak to the resiliency of the chemical engineering discipline.

This curriculum has been aimed at transmitting a body of knowledge that is foundational and translating it into solutions for technological and societal challenges. It has enabled chemical engineers to respond nimbly to the unpredictable nature of these challenges by redirecting fundamental concepts as codified in a historical sequence of core courses: mass and energy balances, transport of mass and energy, chemical kinetics, process design and control, and a capstone undergraduate laboratory. The examples used in transmitting this knowledge and applying it in practice have changed and will continue to do so as chemical engineering finds new applications, even as the foundations have acquired additional complexity through mathematical dexterity fueled by transformational changes in computing power, as well as greater breadth through the growth of the biochemical aspects of the discipline.

The core undergraduate curriculum provides a problem-solving approach to the mastery of concepts in the dynamics and thermodynamics of physical and chemical processes, with a historical evolution from the physical to the chemical; most recently to the biochemical and electrochemical, and at present, toward the photochemical realm. The problems addressed have changed because "engineers solve problems. If they are successful, those problems disappear. Then we find new problems to solve…, but the principles used to solve successive generations of problems change very slowly…" (p. 243, Schowalter, 2003).

The core curriculum as taught represents a method of inquiry and a toolbox for solving problems. It is entirely general in its most abstract mathematical form; perhaps for this reason, it has remained useful for a remarkable breadth of relevant practice. At the same time, however, it can appear to lack merit and utility when first learned, especially if the content is delivered absent context within current modern problems and practice. The consensus of a selected group of graduate students and postdocs is that "process science," their nomenclature for the core curriculum, has proved complete enough and adaptable enough to persist for the next 25 years and longer (Westmoreland and McCabe, 2018). Yet the profession and its undergraduate curriculum do not always succeed in creating the messaging landscape required to attract and retain individuals with the diverse backgrounds and interests that future challenges will demand. As it evolves, then, the curriculum will need to convey with greater purpose and success how "no profession unleashes the spirit of innovation like engineering" and how few other disciplines "have such a direct and positive effect on people's everyday lives" (p. 46, NAE, 2008).

To those embedded within the field at a given time, the evolution of the curriculum has often seemed too slow and the survival of the discipline fraught with perils. This is not a new perception. Decades ago, Denn (p. 565, 1991) observed, "We have been hearing a great deal in recent years about the changing nature of chemical engineering. The emphasis on new fields of research has created the appearance of a fragmented profession…." Nonetheless, as suggested above, the curriculum "has endured, not because it is frozen, but because it has adapted dynamically to new ideas, emphases challenges, and opportunities" (p. 7, Luo et al., 2015). As described later in this chapter, some topics within the curriculum will need to evolve more rapidly, and in some cases, components removed from the canon during earlier cycles of evolution will need to be restored.

Today, as throughout its history, the field of chemical engineering needs to consider what minimal requirements and what set of principles should define the content of its core undergraduate curriculum. This is a question perhaps best posed as: What should an evolving undergraduate curriculum deliver as its product? Nearly 20 years ago, the answer to that question was, an undergraduate chemical engineer trained through "a program of study sufficient for entry-level positions in engineering practice and engineering-related fields," but also exposure to shoulder areas and to professional and personal ethical guidelines and the foundational knowledge required for graduate studies (p. 243, Schowalter, 2003). These requirements have not changed, but chemical engineers function and practice in a much different environment today. They are asked to address more diverse challenges with a body of knowledge and a toolbox that extend beyond what a 4-year core undergraduate curriculum can competently deliver in full. As discussed later in this chapter, master's degrees and continuing education are likely to become increasingly important for working professionals who seek to specialize in one of these shoulder areas.

The toolbox delivered by the undergraduate curriculum provides a mathematical framework for designing and describing (electro-/photo-/bio-) chemical and physical processes across length and time scales spanning many orders of magnitude. It teaches that (1) some quantities are conserved (energy, momentum, atoms, mass); (2) their balances need to be carried out over "control volumes" small enough to be homogeneous but large enough to be described by continuum equations; (3) thermodynamic relations define the point of equilibrium, but also the dynamics by which systems approach such equilibria, whether through chemical or physical changes; and (4) all of this extends, remarkably unperturbed, to molecules and atoms in every state of matter (gas, liquid, solid, supercritical).

A survey of young professionals a few years after they had entered the profession of chemical engineering identified features of the undergraduate curriculum that they considered important to their careers (Figure 9-2); the components of the enduring core curriculum are well represented throughout these features. The four highlighted items represent those in need of revision and strengthening in the face of changes in both the nature of chemical engineering practice and the employment landscape. Two items in particular—process and product safety; and data science and application: design of experiments, statistics, analytics—reside within the core knowledge base and need to become more prominent. Process and product safety will need to become a stronger component within each core undergraduate course. Data science and statistics may be delivered most effectively in a separate course embedded within the core curriculum and taught with specific emphasis on matters of chemistry and engineering. This latter course would also bring a greater emphasis on statistics in the modern context of larger datasets, more powerful computing, and models and methods that are more robust and of greater fidelity.

The other features highlighted in Figure 9-2—business skills, leadership training, management, and economics; and innovation and entrepreneurial skills—represent "softer" skillsets that provide entry-level engineers with significant competitive advantages in today's workplace. Along with other, related skills, such as written and oral

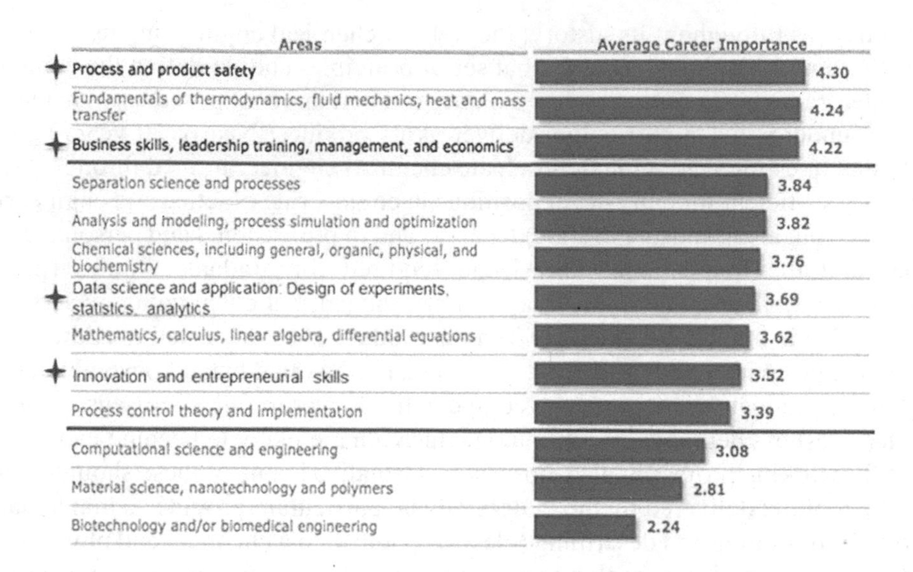

FIGURE 9-2 Career importance of various areas of study, as indicated in a survey of early-career chemical engineers. SOURCE: Modified from Luo et al. (2015).

communication and a baseline understanding of policy and regulatory issues, however, they reside outside the technical core of chemical engineering. In the committee's view, the core undergraduate curriculum is not the most effective vehicle for delivering the necessary foundational knowledge and skills in these areas. Elective courses, postgraduate training in specialized industrial settings, and lifelong learning through professional societies and relevant literature provide more effective routes for acquiring and sharpening these skills, and the application and sharpening of these ancillary skills can be made part of each core course, with emphasis on how they enable and enhance the technical contributions of chemical engineers.

The remainder of this section describes some of the challenges that represent important considerations in the near-term evolution of the undergraduate curriculum, as identified by members of this committee and shared by invited external speakers in discussions and presentations. Three challenges are discussed: the need for experiential learning and greater connectivity among the concepts/tools of the discipline and their application in practice through (1) more effective connections among the individual core courses ("the silos"); (2) experiential learning through virtual or physical laboratory experiences earlier in the undergraduate course of studies; and (3) a more effective and seamless embedding of statistics and of mathematical and computational thinking into the core. The committee emphasizes that these challenges are inextricably connected, and notes that actions suggested in the course of the discussion are meant to be illustrative, and not to prescribe modes of implementation, which will best be identified by experts in discipline-based education research (e.g., NRC, 2012; Paul and Brennan, 2019).

Connecting the Silos

In the core curriculum, concepts of balances (mass, energy, momentum), fluid flow, thermodynamics, kinetics, and process/design control are taught in separate courses that may make them seem disconnected to students. Students are then asked to connect these seemingly disparate concepts with each other, as well as with the bond-making and bond-breaking rules from the chemistry curriculum, the biochemical tenets of the biology curriculum, and the mathematical mechanics taught within the mathematics curriculum. Not surprisingly, students face significant hurdles in bringing these historical repositories of knowledge together to form the problem-solving and reasoning strategies that constitute the practice of chemical engineering. Some of these connections emerge, with varying levels of effectiveness and rigor, in a laboratory or unit operations course near the end of the curriculum. But these connections are better made by anticipating in earlier courses how these concepts will ultimately coalesce at later stages and what kinds of engineering challenges require their combined application—how the balances, the thermodynamics and dynamics, the fluid flow, and the chemistry and biology ultimately merge into the design and control of reaction, separation, biological, and materials synthesis processes. These earlier connections can be made, in the committee's view, by making the boundaries between the silos more porous, highlighting how the individual core concepts first presented within their respective silos ultimately merge into the practice of chemical engineering.

Experiential Learning and a "Laboratory within Each Core Course"

The dense nature of the core undergraduate curriculum leaves few openings for incorporating an additional hands-on laboratory course earlier in the curriculum. In some instances, this has been successfully accomplished, albeit as a broad engineering design course (e.g., the Coffee Lab at the University of California, Davis; see Box 9-2). In other cases, a freshman-level introductory course has attempted to place the curriculum in context at an early stage, but without the rigor of analysis and treatment that will follow later on and with some duplication of the content of subsequent core courses. Such efforts need to continue and expand, but they are likely to miss those students that enter at a later stage of the curriculum through transfer from community college or other majors. An alternative strategy would be to use advances in real-time simulation and demonstration in virtual/digital form to illustrate an "experiment" representing the behavior of a (bio-/photo-/electro-) chemical or physical system as described by a mathematical representation immediately following this "experience." In its interactive mode, this kind of visualization would allow students to design and control the behavior and performance of such systems, to explore how they respond to perturbations in parameters or conditions, and to address and resolve safety and ethical matters in the practice of engineering without risking any direct physical or professional consequences of their actions. Such approaches, currently implemented as more ad hoc strategies, would expose students to issues of design, control,

BOX 9-2
Coffee Lab: An Example of Experiential Learning

At the University of California, Davis an experiential freshman laboratory experience has been developed for both majors and nonmajors.[a] Each version of this course uses the design of a coffee-brewing process, optimizing flavor with respect to energy use, as a vehicle for experiential learning. For nonmajors, the goal is providing a nonmathematical introduction to how chemical engineers think by using the unit operations of coffee brewing and experiments as an illustration. This course for nonmajors has a very large enrollment (>1,000 students/year) and serves the goals of both marketing the major to non–chemical engineering freshmen and providing a basic chemical engineering skillset to other disciplines. The analogous course for majors is mathematical in nature and uses the roasting, grinding, and brewing of coffee to introduce process flow diagrams, mass and energy balances, transport phenomena, separations, and the basics of design. With this experience in hand, students are demonstrably more sophisticated in their understanding of how individual courses and concepts fit into the discipline and various applications as they complete their degrees.

The Coffee Lab also uses relatively inexpensive and readily available household appliances that require minimal training and safety measures. This makes it an accessible introduction to chemical engineering for institutions that do not currently support a chemical engineering program or laboratory and could facilitate interest and preparedness for students who transfer to schools with, or pursue graduate degrees in, this field.

[a] https://coffeecenter.ucdavis.edu/facilities/undergraduate-coffee-lab.

safety, and even economic impact that typically appear in their more formal and foundational contexts later in the curriculum. They could also be used to incorporate sensitivity and statistical analyses, design of experiments, "what if" assessments of realistic scenarios, and a view of mathematical treatments underpinned by their role in analysis and decision making in the practice of the chemical and physical processes that such treatments intend to describe. The application of these approaches will continue to benefit from advances in real-time description and visualization of process (and product) performance, and will sharpen the process synthesis and analysis skills needed for chemical engineering practice. And students will be more effective and feel less intimidated when examining the validity of assumptions made in describing real-world systems that would otherwise require descriptions too complex for analytical solutions or even numerical analysis of more complete equations.

Bringing Mathematics and Statistics into the Core

In most cases, students acquire the mathematical machinery of calculus, differential equations, and linear algebra in courses taught by college math departments or through advanced high school coursework. In such courses, they acquire limited (if any) skills in numerical methods or in the construction of the equations that describe the physical and chemical content they are later asked to implement in chemical engineering courses. Statistical analysis, specifically in the context of acquiring knowledge and analyzing dense datasets, is essentially absent from the curriculum until the capstone laboratory course, where students first encounter the imperfections of the data gathered through their own actions. Previously, such statistical and mathematical methods were more closely integrated into and introduced earlier in the curriculum. The reversal of that pattern seen today in the case of statistics is due to the emergence of data science as a modern catchall for the learning and practice of such methods. The committee believes that training in mathematics and statistics needs to be brought into the core curriculum in a more structured manner, either complementing or replacing some of the education that currently occurs outside the core curriculum. This might take the form of a course in mathematical methods taught within chemical engineering departments focused on illustrating how analytical, numerical, and statistical methods are used in the context of the equations that emerge later within specific core courses. The return of this content to the core needs to occur reasonably early in the course of studies for greatest impact, creating several challenges given the dense nature of the curriculum and the needs of students entering at different stages and with different backgrounds and skillsets in mathematics and statistics.

In summary, the undergraduate chemical engineering curriculum has served the discipline well and will continue to evolve in response to scientific discoveries, technological advances, and societal needs. In this evolution, it will benefit from rapid changes in the ways knowledge is disseminated and transferred within and among disciplines. As part of this evolution, it is also necessary to consider the imperative to attract and retain practitioners of chemical engineering with increasingly diverse backgrounds, as discussed in the next section. Later sections of this chapter address the need to enhance and improve training in business, economics, innovation, and entrepreneurship, as well as lifelong learning. The committee considers these skills to be essential, well suited to being illustrated within the core undergraduate curriculum but entailing foundational knowledge that lies beyond the core.

BECOMING A CHEMICAL ENGINEER: THE IMPORTANCE OF DIVERSITY

As a discipline, chemical engineering is unique in its pervasive contributions to society—in areas ranging from energy; to food, water, and air; to health and medicine; to manufacturing; to materials, as described in earlier chapters of this report. Consequentially, the field is in a strong position to attract a broad range of individuals interested in the many areas associated with chemical engineering who also are seeking a career with the potential for societal impact. Research has shown that members of historically excluded groups are often motivated by the altruistic career goals of making the world better and giving back to their communities (Thoman et al., 2015). Emphasizing the role of chemical engineering in addressing societal issues might therefore help attract more high school students from diverse backgrounds to the undergraduate major.

Women and Black, Indigenous, and People of Color (BIPOC) are underrepresented in chemical engineering relative to the general population, even by comparison with chemical and biological sciences and related fields. While a career in a STEM (science, technology, engineering, and mathematics) field is an attractive path toward altruistic work, persisting barriers impede the entry of women and BIPOC into the field. Some of these barriers affect all historically excluded groups, while others affect students based on their gender or racial identity, and these barriers can be compounded for students who are members of more than one historically excluded group. The National Academies and others have reported extensively on such structural and cultural barriers as unwelcoming and unsupportive cultures and environments; "gatekeeping"; biases; lack of mentors and role models; and inequitable policies and practices that impact recruitment, retention, and career success (see, e.g., McGee, 2020, on barriers for underrepresented and racially minoritized students; NASEM, 2020b, on barriers for Black students; NASEM, 2016, on barriers for women and BIPOC broadly; and NASEM, 2018 and 2020a, on barriers for women).

As a result of such systemic barriers, chemical engineering benefits from the talents of only a fraction of the population. Educational attainment in chemical engineering for women (Figure 9-3) and BIPOC (Figure 9-4) has remained essentially unchanged for more than a decade. The demographics in chemical engineering today reflect the past. Historically, science and engineering have not been welcoming to BIPOC and women and have been particularly harsh to Black Americans. In his essay "The Negro Scientist," published in *The American Scholar* in 1939, W. E. B. Du Bois challenged assumptions held by Whites regarding the propensity of African Americans for science. These types of biases persist, and after starts and stumbles with interventions designed to counter systemic barriers (NASEM, 2016), work still remains to provide clear and inclusive pathways in STEM fields, including chemical engineering, for historically excluded groups.

To fully support members of historically excluded groups in chemical engineering training and education, specialized programs and cultural shifts will be necessary. Interventions and support mechanisms will vary based on which groups are targeted (NASEM, 2021c); the focus may be on supporting women,[2] or on Black[3] or Latinx/Hispanic and Indigenous[4] students. The design of specific support mechanisms will vary as well according to the unique needs and goals of each institution. In this section, and in the later section on graduate education, the committee presents strategies applicable for most chemical engineering departments that are likely to improve the recruitment and retention of and outcomes for multiple underrepresented groups.

Research has illuminated how children's early pathways can be determined, along with some of the critical factors that dictate their future educational options and career trajectories (Akee et al., 2017; Chetty et al., 2016, 2019). These studies have revealed that children with high scores on third-grade math tests who come from high-income families, children who grow up in geographic areas with high rates of invention, and girls who are exposed to women inventors are more likely to become inventors. These findings speak

---

[2] E.g., Society of Women Engineers (https://swe.org/).

[3] E.g., National Organization for the Professional Advancement of Black Chemists and Chemical Engineers (https://www.nobcche.org/).

[4] E.g, Society for the Advancement of Chicanos/Hispanics and Native Americans in Science (https://www.sacnas.org/).

to the social factors that require attention in any technical field seeking to spark creativity and build pathways to include the full pool of talent. The importance of role models and access to opportunities is clear, as is the need for adequate academic preparation for any STEM field, including chemical engineering. In primary and secondary education, studies specific to chemical engineering are lacking; studies that include qualitative data and experiences specific to chemical engineering are lacking even more.

Chemical engineering as a field is not immune from systemic and other barriers to inclusivity, and the field can draw insights and apply the lessons from STEM-wide studies. In short, if chemical engineering is to reach its full potential as a discipline and a major enabler of solutions for societal needs, it will need to address opportunity gaps and ensure that its educational, research, and professional environments support the success of everyone, regardless of their identity.

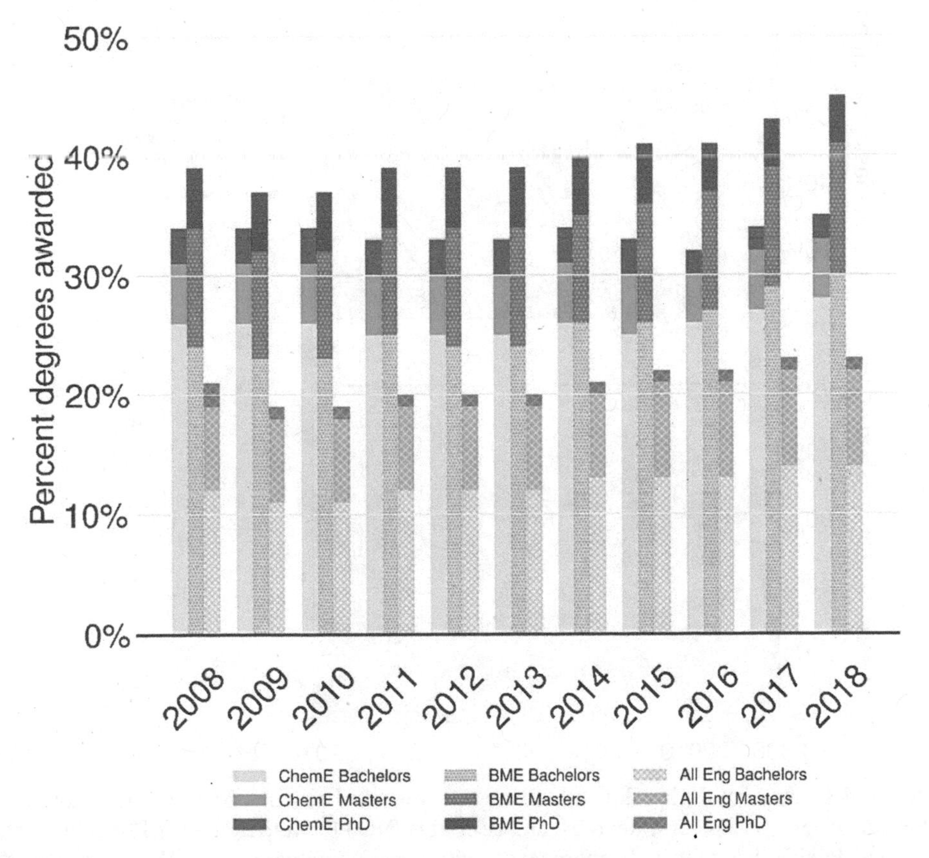

FIGURE 9-3 Percentage of bachelor's, master's, and PhD degrees awarded to women in chemical engineering (ChemE), biomedical engineering (BME), and engineering overall, 2008–2018. Data from the National Center for Science and Engineering Statistics.

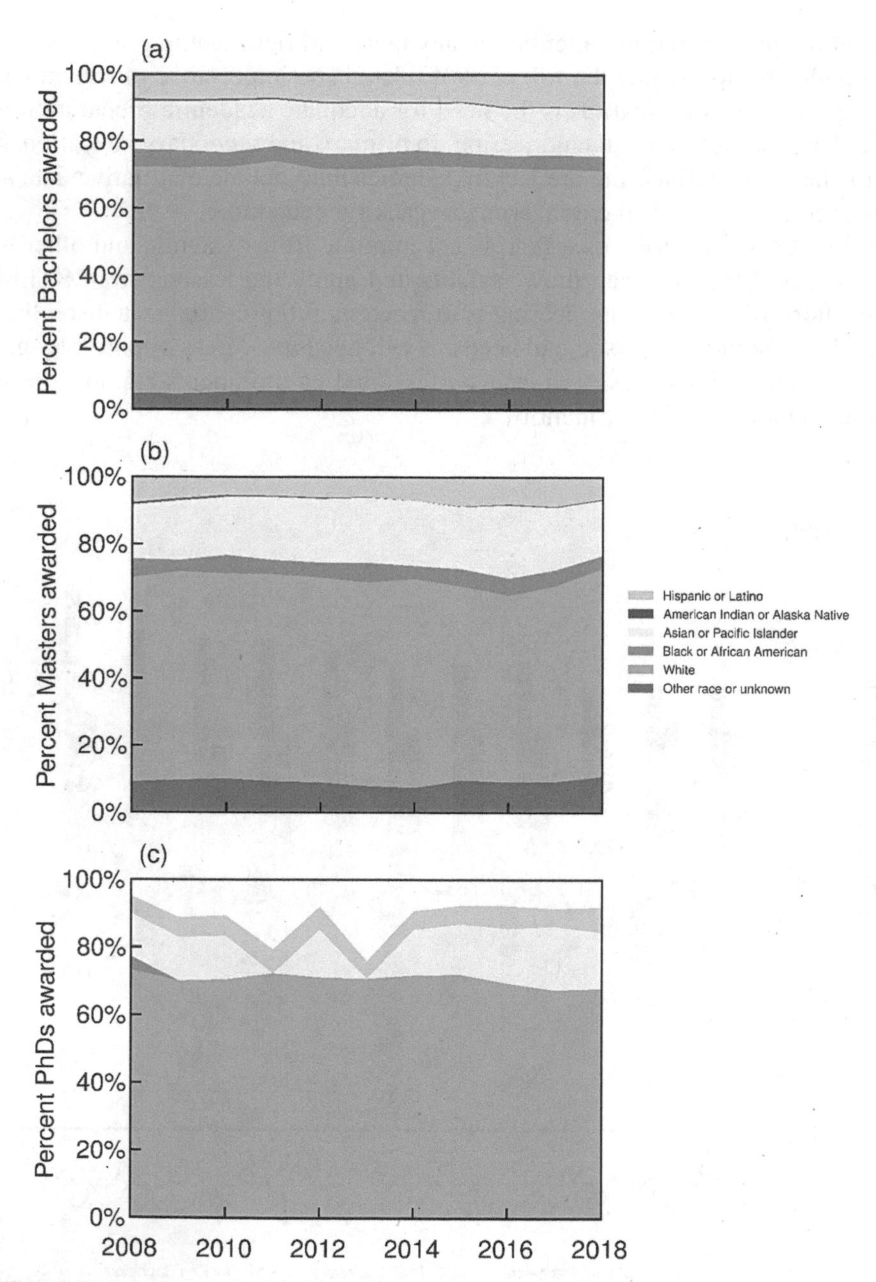

FIGURE 9-4 Demographic breakdown of degrees awarded to chemical engineers by race: (a) bachelor's degrees, (b) master's degrees, and (c) PhDs. NOTE: In the data for PhDs, the category of Asian and Pacific Islander is disaggregated, with separate categories for Asian and for Native Hawaiian and Pacific Islanders; the categories also included an option for "more than one race" rather than "other race or unknown." Therefore, these data do not sum to 100 percent because data were redacted for privacy reasons. Data from the National Center for Science and Engineering Statistics.

Cech (2013) speaks to the (mis)framing of engineering as meritocratic and depoliticized (asocial). This mischaracterization results from the false assumptions that an inherently technical field will inevitably be a meritocracy as though technical fields can operate outside of social influences (Cech, 2013; Cech and Blair-Loy, 2010). This framing has been debunked with research revealing the myriad structural and cultural factors that turn students away from the field regardless of their skills and competencies (Seymour and Hewitt, 1997), such as structural racism (McGee, 2020), gatekeeping, and weeding out through historical exclusion in the education system (Malcom, 1996; Malcom and Malcom-Piquex, 2020), and stereotyping regarding who has innate talent (Leslie et al., 2015). Science and math courses have long been gatekeepers for entry into engineering. Retention data disaggregated by discipline are not readily available, but based on the observations of members of this committee, such courses as the sophomore-level course in mass and energy balances in chemical engineering serve as additional gatekeepers.

Those chemical engineers who are retained in the field play many roles in practice and face significant challenges in retaining relevance and excellence as the field evolves; they do so through diverse educational trajectories and with endpoints in industry, academia, and elsewhere. When considering what draws people to chemical engineering, it is important to acknowledge these different educational and career trajectories and the pressures involved in achieving and maintaining them. Stability, work–life balance, professional support structures and mentorship, and opportunities for growth are important to people in any sector but are not distributed equally across sectors or demographics within them. In academia, professors act as educators, administrators, mentors, researchers, communicators, and fundraisers, but all individuals are not equally suited to all of these tasks, and some can become overburdened by the need to fulfill them all.

At the same time, members of historically excluded groups bear a disproportionate responsibility for promoting diversity, providing representation on committees, and supporting the academic and career progression of other members of underrepresented groups. Institutionalizing the work of diversity requires shared responsibility, not just the efforts of those with a personal stake in improving access to equitable opportunities. Both industry and academia still fail to pay and support women and BIPOC and promote them to executive positions at a proportional rate (Funk and Parker, 2018; Gumpertz et al., 2017; Renzulli, 2019). Promoting and retaining meaningfully diverse talent will have a multiplier effect, engendering greater diversity moving forward as more diverse groups decide who joins the workforce. These issues are pervasive throughout STEM fields and certainly not unique to chemical engineering. In looking to the future, however, chemical engineers have an opportunity to be leaders among STEM fields in increasing diversity and inclusion within their profession.

Greater visibility and connectivity within the broader community can also support diversity, equity, and recruitment, and chemical engineers can make contributions in other areas of public interest beyond diversifying the field. Like all scientists and engineers, they have the opportunity to use effective scientific and popular communication to engage with the general public, as well as improve resources for K-12 educators. Social media and science entertainment have been vital for accessibility and visibility, but given the integral roles of science and engineering in society, scientists and engineers need to be more integrated into policy making. Chemical engineers have a stake in society equal to that of people with law or business backgrounds who commonly take on societal concerns, and are trained in a thought process that would lend itself well to addressing those concerns with respect to both scientific and systemic issues. It is not clear to the average high school senior that chemical engineering will have a critical role in many of the domains that will be central to progress in the coming decades. As demonstrated in earlier chapters, many policy issues in such domains as energy, food security, clean air and water, and health care, and public health involve chemical engineering. Maintaining dialogue among chemical engineers in policy-making positions, communities to be served, and those in academia and industry would also benefit the field, identifying needs and potential priorities.

Science policy is another area in which students are driving growth at universities, founding student groups and advocating for formalized programs. Being responsive to those student interests and facilitating this career path can help chemical engineering programs attract those who want to pursue public-sector work and to develop both the tools and the influence needed to have direct impact in bettering their community.

This section has addressed recruitment of chemical engineers and barriers to entering the field, in particular for women and BIPOC. Recruiting is critical, but so is retention. Mentorship has been shown to have a positive effect on underrepresented students, yet underrepresented individuals enrolled in STEM programs typically receive less mentorship than their well-represented peers (NASEM, 2019e). Institutionalized developmental support needs to evolve from “Are you cut out for this?” to “How can we help you succeed?” Formal support systems for academic success are enhanced by the deliberate formation of peer and mentoring networks. Beyond mentoring, systemic approaches to ensuring success for all individuals along the entire career path will ensure that chemical engineering remains equipped to attract, develop, and retain a diverse cadre of future chemical engineers.

MAKING CHEMICAL ENGINEERING BROADLY ACCESSIBLE

A long-running national dialogue about college affordability and the impact of student loan debt on the overall economy has recently become more visible. The relative affordability of community colleges is a major attraction for a diverse body of students, ranging from talented budget-minded high school seniors to nontraditional students. In 2021, average annual tuition and fees at 2-year public schools was $3,372, versus $9,580 for in-state students at 4-year public schools (Hanson, 2021). In addition, at least 17 states have programs that make community college attendance tuition-free for at least a portion of the student population (Farrington, 2020). Students enrolled at 2-year schools are more racially diverse than those at 4-year schools (NASEM, 2016), and the majority of tribal colleges and universities in the United States are 2-year institutions. Further, many states have existing contractual agreements promising not only admission to public 4-year institutions for students who demonstrate success at a community college, but also the ability to graduate within 2 years after transferring. Indeed, it is possible that 2-year community colleges could become the default choice of the middle class in the relatively near future, requiring a major adaptation of chemical engineering undergraduate programs to become viable options for those students.

This body of transfer students represents an untapped opportunity for chemical engineering to broaden participation in and access to the profession. Over the last decades, science communication and outreach at the K-12 level have resulted in significantly increased interest in STEM fields, but that increase (particularly among diverse groups) has been focused in areas in which high school courses are available—physics; biology; computer science; and, to a lesser degree, mechanical engineering/robotics. Many high school graduates have little exposure to the relevance or potential impact of chemical engineering as a career, and few community colleges offer chemical engineering courses, though many provide transfer pathways to 4-year schools. Building bridges to actively recruit both full-time community college students and nontraditional students and identifying and implementing pathways to support them after they transfer could greatly democratize the profession.

Chemical engineering transfer students face a remarkable challenge beyond the abrupt change from the community college to the larger university environment (sometimes referred to as "transfer shock"; Flaga, 2006). Chemical engineering curricula generally have required courses beginning early in the sophomore year, with many programs offering an introductory course in the first year. Further, education research has underscored the importance of providing early hands-on experiences in engineering to improve students' motivation to complete their degrees (Cui et al., 2011). In fact, such experiences have been shown to be particularly successful in the retention of women and BIPOC (respectively, a 27 percent, 54 percent, and 36 percent retention gain for women, Latinx, and African American students; Hoit and Ohland, 1998; Knight et al., 2003; Napoli et al., 2017; Willson et al., 1995). These gains are attributed not only to increased design, teamwork, and communication skills, but also to the development of a peer support network (Richardson and Dantzler, 2002). Challenges for transfer students are compounded because, in contrast with prerequisites in chemistry, physics, and mathematics, most 2-year community colleges lack chemical engineering departments to offer these courses, much less hands-on experiences. Further, because students at community colleges do not fulfill any major requirements, they do not form a peer support network with other chemical engineering majors. As a result, transfer students are asked to compress most of 3–4 years of chemical engineering curriculum into a 2-year period, and to do so without the same peer support or foundational experience in engineering enjoyed by their nontransfer peers. This is not a recipe for success.

The challenge of accommodating a 2-year path to graduation for community college transfer students is already facing many undergraduate programs at public universities. This is an ideal opportunity for the widespread deployment of online course offerings within the sophomore chemical engineering curriculum (mass and energy balances, a first course in transport phenomena and/or thermodynamics, and likely a course in mathematics for chemical engineering applications). Further, from a student perspective, these courses need to be widely accepted so as to open up options for transfer to a variety of 4-year programs. Despite the obvious administrative hurdles, these courses would be most beneficial if crafted by and offered as a collaboration among leading large universities (public and private), along with the accreditation agencies and the American Institute of Chemical Engineers (AIChE), thereby promoting universal acceptance of these courses as satisfying prerequisites for the junior-level curricula at individual 4-year institutions.

In addition to academic hurdles, community college transfer students may face financial or other challenges that, while unrelated to their academic abilities, affect their performance. For example, the lack of study groups or support systems noted above (Lenaburg et al., 2012) can translate into a sizable drop in their grade point average during the first term, which can have long-term impacts on the future potential for graduate study and career options. In addition, given the relatively short time spent at a 4-year university, such transfer students typically do not participate in campus undergraduate research experiences, thus missing out on important opportunities for professional skills development and resumé building. Universities need to develop systems to support and engage these students early on and establish peer networks that will support their acclimation and academic success (Eris et al., 2010; Litzler and Young, 2012). Importantly, doing so will build students' confidence and teach important workplace skills.

The committee recognizes that adding more experiential learning earlier in a traditional 4-year curriculum (as discussed previously in this chapter) and making that same 4-year curriculum more welcoming to transfer students from 2-year institutions are seemingly at odds with one another. The experience of a transfer student in a 4-year program will never be the same as that of a student who entered as a freshman, but it is the committee's hope that more practicum-like experiences will become standard across all introductory STEM courses, whether offered at a 2-year or 4-year institution. For this reason, the committee also chose to highlight an example of experiential learning (Box 9-2) that does not require expensive or specialized equipment, as well as the development of virtual experiential learning experiences and courses that satisfy the chemical engineering degree requirements typically offered during the sophomore year.

TEACHING UNDERGRADUATE STUDENTS TODAY AND TOMORROW

Delivery Methods

In spring 2020, the COVID-19 pandemic caused almost all U.S. higher education institutions to move abruptly to an online format, greatly accelerating ongoing trends toward efficiency and scale within higher education. The result was the creation of significant online content of widely varying quality and a deeper understanding of what does and does not work in synchronous and asynchronous online modes both for chemical engineering and more generally.

To some degree, chemical engineering courses, regardless of the subject, traditionally start with fundamentals and end with practice (if time permits). This pattern reflects the desire of educators to teach tools that can be adapted throughout a student's career rather than a vocational skill. In practice, however, this approach has resulted in the derivation of fundamental equations in lecture, with application and problem solving occurring in discussion sections and problem sets. The online experience of 2020–2021 amplified existing trends toward classroom delivery that encourages more problem- and project-based and group learning, which appears to be welcomed by students. One mode of implementing this "flipped" approach that became prominent in the online format was including the derivation in an asynchronous recording and then solving the problems live. The obvious weaknesses of this approach are a potential lack of understanding of asynchronous content or noncompliance with requirements to watch it. There are also some indications that such flipped approaches or greater use of team-based learning may disproportionately affect women, BIPOC, and low-income students; however, the results of early research are conflicting in this regard (e.g., Cruz et al., 2021; Deri et al., 2018; Dixon and Wendt, 2021; Raišienė et al., 2021; Sarsons, 2017; Winter et al., 2021). More research is needed in this area, and any move toward more asynchronous learning and/or team- and project-based learning will need to ensure equitable outcomes for all students.

In this regard, it may be more realistic to consider online delivery as the new form of "self-teaching." Historically, the U.S. higher education system (regardless of discipline) has strived to impart to students both concrete knowledge and skills and the ability to learn new skills throughout their career. In this sense, the purpose of courses is to teach content that is critical, but that is either too difficult or not obviously motivating to learn on its own. If viewed through this lens, the curation of online content that is either of a complexity level appropriate for self-teaching or related to subjects for which the student is motivated offers an exciting prospect for lifelong learning, allowing for the uniform distribution of better content at lower cost and democratizing the offering of specialized courses around the world. Augmentation of existing courses (with modules, examples, or alternative explanations), communication and other soft skills, business/entrepreneurship/management, policy and regulatory issues, and extensions of course content are areas in which online and classroom learning may dovetail within a single course or curriculum.

All online content carries the curse of the internet—namely, the varying quality and accuracy of information and content. Within a single course or curriculum, curation of content will become a major faculty responsibility. With respect to extension learning, curation of content may become a major endeavor of professional societies such as AIChE. In the past, these societies have served their membership in terms of information dissemination and continuing professional education via conferences, workshops and seminars, and the publication of journals. Going forward, this role will likely shift not just to one that is online in nature but to one that is more geared toward serving as a trusted curator of outside resources, perhaps via a subscription model rather than content generation and ownership.

Curricular Content Evolution

As discussed earlier in this chapter, the chemical engineering curriculum has in recent years sought to balance the goals of retaining core rigor (mathematical modeling, thermodynamics, kinetics, and design) and incorporating new important topics (most recently, biochemical engineering and data science). With a massive expansion of the number of STEM majors in many institutions (Figure 9-5), enrollment in the introductory sequences of chemistry, physics, mathematics, and biology has grown to the point that chemical engineering students make up a small portion of STEM students. This shift has led to the question, posed differently at every university based on its politics and financial construct, of the degree to which these other departments should teach introductory material relevant to chemical engineering and to what degree a chemical engineering program is responsible for its own introductory material. For example, each chemical engineering major in the United States is required to take a calculus sequence from a mathematics department. Some chemical engineering departments have developed supplemental undergraduate mathematics content to incorporate coverage of relevant partial differential equations, linear algebra, numerical methods, and the data science and statistics tools needed by chemical engineering students. This trend is likely to be limited, however, by the relatively small size of chemical engineering departments and an inability to teach an entire undergraduate curriculum without the aid of sister departments on campus. Further, as discussed above, one major drive toward college affordability and diversity and inclusion is the broadening of a path for transfer students from lower-cost community colleges and students who change their majors. Neither is well served by a curriculum that is monolithically specialized starting in the freshman year.

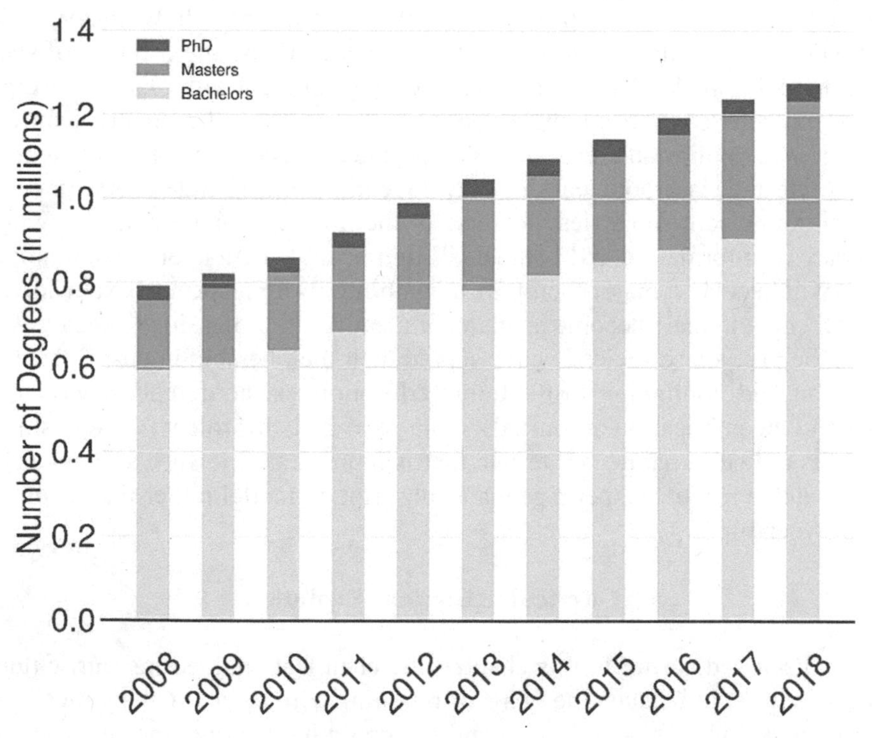

FIGURE 9-5 Total STEM degrees (bachelor's, master's, PhD) awarded between 2008 and 2018. Data from National Center for Science and Engineering Statistics.

TEACHING GRADUATE STUDENTS TODAY AND TOMORROW

Chemical engineering research has expanded considerably in breadth and scope over the past decades, and it is likely to continue doing so. This expansion requires that graduate students acquire deep knowledge in adjacent fields and subfields, including, for example, biology, materials science, and applied physics. At the same time, increasing demands on undergraduate education programs have necessitated reduced coverage of core chemical engineering topics, such as thermodynamics and transport phenomena. The question facing graduate education in the field is whether to compensate for the depth and content that are no longer provided by a chemical engineering undergraduate degree or to focus on giving students the flexibility and opportunities to largely tailor their own graduate program.

A core graduate program consisting of thermodynamics, statistical and quantum mechanics, transport phenomena, chemical kinetics, and applied mathematics will continue to be necessary for chemical engineering research, but that content will have to be delivered in a manner that allows students to apply it in a wider range of contexts. Courses in thermodynamics, for example, will have to rely on approaches and examples that illustrate the general applicability of the underlying concepts, from problems related to issues of protein stability; to general free-energy minimization techniques; to phase transitions in mixtures of solids, liquids, or gases. The core curriculum will need to be limited to foundational concepts, thereby giving students the flexibility to pursue coursework in areas of direct relevance to their research.

While graduate preparation in chemical engineering builds on undergraduate material, this exclusivity comes at the cost of diversity in terms of both the number of women and BIPOC and the breadth of scientific backgrounds in the chemical engineering graduate population. The imperative to recruit talent from a more diverse range of backgrounds will require, in addition to the measures discussed above, the opening up of chemical engineering by providing background content in a manner that creates opportunities for students from other disciplines (e.g., chemists, physicists, biologists) to join a graduate chemical engineering program. That material would include elements from the core subjects covered in the undergraduate curriculum but organized and delivered in a way that is easily accessible to postgraduate scientists or engineers. Interestingly, anecdotal evidence from the members of this committee indicates that while many chemical engineering graduate programs use undergraduate preparation in the field as a major gateway to admission, faculty of their own departments include many members whose training was in related disciplines.

Relative to undergraduate programs in chemical engineering, those in chemistry and biology graduate significantly more women (50 percent and 63 percent, respectively, compared to 35 percent in chemical engineering; NCSES, 2018). Similarly, undergraduate programs in chemistry and biology are more racially diverse (58 percent and 55 percent White, respectively, compared with 64 percent white in chemical engineering; NCSES, 2018). By considering admitting more graduate students with undergraduate degrees in these related disciplines, chemical engineering departments could provide opportunities for more diverse applicant pools. Significantly, a decision not to accept graduate students from other, related fields would generally rule out the enrollment of graduates of many historically Black colleges and universities and other minority-serving institutions that lack undergraduate chemical engineering programs. At the same time, a search for diverse graduate students cannot be limited solely to those institutions. Even when recruiting within undergraduate chemical engineering programs, there is an elitism in some graduate programs that excludes those who may have chosen a bachelor's institution because of its affordability or location. For a graduate program, it is justifiable to want students to understand what it means to do research before committing them to, and supporting them for, a program lasting many years. But how is genuine interest cultivated for those who did not know earlier about or did not have access to undergraduate research opportunities, or for those members of historically excluded groups who did have the chance to participate in such programs as the Research Experience for Undergraduates but are then not actively recruited to the host institutions?

Another vehicle for graduate education that has until recently been largely missing from the graduate chemical engineering curriculum is internships in industry, government, or the nonprofit sector. Experiential learning in the form of graduate internships is currently rare, and providing sufficient opportunities for systematic placement of graduate students will require a conversation among industry, federal funding agencies, universities, and professional organizations such as AIChE to enable the development of suitable frameworks capable of administering effective training programs, perhaps even on a national scale. As with coursework, new opportunities are emerging through remote and virtual access. New models will likely be needed that address issues of equity and inclusion, suitable compensation, intellectual property considerations, and a commitment to the mentoring of graduate interns. Encouraging companies to create educational/internship opportunities by creating model programs would be beneficial.

Master's degrees are likely to play an increasingly important role in graduate education. For the reasons outlined above—whether a need to acquire additional depth in core chemical engineering concepts or to gain breadth in ancillary disciplines such as bioengineering or computing or data science, among many others—master's degrees could offer an attractive solution for chemical engineers needing to adapt and respond to a rapidly changing marketplace. One obstacle that remains to be addressed is the cost of such degrees. However, with the emergence of improved options for remote learning, and with compelling examples of high-quality degrees (e.g., the Online Master of Science in Computer Science from the Georgia Institute of Technology, with more than 5,000 graduates in its 8 years of existence; McMurtrie, 2018; Nietzel, 2021) being offered for less than the cost of the typical undergraduate tuition at a state institution, students and employers alike will benefit from the flexibility offered by master's and graduate certificate programs. The chemical engineering community will have to develop carefully conceived degrees that can not only provide in-depth, topical chemical engineering content for students and practitioners but also attract students from other disciplines. Such programs will also need to provide the flexibility required by working professionals who wish to continue their education and earn an advanced degree through evening and/or weekend programs.

NEW LEARNING AND INNOVATION PRACTICES TO ADDRESS CURRENT CHALLENGES

The preceding sections outline significant steps that could be taken to grow and diversify the field of chemical engineering and deepen its impact on society. These observations about curriculum, experiential learning, approaches to teaching, and ways of building a more diverse and inclusive profession, as well as the broader issue of controlling the costs of higher education, raise the question of whether education in the field can be better designed to deliver on future opportunities. Is it possible to double the quality of an education (including an outcome measure of student success) delivered in half the time and at half the cost? How can education be made globally accessible in real time? What are possible new options for addressing the identified challenges facing chemical engineering education? Answers to these questions could reflect and incorporate the ways in which technology is transforming how work is performed in many professions and how global networks have transformed knowledge management.

In the past, problems were solved based on what an individual or group of individuals knew, acquired from discrete sources (e.g., books, articles, other publications, and their own lived experiences). Today, in contrast, essentially all of the world's public information has been indexed and is quickly available, often at no cost, via internet search engines. When confronted with a problem in almost any setting, those with internet access first search for possible solutions or known information. Their initial findings connect to nearly limitless information about related topics and solutions and opportunities for an individual to find and build upon what is known. An education is necessary to provide sufficient background in the subject matter so the problem can be formulated, to curate and validate information, and then to know how to apply the information to create the needed solutions.

The nature of the education needed to solve problems in this way is evolving. Several companies in the technology arena have eliminated previously held requirements for a 4-year degree for many jobs and are leading the development of more targeted certificate programs to create a pool of talent for the range of jobs that are open. For example, Google has launched "Grow with Google," a program wherein completion of an online certificate program available on Coursera can lead to entry-level jobs with competitive starting salaries.[5] There is no requirement for a 4-year degree or equivalent experience. The Google program engages more than 100 partner companies that also have positions available upon completion of the common certificate programs. The traditional higher education model is linear, with a student moving from K-12 to some amount of university education and then to work, and there are usually limited feedback loops and a lack of integration across the steps. That linear model could be transformed into a general learning model based on a shared platform integrating the educational silos found today.

[5] See http://grow.google.

The degree to which truly new models of skills-based education will penetrate chemical engineering is unclear. The advantages of the integrated approach discussed above are numerous and are likely to remain attractive for many students. On the other hand, there is ample evidence that U.S. research universities are not designed to deliver a low-cost undergraduate education. Their many missions result in high costs (to support faculty and research) and high overhead rates (to pay for people and facilities), and they are not responsive over the time scales of the connected world because of the nature of their research and scholarship. New connected models of learning and innovation that span traditional boundaries could provide solutions more responsive to some of today's needs, although major barriers, including the incorporation of laboratory classes, would have to be overcome.

Learning and innovation could be designed to span the boundary between universities and workplaces. Instead of universities taking on more responsibilities, contributors could build content that would be shared and could be used at all stages of education. That content could be both scalable and used locally in the classroom or as a supplement to classroom learning. Given the accelerating pace of change, such changes would assist learners of all ages in thinking about future career options that can be aligned with their learning and skills development (see Box 9-3).

The move to skills-based hiring opens up other new possibilities. An existing degree is at some basic level a collection of skills. Different disciplines have both unique and common skills, the latter of which enable new mapping of those skills to other disciplines, as well as to different jobs. Skills-based modules offered as complements to existing degree programs could provide a lower-cost path to a first job, along with continued support for additional lifelong learning in response to the evolution of the job market. New learning networks could also help build a more diverse STEM workforce by enhancing access to much-needed career opportunities, background preparation, and support.

As discussed earlier in this chapter, much remains to be learned about the relative advantages of online and classroom teaching and learning. While online programs have gained popularity and were widely used during the COVID-19 pandemic, their advantages clearly come at the cost of the interpersonal interactions and discussions that occur in a traditional classroom. Chemical engineers have an opportunity to lead innovation in STEM fields by building a model that emphasizes scalability as well as human connection. Scalability can come from online content that is curated jointly by companies and universities and made available for local use in the classroom, or from the use of standalone modules to address particular topics that are accessed entirely online. Human interaction can then come from internships or work on extended projects at a company. The possibilities are numerous, even as the existing business model and the set of priorities now in place in universities, companies, and government create barriers to change (e.g., Conn et al., 2021). As with all major disruptions, change will likely be generated first by companies as they struggle to find, hire, and develop future talent.

BOX 9-3
The American Institute of Chemical Engineers (AIChE)
Institute for Learning and Innovation

The AIChE Institute for Learning and Innovation (ILI) provides a "horizontal" connection from university to workplace. In this university/local business model, shared content can be accessed by all stakeholders for use in the classroom. The model also enables sharing across universities and building new collaborations among multiple stakeholders. The ILI recognizes that technology is rapidly transforming most market sectors, that learners of all ages need to keep pace with evolving skill requirements, and that companies will seek contemporary skills and capabilities as their business continues to change. A particular focus of the effort is on providing high-quality and low-cost content for institutions of all sizes.

The initial ILI has four basic modules: Career Discovery, Academy, Practice, and Credential, illustrated below.

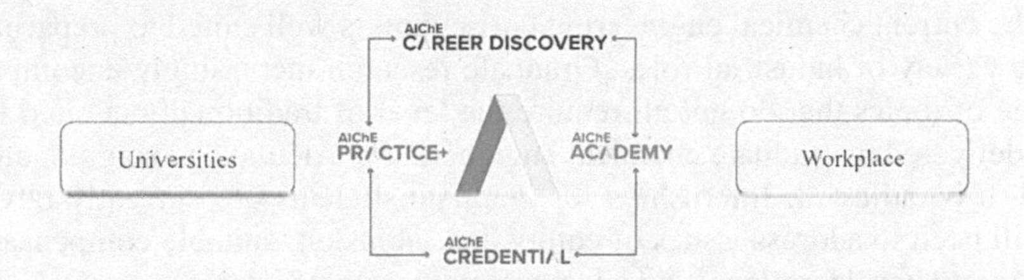

The Career Discovery module, which has been piloted and launched through the ILI, begins with asking students "What might you like to do?" rather than "What do you want to do?" Through a series of exercises, lectures, and group discussions, students develop a personal career plan for working through the university curriculum. Additional skills and experiences are highlighted, and through the broad network of the ILI students have options for engaging in additional skills training, internships, and other outreach activities that will help them get their first job after graduation. The same approach is being used within companies by midcareer professionals seeking new skills.

The Academy, Practice, and Credential modules are all designed cooperatively by industry and universities. The modules provide training in high-priority skills in the marketplace, as well as in specific skills and applications prioritized by companies. AIChE will make shared content available for use in universities, allowing university education to shift in emphasis toward exploring new engineering or business problems. New internship models to engage students with market problems are also under development, and companies can engage with classroom exercises in order to share case studies. Finally, "Lessons from Leaders" is an emerging module in which senior leaders share their career experiences and learning with students considering various career options.

A related scholarship program for Black, Indigenous, and People of Color (BIPOC), the Future of STEM Scholars Initiative (FOSSI), is providing expanded opportunities for students with limited exposure to or understanding of possible market options. Building on studies of disruptive innovation provides a major opportunity to create new education models for historically excluded groups through partnerships among companies, community colleges, and historically Black colleges and universities. Such models would provide a low-cost pathway to high-quality jobs with a direct connection to market need while at the same time broadening outreach to a more diverse talent pool.

CONCLUSION

The core chemical engineering curriculum has contributed to the long-term success and impact of the discipline. The undergraduate curriculum provides a mathematical framework for designing and describing (electro-/photo-/bio-) chemical and physical processes across spatial and temporal scales of many orders of magnitude. Data science and statistics may be delivered most effectively within a separate course embedded within the core curriculum and taught with specific emphasis on matters of chemistry and engineering. At the same time, experiential learning is important, and the majority of industrial and academic chemical engineers interviewed by the committee stressed the importance of internships and other practical experiences. However, far fewer internships are available than the number of students who would benefit from them, and the density of the core undergraduate curriculum leaves few openings for incorporating an additional hands-on laboratory course earlier in the curriculum.

The current chemical engineering curriculum is well suited to preparing students for a wide variety of industrial roles. Graduate research increasingly encompasses a diverse range of topics that do not all require the level of traditionally curated knowledge currently delivered in graduate chemical engineering curricula, and so graduate curricula may need to be adjusted. Internships for graduate students are currently rare, and new models will need to address issues of equity and inclusion, suitable compensation, intellectual property considerations, and adequate mentoring of interns.

Women and members of historically excluded groups are underrepresented in chemical engineering relative to the general population, even by comparison with the chemical and biological sciences and related fields. Diversifying the profession is essential to the field's survival and potential for impact. At all points along their academic path, chemical engineering students need role models and effective, inclusive mentors, including mentors that reflect the diversity of backgrounds needed by the field. Leveraging of professional societies and associated affinity groups could provide valuable support for people of diverse backgrounds entering the field. Strong university support for student chapters of professional organizations would improve access and success.

The general affordability of community colleges is a major attraction for a diverse body of students, ranging from budget-minded high school seniors to nontraditional students. Increased engagement of transfer students is an untapped opportunity for chemical engineering to broaden participation in and access to the profession. Students from 2-year colleges and those who change their major to chemical engineering would benefit from a redesign of the curriculum allowing them to complete the degree in less time. Better academic and social support structures are needed to enable successful pathways for these students. New methods for offering portions of the curriculum in a distributed manner and more general restructuring may require flexibility in curriculum design and changes in university policies, graduation, and accreditation requirements.

Recommendation 9-1: Chemical engineering departments should consider revisions to their undergraduate curricula that would

- **help students understand how individual core concepts merge into the practice of chemical engineering,**
- **include earlier and more frequent experiential learning through physical laboratories and virtual simulations, and**
- **bring mathematics and statistics into the core curriculum in a more structured manner by either complementing or replacing some of the education that currently occurs outside the core curriculum.**

Recommendation 9-2: To provide graduate students with experiential learning opportunities, universities, industry, funding agencies, and the American Institute of Chemical Engineers should coordinate to revise graduate training programs and funding structures to provide opportunities for and remove barriers to systematic placement of graduate students in internships.

Recommendation 9-3: To increase recruitment and retention of women and Black, Indigenous, and People of Color (BIPOC) individuals in undergraduate programs, chemical engineering departments should emphasize opportunities for chemical engineers to make positive societal impacts, and should build effective mentoring and support structures for students who are members of such historically excluded groups. To provide more opportunities for BIPOC students, departments should consider redesigning their undergraduate curricula to allow students from 2-year colleges and those who change their major to chemical engineering to complete their degree without extending their time to degree, and provide the support structures necessary to ensure the retention and success of transfer students.

Recommendation 9-4: To increase the recruitment of students from historically excluded communities into graduate programs, chemical engineering departments should consider revising their admissions criteria to remove barriers faced by, for example, students who attended less prestigious universities or did not participate in undergraduate research. To provide more opportunities for women and Black, Indigenous, and People of Color (BIPOC) individuals, departments should welcome students with degrees in related disciplines and consider additions to their graduate curricula that present the core components of the undergraduate curriculum tailored for postgraduate scientists and engineers.

Recommendation 9-5: A consortium of universities, together with the American Institute of Chemical Engineers, should create incentives and practices for building and sharing curated chemical engineering content for use across universities and industry. Such sharing could reduce costs and advance broad access to high-quality content intended both for students and for professional engineers intending to further their education or change industries later in their careers.

Recommendation 9-6: Universities, industry, federal funding agencies, and professional societies should jointly develop and convene a summit to bring together perspectives represented by existing practices across the ecosystem of stakeholders in chemical engineering professional development. Such a summit would explore the needs, barriers, and opportunities around creating a technology-enabled learning and innovation infrastructure for chemical engineering, extending from university education through to the workplace.

10
International Leadership

- U.S. leadership in chemical engineering has decreased significantly in the past 15 years as measured by number of publications, now about half the number produced by China.
- U.S. leadership in the impact of scholarly research in chemical engineering as measured by citation impact has likewise lost ground to international competitors.
- The United States is in a leadership position in some areas of chemical engineering technology, but lags in many niches compared with various other countries.

This report presents a U.S.-based perspective on the field of chemical engineering, but like most areas of science and engineering, chemical engineering is a globally connected enterprise both commercially and academically. Assigning advances within the field to an individual country is difficult, particularly for commercial innovations developed by multinational companies, but some measures of national scientific (mainly academic) output can be measured quantitatively. To that end, the committee used a methodology developed during a previous National Academies study for benchmarking the competitiveness of U.S. chemical engineering (NRC, 2007). Applying this methodology, the committee examined scientific outputs, defined as papers published in peer-reviewed journals, over 5-year intervals from 1981 to 2020—a period encompassing that covered by the previous study and extending it through the most recent complete calendar year. The committee also estimated the quality and impact of those papers across geographic regions by comparing the field-weighted citation impact of the top 100 papers and the number of highly cited papers.

This chapter provides a perspective on U.S. leadership in chemical engineering that is focused on publications and research output. It should be noted that that the emphases and directions of chemical engineering as a discipline and a profession vary across countries and regions. These variations can result from differences in resource availability, political and economic systems, and chemical engineering education. A more comprehensive view considering national investments in chemical engineering research and development (R&D), training curricula, and impact on gross domestic product, for example, could provide a more nuanced and valuable perspective. Such data, however, are difficult to obtain and are often not disaggregated by field or discipline. Furthermore, such an extensive data analysis was beyond the committee's charge. In many areas of chemical engineering, moreover, there are clear examples in which national or regional challenges, such as water scarcity, have driven research and technology development at a high level, as well as cases in which existing infrastructure, such as refineries, is well matched to such challenges and therefore enables regional advances.

PUBLICATION RATES AND CITATION ANALYSIS

The methodology of the 2007 study was based on analysis of a large base of scientific journals (listed in Appendix 3B of NRC, 2007). For the present analysis, journals on the original list that are no longer being published were removed, and those whose publication began in 2007 or thereafter were incorporated (Appendix B). The total number of papers published was found by searching Scopus for all publications for which any author or coauthor had a chemical engineering affiliation in the address field in each of these journals over each time interval. As indicated in the 2007 report, a chemical engineering affiliation is a reliable measure of a researcher's being involved in chemical engineering in the United States, but likely undercounts authors in other regions where department names do not necessarily include the term "chemical engineering." These data include all countries affiliated with each paper. Over the entire 40-year period, 1,250 papers were published with coauthors from the United States, Asia, and the European Union (EU); 6,619 with coauthors from the EU and Asia; 9,722 with coauthors from the United States and the EU; and 16,153 papers were published with coauthors from the United States and Asia. Note that Asia comprises China, Korea, Taiwan, India, and Japan, and the EU includes the current membership and England.

Publishing by U.S. chemical engineers has continued to grow since 2007, but as was already seen in 2007, the rate of growth of the other regions dramatically exceeds that of the United States. Some of this growth was previously attributed to the increased number of researchers in those other regions rather than increased publication rates by individual chemical engineers; indeed, while China currently has a low number of researchers per capita (Heney, 2020), a larger share of their undergraduates is awarded STEM (science, technology, engineering, and mathematics) degrees (49 percent) relative to the share in the United States (33 percent; NSF, 2016). Publications from Asia have grown nearly 100-fold since 1981, eclipsing other regions and making up 57 percent of the total papers published between 2016 and 2020, compared with 23 percent for the United States (Figure 10-1).

The same trend is visible in the regional shifts in publication in those journals identified as core to chemical engineering in the 2007 study—the *AIChE Journal*, *Industrial & Engineering Chemistry Research*, and *Chemical Engineering Science* (Table 10-1). For these journals, a similar search was performed without specifying chemical engineering affiliation. In the last 5 years for all three journals, Asia has become the dominant source of publications, with 50 percent of the total, while the U.S. share has declined from 42 percent in 1991–1995 to 20 percent in the last 5 years.

While the quantity of articles is easy to determine, their quality and impact are more difficult to ascertain. In the short term, a measure of the impact of a publication is the number of times it is cited (Table 10-2). Asia has recently accounted for the largest share of the 100 most cited articles in the three core chemical engineering journals. For the list of top 100 cited articles, countries within a geographic region that were not included in the original 2007 region search list were binned with that region; for example, Singapore was included in the count for Asia.

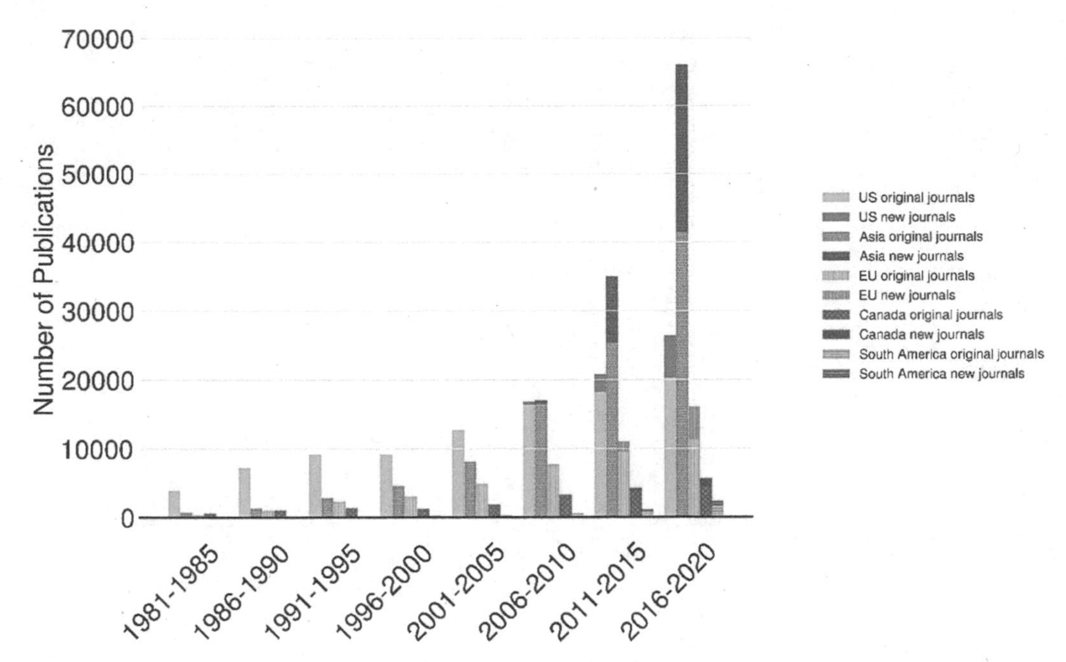

FIGURE 10-1 Number of publications by region from selected journals with at least one author with a chemical engineering department affiliation over 5-year intervals from 1981 to 2020. Asia comprises China, Korea, Taiwan, India, and Japan, and the European Union (EU) includes current membership and England. "Original journals" includes all those used in the 2007 study (NRC, 2007) and "new journals" includes those added for the current analysis whose publication and/or indexing by Scopus began in 2007 or thereafter.

Arguably, the work published by chemical engineers that has the most impact is published in journals whose scope is much broader than chemical engineering. Such journals include *Science*, *Proceedings of the National Academy of Sciences*, and *Nature.* Indeed, the fraction of publications in these three journals with a coauthor having a chemical engineering affiliation has grown over time (Table 10-3), which speaks to a general broadening of the scope and impact of chemical engineering research (or at least research done by chemical engineers). As is the case for the three narrower chemical engineering journals discussed above, contributions from Asia have increased substantially, but in the broader three journals, papers authored by chemical engineers are still primarily of U.S. origin. Interestingly, the number of papers resulting from collaborations across regions has increased substantially in recent decades.

TABLE 10-1 Allocation of All Articles in *AIChE Journal*, *Industrial & Engineering Chemistry (I&EC) Research*, and *Chemical Engineering Science* Published within Previously Defined Regions and Percentages over 5-Year Periods from 1991 to 2020

	Total Articles	**1991–1995**	**1996–2000**	**2001–2005**	**2006–2010**	**2011–2015**	**2016–2020**
AIChE Journal	United States	716 (66%)	749 (52%)	666 (47%)	504 (31%)	637 (35%)	672 (35%)
	Europe	180 (16%)	308 (21%)	409 (29%)	448 (28%)	426 (23%)	382 (20%)
	Asia	117 (11%)	190 (13%)	224 (16%)	429 (26%)	587 (32%)	741 (39%)
	Canada	75 (7%)	80 (6%)	83 (6%)	108 (7%)	146 (8%)	138 (7%)
	S. Am	14 (1%)	23 (2%)	28 (2%)	45 (3%)	24 (1%)	49 (3%)
	Total	1,091	1,440	1,431	1,619	1,825	1,922
I&EC Research	United States	1,049 (45%)	1,183 (38%)	1,463 (33%)	1,489 (24%)	1,604 (18%)	1,615 (19%)
	Europe	495 (21%)	782 (25%)	1,245 (28%)	1,526 (25%)	1,756 (20%)	1,515 (17%)
	Asia	486 (21%)	658 (21%)	1,010 (23%)	2,083 (34%)	4,304 (49%)	4,832 (56%)
	Canada	125 (5%)	206 (7%)	266 (6%)	372 (6%)	434 (5%)	447 (5%)
	S. Am	49 (2%)	86 (3%)	174 (4%)	230 (4%)	316 (4%)	330 (4%)
	Total	2,317	3,094	4,410	6,114	8,765	8,698
Chemical Engineering Science	United States	587 (27%)	553 (23%)	512 (19%)	599 (19%)	541 (17%)	441 (14%)
	Europe	757 (35%)	954 (40%)	990 (37%)	1092 (35%)	1164 (36%)	969 (30%)
	Asia	262 (12%)	362 (15%)	536 (20%)	724 (23%)	938 (29%)	1403 (43%)
	Canada	140 (6%)	153 (6%)	168 (6%)	256 (8%)	208 (6%)	201 (6%)
	S. Am	23 (1%)	65 (3%)	78 (3%)	78 (2%)	60 (2%)	112 (3%)
	Total	2,159	2,377	2,686	3,135	3,232	3,265

TABLE 10-2 Region of Authors of the Top 100 Most Cited Articles Published in a Given 5-Year Period between 1991 and 2020 in Selected Core Chemical Engineering Journals: *AIChE Journal*, *Industrial & Engineering Chemistry Research*, and *Chemical Engineering Science*

	100 most cited	**1991–1995**	**1996–2000**	**2001–2005**	**2006–2010**	**2011–2015**	**2016–2020**
AIChE Journal	United States	69	59	52	35	54	32
	Europe	14	20	28	24	12	15
	Asia	5	9	15	30	27	45
	Canada	10	6	1	8	3	2
	S. Am	1	1	0	0	0	2
	Australia	1	4	1	2	4	2
	Other	0	1	3	1	0	2
I&EC Research	United States	51	41	43	32	14	16
	Europe	22	27	30	25	12	13
	Asia	14	19	17	20	56	57
	Canada	3	5	4	10	7	2
	S. Am	1	0	1	3	0	2
	Australia	4	6	3	9	2	2
	Other	5	2	2	1	9	8
Chemical Engineering Science	United States	30	26	19	20	24	13
	Europe	53	51	48	37	34	19
	Asia	4	14	24	22	24	49
	Canada	10	7	4	11	6	7
	S. Am	0	0	0	2	1	1
	Australia	1	2	3	7	10	5
	Other	2	0	2	1	1	6

NOTE: Individual articles may be attributed to more than one country.

TABLE 10-3 Number and Percentage by Region of Articles in *Science*, *Proceedings of the National Academy of Sciences (PNAS)*, and *Nature* with At Least One Author with a Chemical Engineering Department Affiliation

		1991–1995	1996–2000	2001–2005	2006–2010	2011–2015	2016–2020
Science	Total Papers	11,834	12,816	11,922	11,005	10,347	12,488
	Total ChemE	76 (1%)	56 (0%)	84 (1%)	143 (1%)	174 (2%)	277 (2%)
	U.S. ChemE	60 (79%)	53 (95%)	81 (96%)	134 (94%)	145 (83%)	225 (81%)
	EU ChemE	12 (16%)	5 (9%)	15 (18%)	26 (18%)	41 (24%)	64 (23%)
	Asia ChemE	4 (5%)	3 (5%)	6 (7%)	23 (16%)	37 (21%)	101 (36%)
	Canada ChemE	0 (0%)	0 (0%)	1 (1%)	3 (2%)	5 (3%)	20 (7%)
	S. Am ChemE	0 (0%)	0 (0%)	0 (0%)	2 (1%)	1 (1%)	9 (3%)
	Internationalization	0%	9%	23%	31%	32%	51%
PNAS	Total Papers	12,889	13,994	15,642	19,697	21,065	19,683
	Total ChemE	40 (0%)	66 (0%)	138 (1%)	377 (2%)	528 (3%)	719 (4%)
	U.S. ChemE	38 (95%)	63 (95%)	131 (95%)	336 (89%)	486 (92%)	604 (84%)
	EU ChemE	3 (8%)	2 (3%)	19 (14%)	57 (15%)	100 (19%)	142 (20%)
	Asia ChemE	1 (3%)	3 (5%)	13 (9%)	40 (11%)	109 (21%)	219 (30%)
	Canada ChemE	0 (0%)	2 (3%)	4 (3%)	19 (5%)	22 (4%)	43 (6%)
	S. Am ChemE	0 (0%)	0 (0%)	1 (1%)	2 (1%)	5 (1%)	9 (1%)
	Internationalization	5%	6%	22%	20%	37%	41%

Nature	Total Papers	15,048	14,067	12,871	12,156	12,294	11,938
	Total ChemE	48 (0%)	61 (0%)	51 (0%)	73 (1%)	97 (1%)	186 (2%)
	U.S. ChemE	38 (79%)	52 (85%)	45 (88%)	58 (79%)	76 (78%)	153 (82%)
	EU ChemE	7 (15%)	4 (7%)	10 (20%)	14 (19%)	28 (29%)	58 (31%)
	Asia ChemE	5 (10%)	3 (5%)	4 (8%)	6 (8%)	19 (20%)	75 (40%)
	Canada ChemE	0 (0%)	0 (0%)	2 (4%)	2 (3%)	6 (6%)	17 (9%)
	S. Am ChemE	0 (0%)	0 (0%)	0 (0%)	0 (0%)	4 (4%)	3 (2%)
	Internationalization	4%	–3%	20%	10%	37%	65%

NOTES: "Internationalization" indicates the minimum percentage overlap across regions. Not all countries are included in the defined regions, so this is only an approximate measure of international collaboration.

The committee also made use of the Elsevier SciVal tool,[1] particularly field-weighted citation impact,[2] which indicates how the number of citations of publications from a region compares with the average number of citations of all other similar publications in the data universe, or how the citations of a particular publication compare with the world average. Field-weighted citation impact was averaged for the 100 most cited papers with an author having a chemical engineering affiliation published in each year from each country or region within the comprehensive publication list described in Appendix C (Figure 10-2). In parallel with the other publication metrics reported above, Asia, driven primarily by China, has surpassed other regions and countries in recent years.

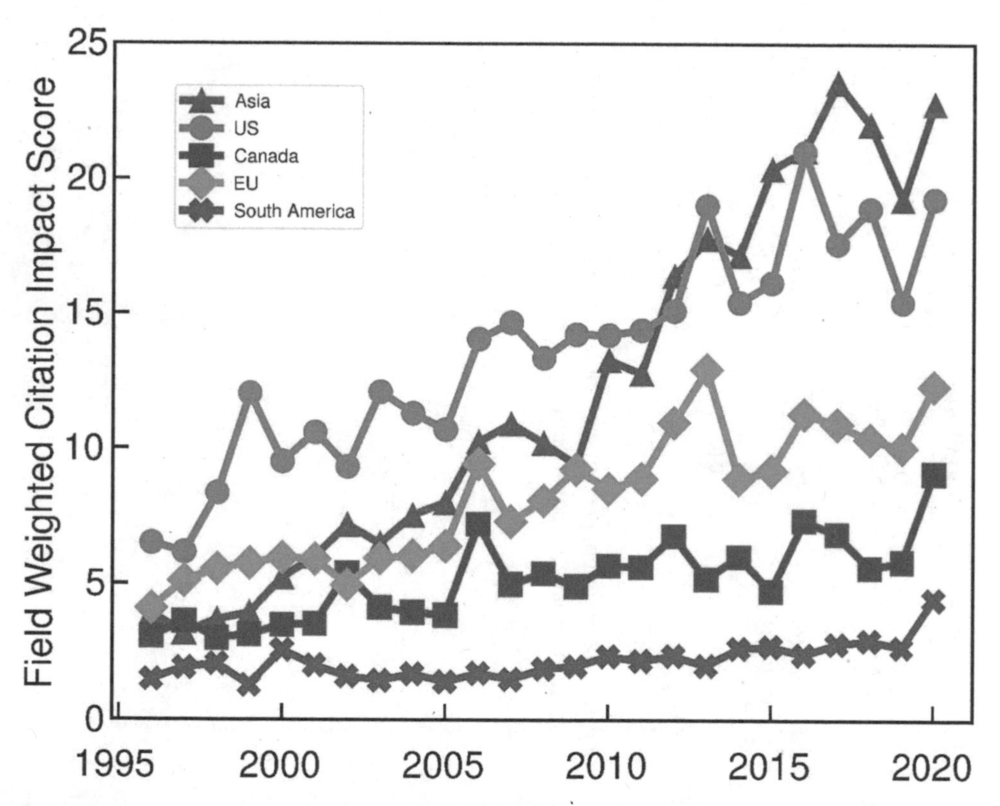

FIGURE 10-2 Comparison of the field-weighted citation impact score for the top 100 cited publications from each region from the full list of journals with an author having a chemical engineering department affiliation. NOTE: 2005 is the first year in which South America published at least 100 papers in these journals, but it did not produce any notable outliers in earlier years.

[1] See https://www.elsevier.com/solutions/scival.

[2] See https://service.elsevier.com/app/answers/detail/a_id/28192/supporthub/scival/p/10961/28192/.

OBSERVATIONS

Beyond the above publication trends, the committee noted differences in chemical engineering advances across specific areas of the world. In energy, for example, the United States is behind both China and Brazil in demonstrating large-scale technologies. The United States remains the leader in catalysis research, which is usually found in chemistry departments in other countries. In the water–energy–food nexus, countries with water scarcity (e.g., Australia, Israel, Saudi Arabia) have advanced R&D much further than has the United States; in fact, in many places in the United States, aspects of the nexus have been used as if they were not independent limited resources.

In the pharmaceutical space, threats from offshore manufacturing include both challenges to the supply chain and concerns about quality control. While the United States still leads the world in the invention and innovation of new therapeutic molecules, India, China, and Singapore are leading the way in the economical production of biosimilar molecules.

In the area of advanced recycling of plastics, the pyrolysis pathway has seen more R&D in Europe than in the United States, a reflection of the difference in refinery capabilities in the two regions. This regional difference has resulted in two trends: many more pyrolysis companies are at pilot and demonstration scales in Europe than in the United States, and more companies in the United States and Canada than in Europe are developing gasification recycling technologies.

In the area of plastics end-of-life management generally, the EU is ahead of the United States in terms of policy drivers toward end-of-life management, as well as such advances in waste conversion as anaerobic digestion and the use of municipal solid waste for energy and other applications. Although there has been a recent increase in funding for and interest in process intensification in the United States, progress in this area has for an extended period been led largely by efforts in Europe.

With respect to the development of tools for future use, the United States leads the world in high-performance computing and the development of new computer architectures. Most codes for molecular simulation were developed in the United States and are open source. The United States also leads in the development of sensors. U.S. leadership is clear as well in the development of tools for medicine (e.g., organs-on-chips), while early work on in vitro models was done in Europe, driven mainly by the region's ban on animal testing of cosmetics.

CONCLUSION

The growth of research output from Asia has been driven mainly by the explosive growth of publications from China (Figure 10-3). This growth reflects a dramatic increase in that country's research intensity and productivity, which was easy to anticipate simply by extrapolating the 2007 study. More concerning for U.S. leadership in chemical engineering and U.S. security is the quality of the work from China as measured both by citations (Figure 10-4) and by its appearance in elite journals (Figure 10-5). Likewise concerning is patent analysis demonstrating that while the United States currently holds an above-average revealed technological advantage in chemical engineering and adjacent industries, that advantage shrank over the period analyzed (2012–2014 compared with 2010–2012; Daiko et al., 2017).

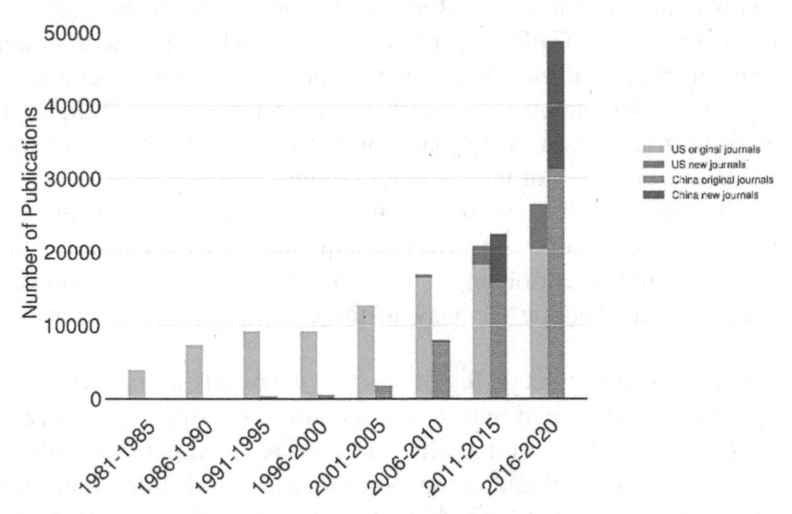

FIGURE 10-3 Number of publications from the United States and China having at least one author with a chemical engineering department affiliation over 5-year intervals from 1981 to 2020. NOTE: "Original journals" includes all those used in the 2007 study (NRC, 2007) and "new journals" includes those added for the current analysis whose publication and/or indexing by Scopus began in 2007 or thereafter.

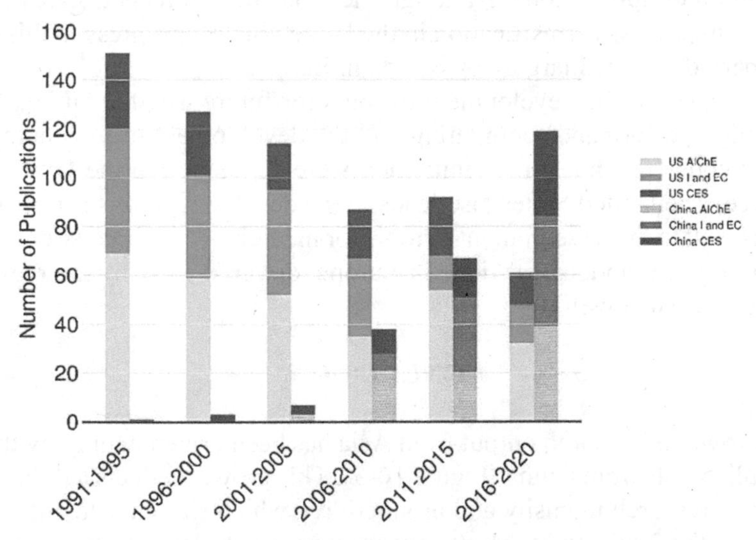

FIGURE 10-4 Number of publications from the United States and China in the top 100 cited articles in a given year over 5-year intervals from 1991 to 2020 within selected core chemical engineering journals: *AIChE Journal*, *Industrial & Engineering Chemistry Research (I and EC)*, and *Chemical Engineering Science (CES)*.

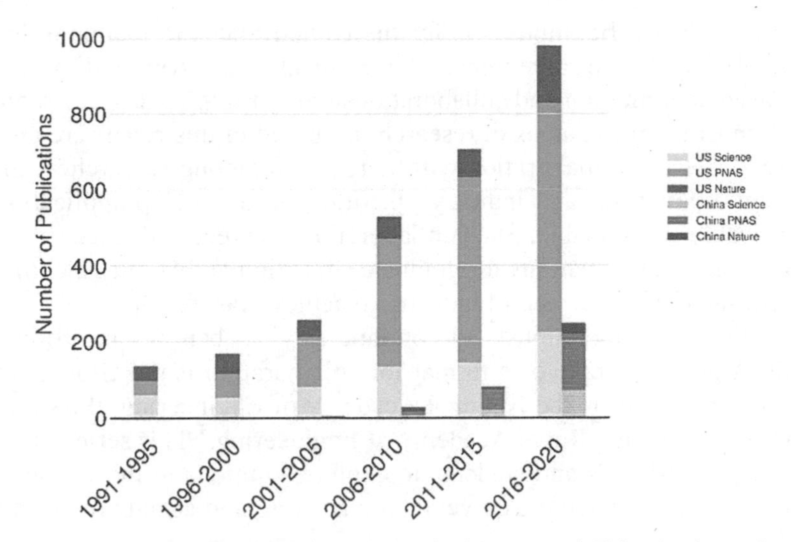

FIGURE 10-5 Number of publications from the United States and China having at least one author with a chemical engineering department affiliation over 5-year intervals from 1981 to 2020 in selected elite journals: *Science*, *Proceedings of the National Academy of Sciences (PNAS)*, and *Nature*.

There is little doubt regarding China's commitment to research in areas central to chemical engineering. Even a casual review of China's Made in China 2025 plan (McBride and Chatzky, 2019) shows a sharp focus on expanding technology and advanced manufacturing—in particular, electric and other alternative-energy vehicles, next-generation information technology, artificial intelligence, new materials, and biomedicine. China also emphasizes semiconductor development and manufacturing. All of these research areas are discussed at various levels of detail in previous chapters of this report. It is clear that anything less than vigorous and sustained investment in these and other areas by the United States will by default cede leadership in technology to China in the next 20 years (Mozur and Myers, 2021).

The increased research output from China is a result of large investments in technology areas, many of which are either central or highly relevant to chemical engineering, investments that are approaching the gross expenditures on R&D of the United States (Heney, 2020).

Similar levels of investment in the U.S. research enterprise are imperative. The committee acknowledges that issues of international leadership are complex, and it is extremely challenging to find a balance between a healthy competitive national aim to lead in a field and national security–driven aspirations of technological superiority, especially given that national needs and priorities differ and can evolve with changing political leadership.

Previous chapters outline the numerous opportunities for chemical engineers to contribute in the areas of energy; food, water, and air; health and medicine; manufacturing; materials; and tools. Without a sustained investment by federal research agencies across these areas, it will be impossible for the United States to maintain leadership in chemical engineering. At the same time, U.S. chemical engineering will be strengthened through increased coordination and collaboration across disciplines, sectors, and political boundaries. Almost all of the areas of research discussed in this report are multidisciplinary in nature, and close collaboration will be required among researchers in academia and government laboratories and industry practitioners to develop applications that are economically viable and scalable. Such collaborations, as recommended throughout this report, will have additional benefits for graduate education and faculty member development while still satisfying the need of industry to achieve fast results.

International collaborations will continue to yield benefits in all the areas discussed herein. A good example of a format for collaboration is the Global Grand Challenges Summits organized by the National Academy of Engineering, the Chinese Academy of Engineering, and the Royal Academy of Engineering.[3] This series of conferences brings together professionals and students to explore solutions to the Grand Challenges for Engineering, and could serve as a venue for creating and sustaining such collaborations.

Recommendation 10-1: Across all areas of chemical engineering, in addition to advancing fundamental understanding, research investments should be set aside for support of interdisciplinary, cross-sector, and international collaborations in the areas of energy; water, food, and air; health and medicine; manufacturing; materials research; tools development; and beyond, with the goal of connecting U.S. research to points of strength in other countries.

[3] See http://www.engineeringchallenges.org/14500.aspx.

References

Abdelkareem, M. A., M. El Haj Assad, E. T. Sayed, and B. Soudan. 2018. Recent progress in the use of renewable energy sources to power water desalination plants. *Desalination* 435:97-113. DOI: 10.1016/j.desal.2017.11.018.

Abejón, R., A. Garea, and A. Irabien. 2010. Ultrapurification of hydrogen peroxide solution from ionic metals impurities to semiconductor grade by reverse osmosis. *Separation and Purification Technology* 76(1):44-51. DOI: 10.1016/j.seppur.2010.09.018.

ACS and RSC (American Chemical Society and Royal Society of Chemistry). 1999. The discovery and development of penicillin 1928-1945. Retrieved August 19, 2021, from https://www.acs.org/content/dam/acsorg/education/whatischemistry/landmarks/flemingpenicillin/the-discovery-and-development-of-penicillin-commemorative-booklet.pdf.

Ahmed, F. 2007. Profile of Mark E. Davis. *Proceedings of the National Academy of Sciences* 104(52). DOI: 10.1073/pnas.0704959105.

Ahmetović, E., Z. Kravanja, N. Ibrić, I. E. Grossmann, and L. E. Savulescu. 2021. State of the art methods for combined water and energy systems optimisation in Kraft pulp mills. *Optimization and Engineering* 22:1831-1852. DOI: 10.1007/s11081-021-09612-4.

AIChE (American Institute of Chemical Engineers). 2017. Achievements in the environment. Retrieved August 17, 2021, from https://www.aiche.org/community/students/career-resources-k-12-students-parents/what-do-chemical-engineers-do/saving-environment/achievements.

AIChE. 2020. Thinking about climate. *CEP Magazine* 116(13). https://www.aiche.org/sites/default/files/cep/20210214.pdf.

AIChE. 2021. 2021 AIChE Salary Survey. *CEP Magazine.* https://www.aiche.org/resources/publications/cep/2021/june/2021-aiche-salary-survey.

Akee, R., M. R. Jones, and S. R. Porter. 2017. Race matters: Income Shares, income inequality, and income mobility for all U.S. Races. *National Bureau of Economic Research Working Paper Series* No. 23733. DOI: 10.3386/w23733.

Albrecht, T. R., A. Crootof, and C. A. Scott. 2018. The water-energy-food nexus: A systematic review of methods for nexus assessment. *Environmental Research Letters* 13(4). DOI: 10.1088/1748-9326/aaa9c6.

Alexandratos, S. D. 2009. Ion-exchange resins: A retrospective from industrial and engineering chemistry research. *Industrial & Engineering Chemistry Research* 48(1):388-398. DOI: 10.1021/ie801242v.

Alger, M., D. Velegol, and R. Shi. 2021. Sustainable energy corps: Building a global collaboration to accelerate transition to a low carbon world. *Chemical Engineering Science: X* 10. DOI: 10.1016/j.cesx.2021.100099.

Alias, A., R. Abhijith, and V. Thankachan. 2019. Review on applications of smart glass in green buildings. Presented at *Green Buildings and Sustainable Engineering*: Singapore.

Allison, G., Y. T. Cain, C. Cooney, T. Garcia, T. G. Bizjak, O. Holte, N. Jagota, B. Komas, E. Korakianiti, D. Kourti, R. Madurawe, E. Morefield, F. Montgomery, M. Nasr, W. Randolph, J. L. Robert, D. Rudd, and D. Zezza. 2015. Regulatory and quality considerations for continuous manufacturing, May 20-21, 2014: Continuous manufacturing symposium. *Journal of Pharmaceutical Sciences* 104(3):803-812. DOI: 10.1002/jps.24324.

Almudever, C. G., L. Lao, X. Fu, N. Khammassi, I. Ashraf, D. Iorga, S. Varsamopoulos, C. Eichler, A. Wallraff, L. Geck, A. Kruth, J. Knoch, H. Bluhm, and K. Bertels. 2017. The engineering challenges in quantum computing. Presented at *Design, Automation & Test in Europe Conference & Exhibition.* Lausanne, Switzerland.

Alonso, C., T. Alig, J. Yoon, F. Bringezu, H. Warriner, and J. A. Zasadzinski. 2004. More than a monolayer: Relating lung surfactant structure and mechanics to composition. *Biophysical Journal* 87(6):4188-4202. DOI: 10.1529/biophysj.104.051201.

Altabet, Y. E., and P. G. Debenedetti. 2017. Communication: Relationship between local structure and the stability of water in hydrophobic confinement. *The Journal of Chemical Physics* 147(24). DOI: 10.1063/1.5013253.

America Makes, AMT, and Deloitte (Association for Manufacturing Technology and Deloitte Consulting). 2021. Assessing the role of additive manufacturing in support of the U.S. COVID-19 response. *Advanced Manufacturing Crisis Production Response.* https://www.fda.gov/media/150615/download.

Anastas, P. T., and J. C. Warner. 1998. *Green chemistry: Theory and practice*. New York: Oxford University Press.

Anastas, P. T., and J. B. Zimmerman. 2003. Design through the 12 principles of green engineering. *Environmental Science & Technology* 37(5):94A-101A. DOI: 10.1021/es032373g.

Andrew, R. M. 2018. Global CO2 emissions from cement production. *Earth System Science Data* 10:195-217. DOI: 10.5194/essd-10-195-2018.

Anna, S. L., N. Bontoux, and H. A. Stone. 2003. Formation of dispersions using "flow focusing" in microchannels. *Applied Physics Letters* 82(3):364-366. DOI: 10.1063/1.1537519.

Anselmo, A. C., and S. Mitragotri. 2014. An overview of clinical and commercial impact of drug delivery systems. *Journal of Controlled Release* 190:15-28. DOI: 10.1016/j.jconrel. 2014.03.053.

API (American Petroleum Institute). 2021. What are alternatives to make fracking less impactful? Retrieved August 17, 2021, from https://www.api.org/oil-and-natural-gas/energy-primers/hydraulic-fracturing/what-are-alternatives-to-make-fracking-less-impactful.

Ardo, S., D. Fernandez Rivas, M. A. Modestino, V. Schulze Greiving, F. F. Abdi, E. Alarcon Llado, V. Artero, K. Ayers, C. Battaglia, J.-P. Becker, D. Bederak, A. Berger, F. Buda, E. Chinello, B. Dam, V. Di Palma, T. Edvinsson, K. Fujii, H. Gardeniers, H. Geerlings, S. M. H. Hashemi, S. Haussener, F. Houle, J. Huskens, B. D. James, K. Konrad, A. Kudo, P. P. Kunturu, D. Lohse, B. Mei, E. L. Miller, G. F. Moore, J. Muller, K. L. Orchard, T. E. Rosser, F. H. Saadi, J.-W. Schüttauf, B. Seger, S. W. Sheehan, W. A. Smith, J. Spurgeon, M. H. Tang, R. van de Krol, P. C. K. Vesborg, and P. Westerik. 2018. Pathways to electrochemical solar-hydrogen technologies. *Energy & Environmental Science* 11(10):2768-2783. DOI: 10.1039/c7ee03639f.

Arent, D. J., S. M. Bragg-Sitton, D. C. Miller, T. J. Tarka, J. A. Engel-Cox, R. D. Boardman, P. C. Balash, M. F. Ruth, J. Cox, and D. J. Garfield. 2021. Multi-input, multi-output hybrid energy systems. *Joule* 5(1):47-58. DOI: 10.1016/j.joule.2020.11.004.

Aresco. 2021. Microwave fracking: The new hydraulic fracturing? Retrieved August 17, 2021, from https://www.arescotx.com/microwave-technology-the-new-fracking/.

Argus. 2020. EU ETS price €32-65/t under 2030 scenarios. Retrieved August 17, 2021, from https://www.argusmedia.com/en/news/2142240-eu-ets-price-3265t-under-2030-scenarios.

Arnold, F. H. 2018. Directed evolution: Bringing new chemistry to life. *Angewandte Chemie International Edition English* 57(16):4143-4148. DOI: 10.1002/anie.201708408.

ASEE (American Society for Engineering Education). 2020. Engineering & engineering technology by the numbers 2019. Retrieved August 17, 2021, from https://ira.asee.org/wp-content/uploads/2021/02/Engineering-by-the-Numbers-FINAL-2021.pdf.

Atkinson, R. D. 2018. Industry funding of university research: Which states lead? Retreived 2021, from https://itif.org/publications/2018/01/08/industry-funding-university-research-which-states-lead.

Auta, H. S., C. U. Emenike, and S. H. Fauziah. 2017. Distribution and importance of microplastics in the marine environment: A review of the sources, fate, effects, and potential solutions. *Environment International* 102:165-176. DOI: 10.1016/j.envint.2017.02.013.

AWEA (American Wind Energy Association). 2020. *Wind powers America annual report 2019*. Retrieved August 18, 2021, from https://www.powermag.com/wp-content/uploads/2020/04/awea_wpa_executivesummary2019.pdf.

Bachler, J., P. H. Handle, N. Giovambattista, and T. Loerting. 2019. Glass polymorphism and liquid-liquid phase transition in aqueous solutions: Experiments and computer simulations. *Physical Chemistry Chemical Physics* 21(42):23238-23268. DOI: 10.1039/c9cp02953b.

Balicka, I. 2020. The food-energy-water nexus: A complex balance. Retrieved July 2020, from https://www.aiche.org/chenected/2020/07/food-energy-water-nexus-complex-balance.

Banholzer, W. F., and M. E. Jones. 2013. Chemical engineers must focus on practical solutions. *AIChE Journal* 59(8):2708-2720. DOI: 10.1002/aic.14172.

Barabino, G. A. 2021. Engineering solutions to COVID-19 and racial and ethnic health disparities. *Journal of Racial and Ethnic Health Disparities* 8(2):277-279. DOI: 10.1007/s40615-020-00953-x.

Bartel, M. A., J. R. Weinstein, and D. V. Schaffer. 2012. Directed evolution of novel adeno-associated viruses for therapeutic gene delivery. *Gene Therapy* 19(6):694-700. DOI: 10.1038/gt.2012.20.

Bates, F. S., P. Brant, G. W. Coates, J. Lipson, C. Osuji, J. de Pablo, S. Rowan, R. Segalman, and K. I. Winey. 2017. Frontiers in polymer science and engineering. NSF Workshop: Frontiers in Polymer Science and Engineering. Arlington, VA. http://nsfpolymerworkshop2016.cems.umn.edu.

Batzner, S., T. Smidt, L. Sun, J. Mailoa, M. Kornbluth, N. Molinari, and B. Kozinsky. 2021. SE(3)-equivariant graph neural networks for data-efficient and accurate interatomic potentials. *arXiv*. DOI: 10.21203/rs.3.rs-244137/v1.

Bausch, M., C. Schultheiss, and J. B. Sieck. 2019. Recommendations for comparison of productivity between fed-batch and perfusion processes. *Biotechnology Journal* 14(2). DOI: 10.1002/biot.201700721.

Bazant, M. Z., and J. W. M. Bush. 2021. A guideline to limit indoor airborne transmission of COVID-19. *Proceedings of the National Academy of Sciences* 118(17). DOI: 10.1073/pnas.2018995118.

Beveridge, G. S. G., and R. S. Schechter. 1975. *Optimization: Theory and practice (chemical engineering)*. New York: McGraw-Hill Education.

Bezek, L. B., J. Pan, C. Harb, C. E. Zawaski, B. Molla, J. R. Kubalak, L. C. Marr, and C. B. Williams. 2021. Additively manufactured respirators: Quantifying particle transmission and identifying system-level challenges for improving filtration efficiency. *Journal of Manufacturing Systems* 60:762-773. DOI: 10.1016/j.jmsy.2021.01.002.

Bhamla, M. S., B. Benson, C. Chai, G. Katsikis, A. Johri, and M. Prakash. 2017. Hand-powered ultralow-cost paper centrifuge. *Nature Biomedical Engineering* 1(1). DOI: 10.1038/s41551-016-0009.

Bhatia, S. N., and D. E. Ingber. 2014. Microfluidic organs-on-chips. *Nature Biotechnology* 32(8):760-772. DOI: 10.1038/nbt.2989.

Bhojwani, S., K. Topolski, R. Mukherjee, D. Sengupta, and M. M. El-Halwagi. 2019. Technology review and data analysis for cost assessment of water treatment systems. *Science of the Total Environment* 651:2749-2761. DOI: 10.1016/j.scitotenv.2018.09.363.

Biddy, M. J. 2016. Chemicals from biomass: A market assessment of bioproducts with near-term potential. Retrieved August 18, 2021, from https://www.energy.gov/sites/prod/files/2016/11/f34/biddy_bioenergy_2016.pdf.

Bielenberg, J., and M. Bryner. 2018. Realize the potential of process intensification. Retrieved March 2018, from https://www.aiche.org/sites/default/files/cep/20180341.pdf.

Biggs, E. M., E. Bruce, B. Boruff, J. M. A. Duncan, J. Horsley, N. Pauli, K. McNeill, A. Neef, F. Van Ogtrop, J. Curnow, B. Haworth, S. Duce, and Y. Imanari. 2015. Sustainable development and the water–energy–food nexus: A perspective on livelihoods. *Environmental Science & Policy* 54:389-397. DOI: 10.1016/j.envsci.2015.08.002.

Billiet, S., and S. R. Trenor. 2020. 100th Anniversary of macromolecular science viewpoint: Needs for plastics packaging circularity. *ACS Macro Letters* 9(9):1376-1390. DOI: 10.1021/acsmacrolett.0c00437.

(BIO) Biotechnology Innovation Organization. 2011. New study shows the rate of drug approvals lower than previously reported. Retrieved August 17, 2021, from https://archive.bio.org/media/press-release/new-study-shows-rate-drug-approvals-lower-previously-reported.

Bioeconomy Capital. 2018. Bioeconomy dashboard: Economic metrics. Retrieved August 17, 2021, from http://www.bioeconomycapital.com/bioeconomy-dashboard.

Bird, R. B., Warren E. Stewart, and Edwin N. Lightfoot. 1960. *Transport phenomena*. New York: John Wiley and Sons, Inc.

Bischoff, K. B. 2015. Pharmacokinetics and cancer chemotherapy. *Journal of Pharmacokinetics and Biopharmaceutics* 1(6):465-480. DOI: 10.1007/bf01059786.

Bjorneholm, O., M. H. Hansen, A. Hodgson, L. M. Liu, D. T. Limmer, A. Michaelides, P. Pedevilla, J. Rossmeisl, H. Shen, G. Tocci, E. Tyrode, M. M. Walz, J. Werner, and H. Bluhm. 2016. Water at interfaces. *Chemical Reviews* 116(13):7698-7726. DOI: 10.1021/acs.chemrev.6b00045.

BloombergNEF. 2020. Battery pack prices cited below $100/kWh for the first time in 2020, while market average sits at $137/kWh. Retrieved August 17, 2021, from https://about.bnef.com/blog/battery-pack-prices-cited-below-100-kwh-for-the-first-time-in-2020-while-market-average-sits-at-137-kwh.

BLS (U.S. Bureau of Labor Statistics). 2021. *Occupational outlook handbook—Chemical engineers*. Retrieved August 16, 2021, from https://www.bls.gov/ooh/architecture-and-engineering/chemical-engineers.htm#tab-1.

Boerner, L. K. 2019. Industrial ammonia production emits more CO_2 than any other chemical-making reaction. Chemists want to change that. *Chemical & Engineering News* 97(24). https://cen.acs.org/environment/green-chemistry/Industrial-ammonia-production-emits-CO2/97/i24.

Boger, T., S. Roy, A. K. Heibel, and O. Borchers. 2003. A monolith loop reactor as an attractive alternative to slurry reactors. *Catalysis Today* 79(2):441-451. DOI: 10.1016/s0920-5861(03)00058-0.

Bozell, J. J., and G. R. Petersen. 2010. Technology development for the production of biobased products from biorefinery carbohydrates—the US Department of Energy's "Top 10" revisited. *Green Chemistry* 12(4). DOI: 10.1039/b922014c.

BP. 2019. BP statistical review of world energy, 68th edition. Retrieved August 17, 2021, from https://www.bp.com/content/dam/bp/business-sites/en/global/corporate/pdfs/energy-economics/statistical-review/bp-stats-review-2019-full-report.pdf.

Brady, J. R., and J. C. Love. 2021. Alternative hosts as the missing link for equitable therapeutic protein production. *Nature Biotechnology* 39(4):404-407. DOI: 10.1038/s41587-021-00884-w.

Branger, C., W. Meouche, and A. Margaillan. 2013. Recent advances on ion-imprinted polymers. *Reactive and Functional Polymers* 73(6):859-875. DOI: 10.1016/j.reactfunctpolym.2013.03.021.

Broekhuis, R. R., R. M. Machado, and A. F. Nordquist. 2001. The ejector-driven monolith loop reactor—Experiments and modeling. *Catalysis Today* 69:87-93. DOI: 10.1016/s0920-5861(01)00358-3.

Brown, K. W., B. Gessesse, L. J. Butler, and D. L. MacIntosh. 2017. Potential effectiveness of point-of-use filtration to address risks to drinking water in the United States. *Environmental Health Insights* 11. DOI: 10.1177/1178630217746997.

Brown, S. A., B. P. Kovatchev, D. Raghinaru, J. W. Lum, B. A. Buckingham, Y. C. Kudva, L. M. Laffel, C. J. Levy, J. E. Pinsker, R. P. Wadwa, E. Dassau, F. J. Doyle, S. M. Anderson, M. M. Church, V. Dadlani, L. Ekhlaspour, G. P. Forlenza, E. Isganaitis, D. W. Lam, C. Kollman, and R. W. Beck. 2019. Six-month randomized, multicenter trial of closed-loop control in type 1 diabetes. *New England Journal of Medicine* 381(18):1707-1717. DOI: 10.1056/NEJMoa1907863.

Brown, T. D., K. A. Whitehead, and S. Mitragotri. 2020. Materials for oral delivery of proteins and peptides. *Nature Reviews Materials* 5(2):127-148. DOI: 10.1038/s41578-019-0156-6.

Buchner, G. A., K. J. Stepputat, A. W. Zimmermann, and R. Schomäcker. 2019. Specifying technology readiness levels for the chemical industry. *Industrial & Engineering Chemistry Research* 58(17):6957-6969. DOI: 10.1021/acs.iecr.8b05693.

Bui, M., C. S. Adjiman, A. Bardow, E. J. Anthony, A. Boston, S. Brown, P. S. Fennell, S. Fuss, A. Galindo, L. A. Hackett, J. P. Hallett, H. J. Herzog, G. Jackson, J. Kemper, S. Krevor, G. C. Maitland, M. Matuszewski, I. S. Metcalfe, C. Petit, G. Puxty, J. Reimer, D. M. Reiner, E. S. Rubin, S. A. Scott, N. Shah, B. Smit, J. P. M. Trusler, P. Webley, J. Wilcox, and N. Mac Dowell. 2018. Carbon capture and storage (CCS): The way forward. *Energy & Environmental Science* 11(5):1062-1176. DOI: 10.1039/C7EE02342A.

Buntz, B. 2021. Pharma's top 20 R&D spenders in 2020. *Drug Discovery and Development*. Retrieved 2021, from https://www.drugdiscoverytrends.com/pharmas-top-20-rd-spenders-in-2020/.

Buzby, J. C., Hodan F. Wells, and J. Hyman. 2014. The estimated amount, value, and calories of postharvest food losses at the retail and consumer levels in the United States. *U.S. Department of Agriculture, Economic Research Service* EIB-121.

Byrne, C., K. M. Zahra, S. Dhaliwal, D. C. Grinter, K. Roy, W. Q. Garzon, G. Held, G. Thornton, and A. S. Walton. 2021. A combined laboratory and synchrotron in-situ photoemission study of the rutile TiO2 (110)/water interface. *Journal of Physics D: Applied Physics* 54(19). DOI: 10.1088/1361-6463/abddfb.

C&EN (*Chemical and Engineering News*). 2021. U.S. top 50 chemical companies of 2021. Retrieved December 3, 2021, from https://cen.acs.org/sections/us-top-50.html.

Caballero, J. A., J. A. Labarta, N. Quirante, A. Carrero-Parreño, and I. E. Grossmann. 2020. Environmental and economic water management in shale gas extraction. *Sustainability* 12(4). DOI: 10.3390/su12041686.

Cao, L., I. K. M. Yu, Y. Liu, X. Ruan, D. C. W. Tsang, A. J. Hunt, Y. S. Ok, H. Song, and S. Zhang. 2018a. Lignin valorization for the production of renewable chemicals: State-of-the-art review and future prospects. *Bioresource Technology* 269:465-475. DOI: 10.1016/j.biortech.2018.08.065.

Cao, Q., M. Huang, T. H. Kuehn, L. Shen, W.-Q. Tao, J. Cao, and D. Y. H. Pui. 2018b. Urban-scale SALSCS, Part II: A Parametric Study of System Performance. *Aerosol and Air Quality Research* 18(11):2879-2894. DOI: 10.4209/aaqr.2018.06.0239.

Cao, Y., J. Romero, J. P. Olson, M. Degroote, P. D. Johnson, M. Kieferova, I. D. Kivlichan, T. Menke, B. Peropadre, N. P. D. Sawaya, S. Sim, L. Veis, and A. Aspuru-Guzik. 2019. Quantum chemistry in the age of quantum computing. *Chemical Review* 119(19):10856-10915. DOI: 10.1021/acs.chemrev.8b00803.

Carayannis, E. G., and J. Alexander. 2004. Strategy, structure, and performance issues of precompetitive R&D consortia: Insights and lessons learned from SEMATECH. *IEEE Transactions on Engineering Management* 51(2):226-232. DOI: 10.1109/TEM.2003.822459.

Carlson, R. 2016. Estimating the biotech sector's contribution to the US economy. *Nature Biotechnology* 34(3):247-255. DOI: 10.1038/nbt.3491.

Carnahan, B., H. A. Luther, and J. O. Wilkes. 1969. *Applied numerical methods*. New York: Wiley.

Carter, P. J. 2011. Introduction to current and future protein therapeutics: A protein engineering perspective. *Experimental Cell Research* 317(9):1261-1269. DOI: 10.1016/j.yexcr.2011.02.013.

Catalanotti, S., V. Cuomo, G. Piro, D. Ruggi, V. Silvestrini, and G. Troise. 1975. The radiative cooling of selective surfaces. *Solar Energy* 17(2):83-89. DOI: 10.1016/0038-092X(75)90062-6.

Caudill, C. L., J. L. Perry, S. Tian, J. C. Luft, and J. M. DeSimone. 2018. Spatially controlled coating of continuous liquid interface production microneedles for transdermal protein delivery. *Journal of Controlled Release* 284:122-132. DOI: 10.1016/j.jconrel.2018.05.042.

Cech, E. 2013. The (mis)framing of social justice: Why ideologies of depoliticization and meritocracy hinder engineers' ability to think about social injustices. In *Engineering education for social justice.* Dordrecht, Netherlands: Springer. DOI 10.1007/978-94-007-6350-0_4.

Cech, E. A., and M. Blair-Loy. 2010. Perceiving glass ceilings? Meritocratic versus structural explanations of gender inequality among women in science and technology. *Social Problems* 57(3):371-397. DOI: 10.1525/sp.2010.57.3.371.

Celik, G., R. M. Kennedy, R. A. Hackler, M. Ferrandon, A. Tennakoon, S. Patnaik, A. M. LaPointe, S. C. Ammal, A. Heyden, F. A. Perras, M. Pruski, S. L. Scott, K. R. Poeppelmeier, A. D. Sadow, and M. Delferro. 2019. Upcycling single-use polyethylene into high-quality liquid products. *ACS Central Science* 5(11):1795-1803. DOI: 10.1021/acscentsci.9b00722.

Centi, G., G. Iaquaniello, and S. Perathoner. 2019. Chemical engineering role in the use of renewable energy and alternative carbon sources in chemical production. *BMC Chemical Engineering* 1. DOI: 10.1186/s42480-019-0006-8.

Cescon, M., S. Deshpande, R. Nimri, I. F. J. Doyle, and E. Dassau. 2021. Using iterative learning for insulin dosage optimization in multiple-daily-injections therapy for people with type 1 diabetes. *IEEE Transactions on Biomedical Engineering* 68(2):482-491. DOI: 10.1109/TBME.2020.3005622.

Chae, D., H. Lim, S. So, S. Son, S. Ju, W. Kim, J. Rho, and H. Lee. 2021. Spectrally selective nanoparticle mixture coating for passive daytime radiative cooling. *ACS Applied Materials & Interfaces* 13(18):21119-21126. DOI: 10.1021/acsami.0c20311.

Chanoca, A., L. de Vries, and W. Boerjan. 2019. Lignin engineering in forest trees. *Frontiers in Plant Science* 10:912. DOI: 10.3389/fpls.2019.00912.

Chen, F., and R. A. Dixon. 2007. Lignin modification improves fermentable sugar yields for biofuel production. *Nature Biotechnology* 25(7):759-761. DOI: 10.1038/nbt1316.

Chen, D., L. Yin, H. Wang, and P. He. 2014. Pyrolysis technologies for municipal solid waste: A review. *Waste Management* 34(12):2466-2486. DOI: 10.1016/j.wasman.2014.08.004.

Cherkasov, A., E. N. Muratov, D. Fourches, A. Varnek, I. I. Baskin, M. Cronin, J. Dearden, P. Gramatica, Y. C. Martin, R. Todeschini, V. Consonni, V. E. Kuz'min, R. Cramer, R. Benigni, C. Yang, J. Rathman, L. Terfloth, J. Gasteiger, A. Richard, and A. Tropsha. 2014. QSAR Modeling: Where have you been? Where are you going to? *Journal of Medicinal Chemistry* 57(12):4977-5010. DOI: 10.1021/jm4004285.

Chetty, R., N. Hendren, F. LIn, J. Majerovitz, and B. Scuderi. 2016. *Childhood environment and gender gaps in adulthood.* Cambridge, MA: National Bureau of Economic Research. DOI: 10.3386/w21936.

Chetty, R., N. Hendren, M. R. Jones, and S. R. Porter. 2019. Race and economic opportunity in the United States: An intergenerational perspective. *The Quarterly Journal of Economics* 135(2):711-783. DOI: 10.1093/qje/qjz042.

CHiMaD (Center for Hierarchical Materials Design). 2021. Polymer property predictor and database. Retrieved August 27, 2021, from https://pppdb.uchicago.edu.

Chmiela, S., H. E. Sauceda, K.-R. Müller, and A. Tkatchenko. 2018. Towards exact molecular dynamics simulations with machine-learned force fields. *Nature Communications* 9, Article 3887. DOI: 10.1038/s41467-018-06169-2.

Chowdhury, S., and S. S. Fong. 2020. Leveraging genome-scale metabolic models for human health applications. *Current Opinion in Biotechnology* 66:267-276. DOI: 10.1016/j.copbio.2020.08.017.

Christopherson, D. A., W. C. Yao, M. Lu, R. Vijayakumar, and A. R. Sedaghat. 2020. High-efficiency particulate air filters in the era of COVID-19: Function and efficacy. *Otolaryngology–Head and Neck Surgery* 163(6):1153-1155. DOI: 10.1177/01945998 20941838.

Chung, M., G. Fortunato, and N. Radacsi. 2019. Wearable flexible sweat sensors for healthcare monitoring: A review. *Journal of the Royal Society Interface* 16(159). DOI: 10.1098/ rsif.2019.0217.

CMS (Centers for Medicare & Medicaid Services). 2020. The National Health Expenditure Accounts (NHEA). Retrieved August 19, 2021, from https://www.cms.gov/Research-Statistics-Data-and-Systems/Statistics-Trends-and-Reports/NationalHealthExpendData/ NationalHealthAccountsHistorical.

Cocchi, M., D. D. Angelis, L. Mazzeo, P. Nardozi, V. Piemonte, R. Tuffi, and S. Vecchio Ciprioti. 2020. Catalytic pyrolysis of a residual plastic waste using zeolites produced by coal fly ash. *Catalysts* 10(10). DOI: 10.3390/catal10101113.

Cohen, C. 1996. The early history of chemical engineering: A reassessment. *British Journal for the History of Science* 29(2):171-194. DOI: 10.1017/s000708740003421x.

Cohen, Y., and J. Glater. 2010. A tribute to Sidney Loeb—The pioneer of reverse osmosis desalination research. *Desalination and Water Treatment* 15:222-227. DOI: 10.5004/ dwt.2010.1762.

Coley, C. W., W. H. Green, and K. F. Jensen. 2018. Machine learning in computer-aided synthesis planning. *Accounts of Chemical Research* 51(5):1281-1289. DOI: 10.1021/acs. accounts.8b00087.

Collias, D. I., A. M. Harris, V. Nagpal, I. W. Cottrell, and M. W. Schultheis. 2014. Biobased terephthalic acid technologies: A literature review. *Industrial Biotechnology* 10(2):91-105. DOI: 10.1089/ind.2014.0002.

Collias, D. I., J. E. Godlewski, and J. E. Velasquez. 2018. Method of making acrylic acid from hydroxypropionic acid. U.S. Patent No. 9,890,102. Washington, DC: U.S. Patent and Trademark Office.

Collias, D. I., M. I. James, and J.M. Jayman. 2021. Circular economy of polymers: Topics in recycling technologies. *ACS Symposium Series* 1391. DOI: 10.1021/bk-2021-1391.ch001.

Colman, R. J., and D. T. Rubin. 2014. Fecal microbiota transplantation as therapy for inflammatory bowel disease: A systematic review and meta-analysis. *Journal of Crohn's and Colitis* 8(12):1569-1581. DOI: 10.1016/j.crohns.2014.08.006.

Colton, C. K., ed. 1991. *Advances in chemical engineering*. San Diego, CA: Academic Press.

Conn, R. W., M. M. Crow, C. M. Friend, and M. McNutt. July 21, 2021. The next 75 years of US science and innovation policy: An introduction. Issues in Science and Technology. https://issues.org/the-next-75-years-of-us-science-and-innovation-policy-an-introduction.

Coronell, D. G., T. H.-L. Hsiung, J. Howard Paul Withers, and A. J. Woytek. 1997. Process for nitrogen trifluoride synthesis. EP Patent No. 0787684B1. Munich, Germany: European Patent Office.

Correa-Baena, J. P., M. Saliba, T. Buonassisi, M. Gratzel, A. Abate, W. Tress, and A. Hagfeldt. 2017. Promises and challenges of perovskite solar cells. *Science* 358(6364):739-744. DOI: 10.1126/science.aam6323.

Cortes Garcia, G. E., J. van der Schaaf, and A. A. Kiss. 2017. A review on process intensification in HiGee distillation. *Journal of Chemical Technology & Biotechnology* 92(6):1136-1156. DOI: 10.1002/jctb.5206.

Coughanowr, D. R., and L. B. Koppel. 1965. *Process systems analysis and control*. New York: McGraw-Hill Book Company, Inc.

Creative Energy. 2007. *European roadmap for process intensification*. Retrieved August 17, 2021, from https://efce.info/efce_media/-p-531.pdf.

Crocker, J. C., M. T. Valentine, E. R. Weeks, T. Gisler, P. D. Kaplan, A. G. Yodh, and D. A. Weitz. 2000. Two-point microrheology of inhomogeneous soft materials. *Physical Review Letters* 85(4):888-891. DOI: 10.1103/PhysRevLett.85.888.

Crone, B. C., T. F. Speth, D. G. Wahman, S. J. Smith, G. Abulikemu, E. J. Kleiner, and J. G. Pressman. 2019. Occurrence of per- and polyfluoroalkyl substances (PFAS) in source water and their treatment in drinking water. *Critical Reviews in Environmental Science and Technology* 49(24):2359-2396. DOI: 10.1080/10643389.2019.1614848.

Cruz, A. C., A. L. Medel, A. C. Bianchi, V. Wong, and M. Danforth. 2021. Impact of flipped classroom model on high-workload and low-income students in upper-division computer science. Presented at *ASEE Virtual Annual Conference*.

Cui, S., Y. Wang, Y. Yang, F. M. Nave, and K. T. Harris. 2011. Connecting incoming freshmen with engineering through hands-on projects. *American Journal of Engineering Education* 2(2):31-42. DOI: 10.19030/ajee.v2i2.6636.

Cybulski, A., and J. A. Moulijn. 1994. Monoliths in heterogeneous catalysis. *Catalysis Reviews* 36(2):179-270. DOI: 10.1080/01614949408013925.

Cywar, R., N. A. Rorrerm, C. A. Hoyt, G. T. Beckham, and E. Chen. 2021. Bio-based polymers with performance-advantaged properties. *Nature Reviews Materials*. In press. DOI: 10.1038/s41578-021-00363-3.

Daglar, H., and S. Keskin. 2020. Recent advances, opportunities, and challenges in high-throughput computational screening of MOFs for gas separations. *Coordination Chemistry Reviews* 422. DOI: 10.1016/j.ccr.2020.213470.

Dahlman, J. E., K. J. Kauffman, Y. Xing, T. E. Shaw, F. F. Mir, C. C. Dlott, R. Langer, D. G. Anderson, and E. T. Wang. 2017. Barcoded nanoparticles for high throughput in vivo discovery of targeted therapeutics. *Proceedings of the National Academy of Sciences* 114(8). DOI: 10.1073/pnas.1620874114.

Daiko, T., H. Dernis, M. Dosso, P. Gkotsis, M. Squicciarini, and A. Vezzani. 2017. World corporate R&D investors: Industrial property strategies in the digital economy. *A JRC and OECD Common Report*. Luxembourg: Publications office of the European Union.

Dan, Y., Y. Zhao, X. Li, S. Li, M. Hu, and J. Hu. 2020. Generative adversarial networks (GAN) based efficient sampling of chemical composition space for inverse design of inorganic materials. *NPJ Computational Materials* 6(1). DOI: 10.1038/s41524-020-00352-0.

Das, T., and H. Cabezas. 2018. Tools and concepts for environmental sustainability in the food-energy-water nexus: Chemical engineering perspective. *Environmental Progress & Sustainable Energy* 37(1):73-81. DOI: 10.1002/ep.12763.

Dassau, E., E. Renard, J. Place, A. Farret, M. J. Pelletier, J. Lee, L. M. Huyett, A. Chakrabarty, F. J. Doyle 3rd, and H. C. Zisser. 2017. Intraperitoneal insulin delivery provides superior glycaemic regulation to subcutaneous insulin delivery in model predictive control-based fully-automated artificial pancreas in patients with type 1 diabetes: A pilot study. *Diabetes, Obesity and Metabolism* 19(12):1698-1705. DOI: 10.1111/dom.12999.

Davis, R., L. Tao, E. C. D. Tan, M. J. Biddy, G. T. Beckham, C. Scarlata, J. Jacobson, K. Cafferty, J. Ross, J. Lukas, D. Knorr, and P. Schoen. 2013. *Process design and economics for the conversion of lignocellulosic biomass to hydrocarbons: Dilute acid and enzymatic deconstruction of biomass to sugars and biological conversion of sugars to hydrocarbons.* Golden, CO: National Renewable Energy Laboratory.

de Beer, M. P., H. L. van der Laan, M. A. Cole, R. J. Whelan, M. A. Burns, and T. F. Scott. 2019. Rapid, continuous additive manufacturing by volumetric polymerization inhibition patterning. *Science Advances* 5(1). DOI: 10.1126/sciadv.aau8723.

de Pee, A., D. Pinner, O. Roelofsen, K. Somers, E. Speelman, and M. Witteveen. 2018. *Decarbonization of industrial sectors: The next frontier.* Amsterdam: McKinsey.

Debenedetti, P. G. 2003. Supercooled and glassy water. *Journal of Physics: Condensed Matter* 15(45):R1669-1726. DOI: 10.1088/0953-8984/15/45/r01.

Debenedetti, P. G., and M. L. Klein. 2017. Chemical physics of water. *Proceedings of the National Academy of Sciences* 114(51):13325-13326. DOI: 10.1073/pnas.1719350115.

Decante, G., J. B. Costa, J. Silva-Correia, M. N. Collins, R. L. Reis, and J. M. Oliveira. 2021. Engineering bioinks for 3D bioprinting. *Biofabrication* 13(3). DOI: 10.1088/1758-5090/abec2c.

Denn, M. M. 1991. The identity of our profession. In *Advances in chemical engineering*. C. K. Colton, ed. San Diego, CA: Academic Press.

Deri, M. A., P. Mills, and D. McGregor. 2018. Structure and evaluation of a flipped general chemistry course as a model for small and large gateway science courses at an urban public institution. *Journal of College Science Teaching* 47(3):68-77.

DHS (Department of Homeland Security). 2019. Chemical sector profile. Retrieved August 16, 2021, from https://www.cisa.gov/sites/default/files/publications/Chemical-Sector-Profile_Final%20508.pdf.

DiMasi, J. A., H. G. Grabowski, and R. W. Hansen. 2016. Innovation in the pharmaceutical industry: New estimates of R&D costs. *Journal of Health Economics* 47:20-33. DOI: 10.1016/j.jhealeco.2016.01.012.

Dixon, K., and J. L. Wendt. 2021. Science motivation and achievement among minority urban high school students: An examination of the flipped classroom model. *Journal of Science Education and Technology* 30(5):642-657. DOI: 10.1007/s10956-021-09909-0.

Dobbelaere, M. R., P. P. Plehiers, R. Van de Vijver, C. V. Stevens, and K. M. Van Geem. 2021. Machine learning in chemical engineering: Strengths, weaknesses, opportunities, and threats. *Engineering* 7(9):1201-1211. DOI: 10.1016/j.eng.2021.03.019.

D'Odorico, P., K. F. Davis, L. Rosa, J. A. Carr, D. Chiarelli, J. Dell'Angelo, J. Gephart, G. K. MacDonald, D. A. Seekell, S. Suweis, and M. C. Rulli. 2018. The global food-energy-water nexus. *Reviews of Geophysics* 56(3):456-531. DOI: 10.1029/2017rg000591.

DOE (U.S. Department of Energy). 2014. The water-energy nexus: Challenges and opportunities. Retrieved August 16, 2021, from https://www.energy.gov/articles/water-energy-nexus-challenges-and-opportunities.

DOE. 2015. Revolution...now—The future arrives for five clean energy technologies. Update retrieved August 16, 2021, from https://www.energy.gov/sites/prod/files/2015/11/f27/Revolution-Now-11132015.pdf.

DOE. 2016. 2016 Billion-ton report—Advancing domestic resources for a thriving bioeconomy. Retrieved August 18, 2021, from https://www.energy.gov/sites/default/files/2016/12/f34/2016_billion_ton_report_12.2.16_0.pdf.

DOE. 2018a. Moving beyond drop-in replacements: Performance-advantaged biobased chemicals. Retrieved August 16, 2021, from https://www.energy.gov/sites/prod/files/2018/06/f53/Performance-Advantaged%20Biobased%20Chemicals%20Workshop%20Report.pdf.

DOE. 2018b. R&D Opportunities for natural gas technologies in building applications. Retrieved August 16, 2021, from https://www.energy.gov/sites/prod/files/2018/08/f55/bto-Natural-Gas-RD-Opportunities-082918.pdf.

DOE. 2018c. Accelerating breakthrough innovation in carbon capture, utilization, and storage. Retrieved 2021, https://www.energy.gov/fe/downloads/accelerating-breakthrough-innovation-carbon-capture-utilization-and-storage.

DOE. 2021a. Solar energy technologies office multi-year program plan. Retrieved August 16, 2021, from https://www.energy.gov/eere/solar/articles/solar-energy-technologies-office-multi-year-program-plan.

DOE. 2021b. Shale research & development. Retrieved August 16, 2021, from https://www.energy.gov/fe/science-innovation/oil-gas-research/shale-gas-rd.

DOE. 2021c. Nuclear reactor technologies. Retrieved August 16, 2021, from https://www.energy.gov/ne/nuclear-reactor-technologies.

Doyle, F. J., 3rd, L. M. Huyett, J. B. Lee, H. C. Zisser, and E. Dassau. 2014. Closed-loop artificial pancreas systems: Engineering the algorithms. *Diabetes Care* 37(5):1191-1197. DOI: 10.2337/dc13-2108.

Du Bois, W. E. B. 1939. The negro scientist. *The American Scholar* 8(3):309-320. http://www.jstor.org/stable/41204425.

Eberle, A., A. Bhatt, Y. Zhang, and G. Heath. 2017. Potential air pollutant emissions and permitting classifications for two biorefinery process designs in the United States. *Environmental Science & Technology* 51(11):5879-5888. DOI: 10.1021/acs.est.7b00229.

Edwards, D. A., D. Ausiello, J. Salzman, T. Devlin, R. Langer, B. J. Beddingfield, A. C. Fears, L. A. Doyle-Meyers, R. K. Redmann, S. Z. Killeen, N. J. Maness, and C. J. Roy. 2021. Exhaled aerosol increases with COVID-19 infection, age, and obesity. *Proceedings of the National Academy of Sciences* 118(8). DOI: 10.1073/pnas.2021830118.

Eggleton, C. D., T.-M. Tsai, and K. J. Stebe. 2001. Tip streaming from a drop in the presence of surfactants. *Physical Review Letters* 87(4). DOI: 10.1103/PhysRevLett.87.048302.

EIA (U.S. Energy Information Administration). 2013. Few transportation fuels surpass the energy densities of gasoline and diesel. Retrieved July 18, 2021, from https://www.eia.gov/todayinenergy/detail.php?id=9991.

EIA. 2020a. U.S. fuel ethanol production capacity increased by 3% in 2019. Retrieved August 16, 2021, from https://www.eia.gov/todayinenergy/detail.php?id=45316.

EIA. 2020b. International energy outlook 2020. Retrieved August 16, 2021, from https://www.eia.gov/outlooks/ieo/.

EIA. 2020c. Biomass explained: Waste-to-energy (municipal solid waste). Retrieved August 16, 2021, from https://www.eia.gov/energyexplained/biomass/waste-to-energy.php.

EIA. 2020d. U.S. coal-fired electricity generation in 2019 falls to 42-year low. Retrieved August 16, 2021, from https://www.eia.gov/todayinenergy/detail.php?id=43675#.

EIA. 2021a. Global ethanol production by country or region. Retrieved August 16, 2021, from https://afdc.energy.gov/data/10331.

EIA. 2021b. *Annual energy outlook 2021*. U.S. Department of Energy, Washington, DC.

EIA. 2021c. How much shale (tight) oil is produced in the United States? Retrieved August 16, 2021, from https://www.eia.gov/tools/faqs/faq.php?id=847&t=6.

EIA. 2021d. Short-term energy outlook: Global liquid fuels. Retrieved August 9, 2021, from https://www.eia.gov/outlooks/steo/report/global_oil.php.

EIA. 2021e. Natural gas explained: Use of natural gas. Retrieved August 31, 2021, from https://www.eia.gov/energyexplained/natural-gas/use-of-natural-gas.php.

EIA. 2021f. Natural gas explained: Delivery and storage of natural gas. Retrieved August 31, 2021, from https://www.eia.gov/energyexplained/natural-gas/delivery-and-storage.php.

EIA. 2021g. Natural gas explained. Retrieved September 11, 2021, from https://www.eia.gov/energyexplained/natural-gas.

EIA. 2021h. *Monthly energy review, August 2021*. Washington, DC: Office of Energy Statistics, U.S. Department of Energy.

EIA. 2021i. Frequently asked questions: How much carbon dioxide is produced when different fuels are burned? Retrieved January 31, 2022 from https://www.eia.gov/tools/faqs/faq.php?id=73&t=11.

EIA. 2021j. Oil and petroleum products explained: Where our oil comes from. Retrieved August 16, 2021, from https://www.eia.gov/energyexplained/oil-and-petroleum-products/where-our-oil-comes-from.php.

EIA. 2021k. Oil and petroleum products explained: Use of oil. Retrieved August 16, 2021, from https://www.eia.gov/energyexplained/oil-and-petroleum-products/use-of-oil.php.

Elbashier, E., A. Mussa, M. Hafiz, and A. H. Hawari. 2021. Recovery of rare earth elements from waste streams using membrane processes: An overview. *Hydrometallurgy* 204. DOI: 10.1016/j.hydromet.2021.105706.

Elhacham, E., L. Ben-Uri, J. Grozovski, Y. M. Bar-On, and R. Milo. 2020. Global human-made mass exceeds all living biomass. *Nature* 588(7838):442-444. DOI: 10.1038/s41586-020-3010-5.

Elton, D. C., Z. Boukouvalas, M. D. Fuge, and P. W. Chung. 2019. Deep learning for molecular design—A review of the state of the art. *Molecular Systems Design & Engineering* 4(4):828-849. DOI: 10.1039/c9me00039a.

EMF (Ellen MacArthur Foundation). 2013a. *Towards the circular economy, volume 2: Opportunities for the consumer goods sector.* Cowes, United Kingdom. https://emf.thirdlight.com/link/coj8yt1jogq8-hkhkq2/@/preview/1?o.

EMF. 2013b. *Towards the circular economy, volume 1: Economic and business rationale for an accelerated transition.* Cowes, United Kingdom. https://emf.thirdlight.com/link/x8ay372a3r11-k6775n/@/preview/1?o.

EMF. 2014. *Towards the circular economy, volume 3: Accelerating the scale up across global supply chains*. Cowes, United Kingdom. https://emf.thirdlight.com/link/t4gb0fs4knot-n8nz6f/@/preview/1?o.

EMF. 2016. The new plastics economy: Rethinking the future of plastics. Retrieved September 14, 2021, 2021, from https://emf.thirdlight.com/link/faarmdpz93ds-5vmvdf/@/preview/1?o.

EMF. 2017a. What is the circular economy? Retrieved August 17, 2021, from https://archive.ellenmacarthurfoundation.org/circular-economy/what-is-the-circular-economy.

EMF. 2017b. The circular economy in detail. Retrieved August 17, 2021, from https://archive.ellenmacarthurfoundation.org/explore/the-circular-economy-in-detail.

EMF. 2019. Circular economy diagram. Retrieved 2021, from https://ellenmacarthurfoundation.org/circular-economy-diagram.

EMF. 2021. Circular economy introduction. Retrieved August 17, 2021, from https://www.ellenmacarthurfoundation.org/circular-economy/concept.

EPA (U.S. Environmental Protection Agency). 2018. Indoor air quality—What are the trends in indoor air quality and their effects on human health? Retrieved August 16, 2021, from https://www.epa.gov/report-environment/indoor-air-quality#note1.

EPA. 2020. Health, energy efficiency and climate change. Retrieved August 16, 2021, from https://www.epa.gov/indoor-air-quality-iaq/health-energy-efficiency-and-climate-change.

EPA. 2021a. Air pollution: Current and future challenges. Retrieved August 16, 2021, from https://www.epa.gov/clean-air-act-overview/air-pollution-current-and-future-challenges.

EPA. 2021b. The process of unconventional natural gas production. Retrieved September 11, 2021, 2021, from https://www.epa.gov/uog/process-unconventional-natural-gas-production.

Eris, O., D. Chachra, H. L. Chen, S. Sheppard, L. Ludlow, C. Rosca, T. Bailey, and G. Toye. 2010. Outcomes of a longitudinal administration of the persistence in engineering survey. *Journal of Engineering Education* 99(4):371-395. DOI: 10.1002/j.2168-9830.2010.tb01069.x.

Eun, C., and M. L. Berkowitz. 2011. Molecular dynamics simulation study of interaction between model rough hydrophobic surfaces. *The Journal of Physical Chemistry A* 115(23):6059-6067. DOI: 10.1021/jp110608p.

European Bioplastics. 2021. Bioplastic materials. Retrieved August 17, 2021, from https://www.european-bioplastics.org/bioplastics/materials.

Extance, A. 2020. IBM seeks to simplify robotic chemistry. *Chemistry World*. From https://www.chemistryworld.com/news/ibm-seeks-to-simplify-robotic-chemistry/4012359.article.

ExxonMobil. 2019. 2019 Outlook for energy: A perspective to 2040. Retrieved August 17, 2021, from https://corporate.exxonmobil.com/-/media/Global/Files/outlook-for-energy/2019-Outlook-for-Energy_v4.pdf.

Faller, C., M. Bracher, N. Dami, and R. Roguet. 2002. Predictive ability of reconstructed human epidermis equivalents for the assessment of skin irritation of cosmetics. *Toxicology in Vitro* 16(5):557-572. DOI: 10.1016/s0887-2333(02)00053-x.

FAO (Food and Agriculture Organization of the United Nations). 2020. Land use in agriculture by the numbers. Retrieved August 16, 2021, from http://www.fao.org/sustainability/news/detail/en/c/1274219.

FAO. 2021. Key facts and findings. Retrieved August 16, 2021, from http://www.fao.org/news/story/en/item/197623/icode.

Farrington, R. 2020. These states offer tuition-free community college. *Forbes*. https://www.forbes.com/sites/robertfarrington/2020/03/25/these-states-offer-tuition-free-community-college.

FDA (U.S. Food and Drug Administration). 2021. Approved cellular and gene therapy products. Retrieved August 16, 2021, from https://www.fda.gov/vaccines-blood-biologics/cellular-gene-therapy-products/approved-cellular-and-gene-therapy-products.

Fenton, O. S., K. N. Olafson, P. S. Pillai, M. J. Mitchell, and R. Langer. 2018. Advances in biomaterials for drug delivery. *Advanced Materials* 30(29). DOI: 10.1002/adma.201705328.

Finley, J. W., and J. N. Seiber. 2014. The nexus of food, energy, and water. *Journal of Agricultural and Food Chemistry* 62(27):6255-6262. DOI: 10.1021/jf501496r.

Flaga, C. T. 2006. The process of transition for community college transfer students. *Community College Journal of Research and Practice* 30(1):3-19. DOI: 10.1080/10668920500248845.

Flavell-While, C. 2011. Arthur D Little—Dedicated to industrial progress. *The Chemical Engineer*. https://www.thechemicalengineer.com/features/cewctw-arthur-d-little-dedicated-to-industrial-progress.

Florey, H. W. 1949. *Antibiotics: A survey of penicillin, streptomycin, and other antimicrobial substances from fungi, actinomycetes, bacteria, and plants*. New York: Oxford University Press.

Ford, M. A. 1982. Computer control of equipment and data handling. *Philosophical Transactions of the Royal Society of London. Series A, Mathematical and Physical Sciences* 307(1500):491-501. DOI: https://www.jstor.org/stable/37280.

Fortune. 1985. A database of 50 years of FORTUNE's list of America's largest corporations. Retrieved August 16, 2021, from https://archive.fortune.com/magazines/fortune/fortune500_archive/full/1985/.

Fortune. 2020. Fortune 500. Retrieved August 17, 2021, from https://fortune.com/fortune500/2020.

Franklin Associates. 2018. Life cycle impacts for postconsumer recycled resins: PET, HDPE, and PP. Retrieved August 17, 2021, from https://plasticsrecycling.org/images/library/2018-APR-LCI-report.pdf.

Franks, R. G. E. 1972. *Modeling and simulation in chemical engineering*. Hoboken, New Jersey: Wiley-Interscience.

Frenkel, D., and A. J. C. Ladd. 1984. New Monte Carlo method to compute the free energy of arbitrary solids. Application to the FCC and HCP phases of hard spheres. *The Journal of Chemical Physics* 81(7):3188-3193. DOI: 10.1063/1.448024.

Frueh, S. 2020. Engineering a response to the COVID-19 pandemic. *National Academies of Sciences, Engineering, and Medicine*. https://www.nationalacademies.org/news/2020/09/engineering-a-response-to-the-covid-19-pandemic.

Funk, C., and K. Parker. 2018. Women in STEM see more gender disparities at work, especially those in computer jobs, majority-male workplaces. In women and men in STEM often at odds over workplace equity. Washington, DC: Pew Research Center.

Furter, W., F., ed. 1983. *History of chemical engineering: Advances in chemistry, series 190*. Washington, DC: American Chemical Society.

Fuss, S., W. F. Lamb, M. W. Callaghan, J. Hilaire, F. Creutzig, T. Amann, T. Beringer, W. De Oliveira Garcia, J. Hartmann, T. Khanna, G. Luderer, G. F. Nemet, J. Rogelj, P. Smith, J. V. Vicente, J. Wilcox, M. Del Mar Zamora Dominguez, and J. C. Minx. 2018. Negative emissions—Part 2: Costs, potentials and side effects. *Environmental Research Letters* 13(6). DOI: 10.1088/1748-9326/aabf9f.

Gagliano, E., M. Sgroi, P. P. Falciglia, F. G. A. Vagliasindi, and P. Roccaro. 2020. Removal of poly- and perfluoroalkyl substances (PFAS) from water by adsorption: Role of PFAS chain length, effect of organic matter and challenges in adsorbent regeneration. *Water Research* 171. DOI: 10.1016/j.watres.2019.115381.

Galán, G., M. Martín, and I. E. Grossmann. 2021. Integrated renewable production of sorbitol and xylitol from switchgrass. *Industrial & Engineering Chemistry Research* 60(15):5558-5573. DOI: 10.1021/acs.iecr.1c00397.

Gallo, P., K. Amann-Winkel, C. A. Angell, M. A. Anisimov, F. Caupin, C. Chakravarty, E. Lascaris, T. Loerting, A. Z. Panagiotopoulos, J. Russo, J. A. Sellberg, H. E. Stanley, H. Tanaka, C. Vega, L. Xu, and L. G. Pettersson. 2016. Water: A tale of two liquids. *Chemical Reviews* 116(13):7463-7500. DOI: 10.1021/acs.chemrev.5b00750.

Galloway, J. N., A. R. Townsend, J. W. Erisman, M. Bekunda, Z. Cai, J. R. Freney, L. A. Martinelli, S. P. Seitzinger, and M. A. Sutton. 2008. Transformation of the nitrogen cycle: Recent trends, questions, and potential solutions. *Science* 320(5878):889-892. DOI: 10.1126/science.1136674.

Gao, Z., S. Rohani, J. Gong, and J. Wang. 2017. Recent developments in the crystallization process: Toward the pharmaceutical industry. *Engineering* 3(3):343-353. DOI: 10.1016/J.ENG.2017.03.022.

Gao, W., H. Ota, D. Kiriya, K. Takei, and A. Javey. 2019. Flexible electronics toward wearable sensing. *Accounts of Chemical Research* 52(3):523-533. DOI: 10.1021/acs.accounts.8b00500.

Garcia, D. J., and F. You. 2016. The water-energy-food nexus and process systems engineering: A new focus. *Computers & Chemical Engineering* 91:49-67. DOI: 10.1016/j.compchemeng.2016.03.003.

Gargini, P., F. Balestra, and Y. Hayashi. 2020. The international roadmap for devices and systems (IRDS). *IEEE Electron Devices Society Newsletter* 27(3).

Garnier, G. 2014. Grand challenges in chemical engineering. *Frontiers in Chemistry* 2, Article 17. DOI: 10.3389/fchem.2014.00017.

Gartner. 2020. Gartner says worldwide PC shipments grew 2.3% in 4Q19 and 0.6% for the year. *Gartner Newsroom*. Press release https://www.gartner.com/en/newsroom/press-releases/2020-01-13-gartner-says-worldwide-pc-shipments-grew-2-point-3-percent-in-4q19-and-point-6-per
cent-for-the-year.

Gaynes, R. 2017. The discovery of penicillin—New insights after more than 75 years of clinical use. *Emerging Infectious Diseases* 23(5):849-853. DOI: 10.3201/eid2305.161556.

Geoscience News and Information. 2021. Hydraulic fracturing fluids—Composition and additives. Retrieved August 17, 2021, from https://geology.com/energy/hydraulic-fracturing-fluids.

Gerber, C., R. Vaikmäe, W. Aeschbach, A. Babre, W. Jiang, M. Leuenberger, Z.-T. Lu, R. Mokrik, P. Müller, V. Raidla, T. Saks, H. N. Waber, T. Weissbach, J. C. Zappala, and R. Purtschert. 2017. Using 81Kr and noble gases to characterize and date groundwater and brines in the Baltic Artesian Basin on the one-million-year timescale. *Geochimica et Cosmochimica Acta* 205:187-210. DOI: 10.1016/j.gca.2017.01.033.

Geyer, R. 2021. Plastic: Too much of a good thing? Presented at *Wallace Stegner Center 26th Annual Symposium—The Plastics Paradox: Societal Boon or Environmental Bane?*. Salt Lake City, UT.

Geyer, R., J. R. Jambeck, and K. L. Law. 2017. Production, use, and fate of all plastics ever made. *Science Advances* 3(7). DOI: 10.1126/sciadv.1700782.

Ghaemmaghami, A. M., M. J. Hancock, H. Harrington, H. Kaji, and A. Khademhosseini. 2012. Biomimetic tissues on a chip for drug discovery. *Drug Discovery Today* 17:173-181. DOI: 10.1016/j.drudis.2011.10.029.

Ghosal, S. 2005. Electron spectroscopy of aqueous solution interfaces reveals surface enhancement of halides. *Science* 307(5709):563-566. DOI: 10.1126/science.1106525.

Glasser, W. G. 2019. About making lignin great again—Some lessons from the past. *Frontiers in Chemistry* 7, Article 565. DOI: 10.3389/fchem.2019.00565.

Glotzer, S. C., and M. J. Solomon. 2007. Anisotropy of building blocks and their assembly into complex structures. *Nature Materials* 6(8):557-562. DOI: 10.1038/nmat1949.

Godlewski, J. E., J. Villalobos, D. I. Collias, J. E. and Velasquez. 2014. Process for production of acrylic acid or its derivatives from hydroxypropionic acid or its derivatives. U.S. Patent No. 8,884,050. Washington, DC: U.S. Patent and Trademark Office.

Gong, K., Y. Cheng, L. L. Daemen, and C. E. White. 2019. In situ quasi-elastic neutron scattering study on the water dynamics and reaction mechanisms in alkali-activated slags. *Physical Chemistry Chemical Physics* 21(20):10277-10292. DOI: 10.1039/c9cp00889f.

Gonzalez-Portillo, L. F., K. Albrecht, and C. K. Ho. 2021. Techno-economic optimization of CSP plants with free-falling particle receivers. *Entropy* 23(1). DOI: 10.3390/e23010076.

Goto, Y., T. Hisatomi, Q. Wang, T. Higashi, K. Ishikiriyama, T. Maeda, Y. Sakata, S. Okunaka, H. Tokudome, M. Katayama, S. Akiyama, H. Nishiyama, Y. Inoue, T. Takewaki, T. Setoyama, T. Minegishi, T. Takata, T. Yamada, and K. Domen. 2018. A particulate photocatalyst water-splitting panel for large-scale solar hydrogen generation. *Joule* 2(3):509-520. DOI: 10.1016/j.joule.2017.12.009.

Gove, J. M., J. L. Whitney, M. A. McManus, J. Lecky, F. C. Carvalho, J. M. Lynch, J. Li, P. Neubauer, K. A. Smith, J. E. Phipps, D. R. Kobayashi, K. B. Balagso, E. A. Contreras, M. E. Manuel, M. A. Merrifield, J. J. Polovina, G. P. Asner, J. A. Maynard, and G. J. Williams. 2019. Prey-size plastics are invading larval fish nurseries. *Proceedings of the National Academy of Sciences* 116(48):24143-24149. DOI: 10.1073/pnas.1907496116.

Goyal, H., A. Mehdad, R. F. Lobo, G. D. Stefanidis, and D. G. Vlachos. 2019. Scaleup of a single-mode microwave reactor. *Industrial & Engineering Chemistry Research* 59(6):2516-2523. DOI: 10.1021/acs.iecr.9b04491.

Grand View Research. 2021. Artificial intelligence market size, share & trends analysis report by solution, by technology (deep learning, machine learning, natural language processing, machine vision), by end use, by region, and segment forecasts, 2021-2028. Retrieved 2021, from https://www.researchandmarkets.com/reports/4375395/global-artificial-intelligence-market-size-share.

Grätzel, M., and J. Milić. 2019. The advent of molecular photovoltaics and hybrid perovskite solar cells. *Substantia* 3(2):27-43. DOI: 10.13128/Substantia-697.

Green, D. W., and M. Z. Southard, eds. 2019. *Perry's chemical engineers' handbook*. New York: McGraw Hill Education.

Greenstein, K. E., N. V. Myung, G. F. Parkin, and D. M. Cwiertny. 2019. Performance comparison of hematite (α-Fe2O3)-polymer composite and core-shell nanofibers as point-of-use filtration platforms for metal sequestration. *Water Research* 148:492-503. DOI: 10.1016/j.watres.2018.10.048.

Gumpertz, M., R. Durodoye, E. Griffith, and A. Wilson. 2017. Retention and promotion of women and underrepresented minority faculty in science and engineering at four large land grant institutions. *PLoS One* 12(11). DOI: 10.1371/journal.pone.0187285.

Gupta, A. 2018. Introduction to deep learning: Part 1. *CEP Magazine*. From https://www.aiche.org/resources/publications/cep/2018/june/introduction-deep-learning-part-1.

Gutowski, T. G., S. Sahni, J. M. Allwood, M. F. Ashby, and E. Worrell. 2013. The energy required to produce materials: Constraints on energy-intensity improvements, parameters of demand. *Philosophical Transactions of the Royal Society A* 371(1986):20120003. DOI: 10.1098/rsta.2012.0003.

Haegel, N. M., H. Atwater, Jr., T. Barnes, C. Breyer, A. Burrell, Y. M. Chiang, S. De Wolf, B. Dimmler, D. Feldman, S. Glunz, J. C. Goldschmidt, D. Hochschild, R. Inzunza, I. Kaizuka, B. Kroposki, S. Kurtz, S. Leu, R. Margolis, K. Matsubara, A. Metz, W. K. Metzger, M. Morjaria, S. Niki, S. Nowak, I. M. Peters, S. Philipps, T. Reindl, A. Richter, D. Rose, K. Sakurai, R. Schlatmann, M. Shikano, W. Sinke, R. Sinton, B. J. Stanbery, M. Topic, W. Tumas, Y. Ueda, J. van de Lagemaat, P. Verlinden, M. Vetter, E. Warren, M. Werner, M. Yamaguchi, and A. W. Bett. 2019. Terawatt-scale photovoltaics: Transform global energy. *Science* 364(6443):836-838. DOI: 10.1126/science.aaw1845.

Hafner, J., C. Wolverton, and G. Ceder. 2006. Toward computational materials design: The impact of density functional theory on materials research. *MRS Bulletin* 31(9):659-668. DOI: 10.1557/mrs2006.174.

Hamilton, S. R., R. C. Davidson, N. Sethuraman, J. H. Nett, Y. Jiang, S. Rios, P. Bobrowicz, T. A. Stadheim, H. Li, B.-K. Choi, D. Hopkins, H. Wischnewski, J. Roser, T. Mitchell, R. R. Strawbridge, J. Hoopes, S. Wildt, and T. U. Gerngross. 2006. Humanization of yeast to produce complex terminally sialylated glycoproteins. *Science* 313(5792):1441-1443. DOI: 10.1126/science.1130256.

Han, K., C. W. Shields Iv, and O. D. Velev. 2018. Engineering of self-propelling microbots and microdevices powered by magnetic and electric fields. *Advanced Functional Materials* 28(25). DOI: 10.1002/adfm.201705953.

Handle, P. H., T. Loerting, and F. Sciortino. 2017. Supercooled and glassy water: Metastable liquid(s), amorphous solid(s), and a no-man's land. *Proceedings of the National Academy of Sciences* 114(51):13336-13344. DOI: 10.1073/pnas.1700103114.

Hanson, M. 2021. Average cost of college & tuition. Retreived 2021, from https://educationdata.org/average-cost-of-college.

Hao, Y., W. Li, X. Zhou, F. Yang, and Z. Qian. 2017. Microneedles-based transdermal drug delivery systems: A review. *Journal of Biomedical Nanotechnology* 13(12):1581-1597. DOI: 10.1166/jbn.2017.2474.

Hardin, B. E., H. J. Snaith, and M. D. McGehee. 2012. The renaissance of dye-sensitized solar cells. *Nature Photonics* 6(3):162-169. DOI: 10.1038/nphoton.2012.22.

Harimoto, T., and T. Danino. 2019. Engineering bacteria for cancer therapy. *Emerging Topics in Life Sciences* 3(5):623-629. DOI: 10.1042/etls20190096.

Harmsen, G. J. 2007. Reactive distillation: The front-runner of industrial process intensification. *Chemical Engineering and Processing: Process Intensification* 46(9):774-780. DOI: 10.1016/j.cep.2007.06.005.

Hart, J., R. M. Machado, H. Withers Jr., S. Lo, E. Cialkowski, K. Jambunathan. 2015. Electrolytic apparatus, system and method for the safe production of nitrogen trifluoride. U.S. Patent No. 8945367. Washington, DC: U.S. Patent and Trademark Office.

Hasanbeigi, A., and C. Springer. 2019. *Deep decarbonization roadmap for the cement and concrete industries in California.* San Fransisco, CA: Global Efficiency Intelligence. https://www.climateworks.org/wp-content/uploads/2019/09/Decarbonization-Roadmap-CA-Cement-Final.pdf.

Häußler, M., M. Eck, D. Rothauer, and S. Mecking. 2021. Closed-loop recycling of polyethylene-like materials. *Nature* 590(7846):423-427. DOI: 10.1038/s41586-020-03149-9.

Hautier, G., A. Jain, and S. P. Ong. 2012. From the computer to the laboratory: Materials discovery and design using first-principles calculations. *Journal of Materials Science* 47(21):7317-7340. DOI: 10.1007/s10853-012-6424-0.

Haywood, J. 2016. Chapter 27—Atmospheric aerosols and their role in climate change. In *Climate Change (Second Edition)*. T. M. Letcher, ed. Boston: Elsevier.

Heinken, A., G. Acharya, D. A. Ravcheev, J. Hertel, M. Nyga, O. E. Okpala, M. Hogan, S. Magnúsdóttir, F. Martinelli, G. Preciat, J. N. Edirisinghe, C. S. Henry, R. M. T. Fleming, and I. Thiele. 2020. AGORA2: Large scale reconstruction of the microbiome highlights wide-spread drug-metabolising capacities. *bioRxiv*. DOI: 10.1101/2020.11.09.375451.

Helgeson, M. E. 2016. Colloidal behavior of nanoemulsions: Interactions, structure, and rheology. *Current Opinion in Colloid & Interface Science* 25:39-50. DOI: 10.1016/j.cocis.2016.06.006.

Heney, P. 2020. Global R&D investments unabated in spending growth. *Global funding forecast.* Cleveland, OH: R&D World. From https://www.rdworldonline.com/global-rd-investments-unabated-in-spending-growth.

Hepburn, C., E. Adlen, J. Beddington, E. A. Carter, S. Fuss, N. Mac Dowell, J. C. Minx, P. Smith, and C. K. Williams. 2019. The technological and economic prospects for CO2 utilization and removal. *Nature* 575(7781):87-97. DOI: 10.1038/s41586-019-1681-6.

Higman, C., and M. van der Burgt. 2008. *Gasification.* Oxford, UK: Gulf Professional Publishing.

Himmelblau, D. M., and K. B. Bischoff. 1968. *Process analysis and simulation: Deterministic systems.* New York: John Wiley & Sons, Inc.

Hoit, M., and M. Ohland. 1998. The impact of a discipline-based introduction to engineering course on improving retention. *Journal of Engineering Education* 87(1):79-85. DOI: 10.1002/j.2168-9830.1998.tb00325.x.

Holt, J. K., H. G. Park, Y. Wang, M. Stadermann, A. B. Artyukhin, C. P. Grigoropoulos, A. Noy, and O. Bakajin. 2006. Fast mass transport through sub-2-nanometer carbon nanotubes. *Science* 312(5776):1034-1037. DOI: 10.1126/science.1126298.

Hoornweg, D., P. Bhada-Tata, and C. Kennedy. 2013. Environment: Waste production must peak this century. *Nature* 502(7473):615-617. DOI: 10.1038/502615a.

Hori, Y. 2008. Electrochemical CO2 Reduction on metal electrodes. In *Modern aspects of electrochemistry*. C. G., Vayenas, R. E. White, and M. E. Gamboa-Aldeco, eds. New York: Springer.

Hougen, O. A., and K. M. Watson. 1943. *Chemical process principles. Part I: Material and energy balances*. New York: John Wiley and Sons, Inc.

Hougen, O. A., and K. M. Watson. 1947a. *Chemical process principles. Part III: Kinetics and catalysis*. New York: John Wiley and Sons, Inc.

Hougen, O. A., and K. M. Watson. 1947b. *Chemical process principles. Part II: Thermodynamics.* New York: John Wiley and Sons, Inc.

Hsieh, C.-T., M.-J. Lee, and H.-m. Lin. 2006. Multiphase equilibria for mixtures containing acetic acid, water, propylene glycol monomethyl ether, and propylene glycol methyl ether acetate. *Industrial & Engineering Chemistry Research* 45(6):2123-2130. DOI: 10.1021/ie051245t.

Hua, L., R. Zangi, and B. J. Berne. 2009. Hydrophobic interactions and dewetting between plates with hydrophobic and hydrophilic domains. *The Journal of Physical Chemistry C* 113(13):5244-5253. DOI: 10.1021/jp8088758.

Hubbs, C. D., C. Li, N. V. Sahinidis, I. E. Grossmann, and J. M. Wassick. 2020. A deep reinforcement learning approach for chemical production scheduling. *Computers & Chemical Engineering* 141. DOI: 10.1016/j.compchemeng.2020.106982.

Hussain, A., Y. D. Chaniago, A. Riaz, and M. Lee. 2019. Process design alternatives for producing ultra-high-purity electronic-grade propylene glycol monomethyl ether acetate. *Industrial & Engineering Chemistry Research* 58(6):2246-2257. DOI: 10.1021/acs.iecr.8b04052.

Hwang, I. Y., H. L. Lee, J. G. Huang, Y. Y. Lim, W. S. Yew, Y. S. Lee, and M. W. Chang. 2018. Engineering microbes for targeted strikes against human pathogens. *Cellular and Molecular Life Sciences* 75(15):2719-2733. DOI: 10.1007/s00018-018-2827-7.

Hydrogen Council. 2019. Hydrogen decarbonization pathways. Retrieved August 17, 2021, from https://hydrogencouncil.com/en/hydrogen-decarbonization-pathways.

IChemE (Institution of Chemical Engineers). 2015. Ten ways chemical engineers can save the world from climate change #COP21. Retrieved January 18, 2021, from https://ichemeblog.org/2015/12/21/ten-ways-chemical-engineers-can-save-the-world-from-climate-change-cop21.

IEA (International Energy Agency). 2012. Technology roadmap—High-efficiency, low-emissions coal-fired power generation. Retrieved August 18, 2021, from https://www.iea.org/reports/technology-roadmap-high-efficiency-low-emissions-coal-fired-power-generation.

IEA. 2019. The future of hydrogen. Retrieved August 16, 2021, from https://www.iea.org/reports/the-future-of-hydrogen.

IEA. 2020a. Global EV outlook 2020. Retrieved August 16, 2021, from https://www.iea.org/reports/global-ev-outlook-2020.

IEA. 2020b. Bio-based chemicals—A 2020 update. Retrieved August 16, 2021, from https://www.ieabioenergy.com/blog/publications/new-publication-bio-based-chemicals-a-2020-update.

IEA. 2020c. Coal-fired power. https://prod.iea.org/reports/coal-fired-power.

IEA. 2020d. Energy technology perspectives 2020. Retrieved August 16, 2021, from https://www.iea.org/reports/energy-technology-perspectives-2020.

IEA. 2020e. Global hydrogen demand by sector in the Sustainable Development Scenario, 2019-2070. Retrieved August 16, 2021, from https://www.iea.org/data-and-statistics/charts/global-hydrogen-demand-by-sector-in-the-sustainable-development-scenario-2019-2070.

IEA. 2021a. Transport. Retrieved August 16, 2021, from https://www.iea.org/topics/transport.

IEA. 2021b. Net zero by 2050: A roadmap for the global energy sector. Retrieved August 16, 2021, from https://www.iea.org/reports/net-zero-by-2050.

IEA. 2021c. Coal Information: Overview. Retrieved August 9, 2021, from https://www.iea.org/reports/coal-information-overview.

IEA. 2021d. Natural gas information: Overview. Retrieved August 9, 2021, from https://www.iea.org/reports/natural-gas-information-overview.

IEA. 2021e. World energy balances: Overview. Retrieved August 16, 2021, from https://www.iea.org/reports/world-energy-balances-overview.

IEA, ICCA, and DECHEMA (International Council of Chemical Associations, and German Society for Chemical Engineering and Biotechnology). 2013. Technology roadmap: Energy and GHG Reductions in the chemical industry via catalytic processes. https://iea.blob.core.windows.net/assets/d0f7ff3a-0612-422d-ad7d-a682091cb500/TechnologyRoadmapEnergyandGHGReductionsintheChemicalIndustryviaCatalyticProcesses.pdf.

IndustryARC. 2021. Protein therapeutics market—Forecast (2021–2026). Retrieved August 17, 2021, from https://www.industryarc.com/Report/16207/protein-therapeutics-market.html.

IRENA (International Renewable Energy Agency). 2018. Hydrogen from renewable power technology outlook for the energy transition. Retrieved August 18, 2021, from https://www.irena.org/publications/2018/sep/hydrogen-from-renewable-power.

IRENA. 2020. Innovation outlook: Ocean energy technologies. Retrieved August 18, 2021, from https://www.irena.org/publications/2020/Dec/Innovation-Outlook-Ocean-Energy-Technologies.

IRENA. 2021. World energy transitions outlook: 1.5°C pathway. Retrieved August 16, 2021, from https://irena.org/publications/2021/Jun/World-Energy-Transitions-Outlook.

Irwin, D. A., and P. J. Klenow. 1996. High-tech R&D subsidies estimating the effects of Sematech. *Journal of International Economics* 40(3):323-344. DOI: 10.1016/0022-1996(95)01408-X.

ISO (International Standards Organization). 2006. Environmental management—Life cycle assessment—Principles and framework. https://www.iso.org/standard/37456.html.

Iulianelli, A., S. Liguori, J. Wilcox, and A. Basile. 2016. Advances on methane steam reforming to produce hydrogen through membrane reactors technology: A review. *Catalysis Reviews* 58(1):1-35. DOI: 10.1080/01614940.2015.1099882.

Iyare, P. U., S. K. Ouki, and T. Bond. 2020. Microplastics removal in wastewater treatment plants: A critical review. *Environmental Science: Water Research & Technology* 6(10):2664-2675. DOI: 10.1039/d0ew00397b.

Jagschies, G. 2020. Hierarchy of high impact improvements in biomanufacturing. Presented at *Workshop on Innovations in Pharmaceutical Manufacturing, National Academies of Sciences, Engineering, and Medicine*. Washington, DC.

Jain, A., S. P. Ong, G. Hautier, W. Chen, W. D. Richards, S. Dacek, S. Cholia, D. Gunter, D. Skinner, G. Ceder, and K. A. Persson. 2013. The materials project: A materials genome approach to accelerating materials innovation. *APL Materials* 1. DOI: 10.1063/1.481 2323.

Jama-Rodzenska, A., A. Bialowiec, J. A. Koziel, and J. Sowinski. 2021. Waste to phosphorus: A transdisciplinary solution to P recovery from wastewater based on the TRIZ approach. *Journal of Environmental Economics and Management* 287. DOI: 10.1016/j.jenvman.2021.112235.

Jambeck, J. R., R. Geyer, C. Wilcox, T. R. Siegler, M. Perryman, A. Andrady, R. Narayan, and K. L. Law. 2015. Plastic waste inputs from land into the ocean. *Science* 347(6223):768-771. DOI: 10.1126/science.1260352.

Janes, K. A., P. L. Chandran, R. M. Ford, M. J. Lazzara, J. A. Papin, S. M. Peirce, J. J. Saucerman, and D. A. Lauffenburger. 2017. An engineering design approach to systems biology. *Integrative Biology* 9(7):574-583. DOI: 10.1039/c7ib00014f.

Jayapal, K. P., K. F. Wlaschin, W. S. Hu, and M. G. S. Yap. 2007. Recombinant protein therapeutics from CHO cells—20 years and counting. *Chemical Engineering Progress* 103(10):40-47.

Jiang, H., A. A. Horwitz, C. Wright, A. Tai, E. A. Znameroski, Y. Tsegaye, H. Warbington, B. S. Bower, C. Alves, C. Co, K. Jonnalagadda, D. Platt, J. M. Walter, V. Natarajan, J. A. Ubersax, J. R. Cherry, and J. C. Love. 2019. Challenging the workhorse: Comparative analysis of eukaryotic micro-organisms for expressing monoclonal antibodies. *Biotechnology and Bioengineering* 116(6):1449-1462. DOI: 10.1002/bit.26951.

Johnson, E. F. 1967. *Automatic process control*. New York: McGraw Hill.

Jones, P. T., P. H. Dear, J. Foote, M. S. Neuberger, and G. Winter. 1986. Replacing the complementarity-determining regions in a human antibody with those from a mouse. *Nature* 321(6069):522-525. DOI: 10.1038/321522a0.

Jungwirth, P., and D. J. Tobias. 2006. Specific ion effects at the air/water interface. *Chemical Reviews* 106(4):1259-1281. DOI: 10.1021/cr0403741.

Jyothi, R. K., T. Thenepalli, J. W. Ahn, P. K. Parhi, K. W. Chung, and J.-Y. Lee. 2020. Review of rare earth elements recovery from secondary resources for clean energy technologies: Grand opportunities to create wealth from waste. *Journal of Cleaner Production* 267. DOI: 10.1016/j.jclepro.2020.122048.

Kamm, R. D., R. Bashir, N. Arora, R. D. Dar, M. U. Gillette, L. G. Griffith, M. L. Kemp, K. Kinlaw, M. Levin, A. C. Martin, T. C. McDevitt, R. M. Nerem, M. J. Powers, T. A. Saif, J. Sharpe, S. Takayama, S. Takeuchi, R. Weiss, K. Ye, H. G. Yevick, and M. H. Zaman.

2018. Perspective: The promise of multi-cellular engineered living systems. *APL Bioengineering* 2(4). DOI: 10.1063/1.5038337.

Kang, F., D. Wang, Y. Pu, X.-F. Zeng, J.-X. Wang, and J.-F. Chen. 2018. Efficient preparation of monodisperse CaCO3 nanoparticles as overbased nanodetergents in a high-gravity rotating packed bed reactor. *Powder Technology* 325:405-411. DOI: 10.1016/j.powtec.2017.11.036.

Kantor, E. D., C. D. Rehm, J. S. Haas, A. T. Chan, and E. L. Giovannucci. 2015. Trends in prescription drug use among adults in the United States From 1999-2012. *The Journal of the American Medical Association* 314(17):1818-1830. DOI: 10.1001/jama.2015.13766.

Kapoor, R. 2012. Collaborative innovation in the global semiconductor industry: A report on the findings from the 2010 Wharton-GSA semiconductor ecosystem survey. https://faculty.wharton.upenn.edu/wp-content/uploads/2012/05/SemiconductorEcosystemStudyFinal.pdf.

Kapteijn, F., and J. A. Moulijn. 2020. Structured catalysts and reactors—Perspectives for demanding applications. *Catalysis Today* 383:5-14. DOI: 10.1016/j.cattod.2020.09.026.

Karabasz, A., M. Bzowska, and K. Szczepanowicz. 2020. Biomedical applications of multifunctional polymeric nanocarriers: A review of current literature. *International Journal of Nanomedicine* 15:8673-8696. DOI: 10.2147/IJN.S231477.

Karsch-Mizrachi, I., Y. Nakamura, and G. Cochrane. 2012. The international nucleotide sequence database collaboration. *Nucleic Acids Research* 40(D1):D33-37. DOI: 10.1093/nar/gkr1006.

Kass, I., C. F. Reboul, and A. M. Buckle. 2011. Chapter 14—Computational methods for studying serpin conformational change and structural plasticity. In *Methods in enzymology*. J.C. Whisstock and P. I. Bird, eds. San Diego, CA: Academic Press.

Kastner, M. 2018. Philanthropy: A critical player in supporting scientific research. *Science Philanthropy Alliance News.* https://sciencephilanthropyalliance.org/philanthropy-a-critical-player-in-supporting-scientific-research-alliance-blog.

Katz, D. L. 1966. *Computers in engineering design education.* Ann Arbor, MI: University of Michigan.

Kaya, M. 2016. Recovery of metals and nonmetals from electronic waste by physical and chemical recycling processes. *Waste Management* 57:64-90. DOI: 10.1016/j.wasman.2016.08.004.

Kaz, D. M., R. McGorty, M. Mani, M. P. Brenner, and V. N. Manoharan. 2012. Physical ageing of the contact line on colloidal particles at liquid interfaces. *Nature Materials* 11(2):138-142. DOI: 10.1038/nmat3190.

Kaza, S., L. Yao, P. Bhada-Tata, and F. Van Woerden. 2018. *What a waste 2.0—A global snapshot of solid waste management to 2050*. Washington, DC: The World Bank.

Kekicheff, P., J. Iss, P. Fontaine, and A. Johner. 2018. Direct measurement of lateral correlations under controlled nanoconfinement. *Physical Review Letters* 120(11). DOI: 10.1103/PhysRevLett.120.118001.

Kellett, P. J., and D. I. Collias. 2016. Catalysts and processes for the production of aromatic compounds from lignin. U.S. Patent No. 9452422. Washington, DC: U.S. Patent and Trademark Office.

Kevrekidis, I. G., and G. Samaey. 2009. Equation-free multiscale computation: Algorithms and applications. *Annual Review of Physical Chemistry* 60(1):321-344. DOI: 10.1146/annurev.physchem.59.032607.093610.

Khalilpour, R., and I. A. Karimi. 2012. Evaluation of utilization alternatives for stranded natural gas. *Energy* 40(1):317-328. DOI: 10.1016/j.energy.2012.01.068.

Kim, A. J., P. L. Biancaniello, and J. C. Crocker. 2006. Engineering DNA-mediated colloidal crystallization. *Langmuir* 22(5):1991-2001. DOI: 10.1021/la0528955.

Kim, S. B., J. C. Palmer, and P. G. Debenedetti. 2015. A computational study of the effect of matrix structural order on water sorption by TRP-cage miniproteins. *The Journal of Physical Chemistry B* 119(5):1847-1856. DOI: 10.1021/jp510172w.

Kim, T. H., Y. Wang, C. R. Oliver, D. H. Thamm, L. Cooling, C. Paoletti, K. J. Smith, S. Nagrath, and D. F. Hayes. 2019. A temporary indwelling intravascular aphaeretic system for in vivo enrichment of circulating tumor cells. *Nature Communications* 10(1). DOI: 10.1038/s41467-019-09439-9.

Kinney, M. A., L. T. Vo, J. M. Frame, J. Barragan, A. J. Conway, S. Li, K. K. Wong, J. J. Collins, P. Cahan, T. E. North, D. A. Lauffenburger, and G. Q. Daley. 2019. A systems biology pipeline identifies regulatory networks for stem cell engineering. *Nature Biotechnology* 37(7):810-818. DOI: 10.1038/s41587-019-0159-2.

Kleiner, L. W., J. C. Wright, and Y. Wang. 2014. Evolution of implantable and insertable drug delivery systems. *Journal of Controlled Release* 181:1-10. DOI: 10.1016/j.jconrel.2014.02.006.

Knight, D., J. Sullivan, and L. Carlson. 2003. Staying in engineering: Effects of a hands on, team based, first year projects course on student retention. Presented at *American Society for Engineering Education Annual Conference, 2003*. Nashville, TN.

Kofke, D. A. 1993. Direct evaluation of phase coexistence by molecular simulation via integration along the saturation line. *The Journal of Chemical Physics* 98(5):4149-4162. DOI: 10.1063/1.465023.

König, A., W. Marquardt, A. Mitsos, J. Viell, and M. Dahmen. 2020. Integrated design of renewable fuels and their production processes: Recent advances and challenges. *Current Opinion in Chemical Engineering* 27:45-50. DOI: 10.1016/j.coche.2019.11.001.

Koros, W. J., and R. P. Lively. 2012. Water and beyond: Expanding the spectrum of large-scale energy efficient separation processes. *American Institute of Chemical Engineers Journal* 58(9):2624-2633. DOI: 10.1002/aic.13888.

Koros, W. J., and C. Zhang. 2017. Materials for next-generation molecularly selective synthetic membranes. *Nature Materials* 16(3):289-297. DOI: 10.1038/nmat4805.

Kriegler, E., J. P. Weyant, G. J. Blanford, V. Krey, L. Clarke, J. Edmonds, A. Fawcett, G. Luderer, K. Riahi, R. Richels, S. K. Rose, M. Tavoni, and D. P. van Vuuren. 2014. The role of technology for achieving climate policy objectives: Overview of the EMF 27 study on global technology and climate policy strategies. *Climatic Change* 123(3):353-367. DOI: 10.1007/s10584-013-0953-7.

Kromdijk, J., K. Głowacka, L. Leonelli, S. T. Gabilly, M. Iwai, K. K. Niyogi, and S. P. Long. 2016. Improving photosynthesis and crop productivity by accelerating recovery from photoprotection. *Science* 354(6314):857-861. DOI: 10.1126/science.aai8878.

Krouse, S., R. M. Machado, J. Hart, and J. Nehlsen. 2016. Electrolytic apparatus, system, and method for the efficient production of nitrogen trifluoride. U.S. Patent 9528191. Washington, DC: U.S. Patent and Trademark Office.

Kumar, A., H. Fukuda, T. A. Hatton, and J. H. Lienhard. 2019. Lithium recovery from oil and gas produced water: A need for a growing energy industry. *ACS Energy Letters* 4(6):1471-1474. DOI: 10.1021/acsenergylett.9b00779.

Lachance, J. C., D. Matteau, J. Brodeur, C. J. Lloyd, N. Mih, Z. A. King, T. F. Knight, A. M. Feist, J. M. Monk, B. O. Palsson, P. E. Jacques, and S. Rodrigue. 2021. Genome-scale metabolic modeling reveals key features of a minimal gene set. *Molecular Systems Biology* 17(7). DOI: 10.15252/msb.202010099.

Landsman, M. R., R. Sujanani, S. H. Brodfuehrer, C. M. Cooper, A. G. Darr, R. J. Davis, K. Kim, S. Kum, L. K. Nalley, S. M. Nomaan, C. P. Oden, A. Paspureddi, K. K. Reimund, L. S. Rowles 3rd, S. Yeo, D. F. Lawler, B. D. Freeman, and L. E. Katz. 2020. Water treatment:

Are membranes the panacea? *Annual Review of Chemical and Biomolecular Engineering* 11:559-585. DOI: 10.1146/annurev-chembioeng-111919-091940.

Langhoff, S. R., ed. 1995. *Quantum mechanical electronic structure calculations with chemical accuracy*. Dordrecht, Netherlands: Springer.

Lapidus, L. 1962. *Digital computation for chemical engineers*. New York: McGraw-Hill.

Lapointe, M., J. M. Farner, L. M. Hernandez, and N. Tufenkji. 2020. Understanding and improving microplastic removal during water treatment: Impact of coagulation and flocculation. *Environmental Science & Technology* 54(14):8719-8727. DOI: 10.1021/acs.est.0c00712.

Laramy, C. R., M. N. O'Brien, and C. A. Mirkin. 2019. Crystal engineering with DNA. *Nature Reviews Materials* 4(3):201-224. DOI: 10.1038/s41578-019-0087-2.

Larson, R. G. 1999. *The structure and rheology of complex fluids*. New York: Oxford University Press.

Layman, J. M., D. I. Collias, H. Schonemann, and K. Williams. 2019a. Method for purifying reclaimed polypropylene. U.S. Patent No. 10450436. Washington, DC: U.S. Patent and Trademark Office.

Layman, J. M., D. I. Collias, H. Schonemann, and K. Williams. 2019b. Method for purifying reclaimed polymers. U.S. Patent No. 10465058. Washington, DC:U.S. Patent and Trademark Office.

Lazard. 2019. Lazard's levelized cost of energy analysis—Version 13.0. Retrieved August 18, 2021, from https://www.lazard.com/media/451086/lazards-levelized-cost-of-energy-version-130-vf.pdf.

Leader, B., Q. J. Baca, and D. E. Golan. 2008. Protein therapeutics: A summary and pharmacological classification. *Nature Reviews Drug Discovery* 7(1):21-39. DOI: 10. 1038/nrd2399.

Lebreton, L., B. Slat, F. Ferrari, B. Sainte-Rose, J. Aitken, R. Marthouse, S. Hajbane, S. Cunsolo, A. Schwarz, A. Levivier, K. Noble, P. Debeljak, H. Maral, R. Schoeneich-Argent, R. Brambini, and J. Reisser. 2018. Evidence that the Great Pacific Garbage Patch is rapidly accumulating plastic. *Scientific Reports* 8(1). DOI: 10.1038/s41598-018-22939-w.

Lenaburg, L., O. Aguirre, F. Goodchild, and J.-U. Kuhn. 2012. Expanding pathways: A summer bridge program for community college STEM students. *Community College Journal of Research and Practice* 36(3):153-168. DOI: 10.1080/10668921003609210.

Leslie, S.-J., A. Cimpian, M. Meyer, and E. Freeland. 2015. Expectations of brilliance underlie gender distributions across academic disciplines. *Science* 347(6219):262-265. DOI: 10.1126/science.1261375.

Lewis, N. S. 2016. Research opportunities to advance solar energy utilization. *Science* 351(6271). DOI: 10.1126/science.aad1920.

Li, Z., T. R. Klein, D. H. Kim, M. Yang, J. J. Berry, M. F. A. M. van Hest, and K. Zhu. 2018. Scalable fabrication of perovskite solar cells. *Nature Reviews Materials* 3(4). DOI: 10.1038/natrevmats.2018.17.

Li, L., J. Zhong, Y. Yan, J. Zhang, J. Xu, J. S. Francisco, and X. C. Zeng. 2020a. Unraveling nucleation pathway in methane clathrate formation. *Proceedings of the National Academy of Sciences* 117(40):24701-24708. DOI: 10.1073/pnas.2011755117.

Li, Y., Y. Sun, Y. Qin, W. Zhang, L. Wang, M. Luo, H. Yang, and S. Guo. 2020b. Recent advances on water-splitting electrocatalysis mediated by noble-metal-based nanostructured materials. *Advanced Energy Materials* 10(11). DOI: 10.1002/aenm.201903120.

Li, X., J. Peoples, P. Yao, and X. Ruan. 2021a. Ultrawhite BaSO4 paints and films for remarkable daytime subambient radiative cooling. *ACS Applied Materials & Interfaces* 13(18): 21733-21739. DOI: 10.1021/acsami.1c02368.

Li, C., D. L. Ramasamy, M. Sillanpää, and E. Repo. 2021b. Separation and concentration of rare earth elements from wastewater using electrodialysis technology. *Separation and Purification Technology* 254, Article 117442. DOI: 10.1016/j.seppur.2020.117442.

Liang, R., H. Xu, Y. Shen, S. Sun, J. Xu, S. Meng, Y. R. Shen, and C. Tian. 2019. Nucleation and dissociation of methane clathrate embryo at the gas-water interface. *Proceedings of the National Academy of Sciences* 116(47):23410-23415. DOI: 10.1073/pnas.1912592116.

Lin, H., S. Lee, L. Sun, M. Spellings, M. Engel, C. Glotzer Sharon, and A. Mirkin Chad. 2017. Clathrate colloidal crystals. *Science* 355(6328):931-935. DOI: 10.1126/science.aal3919.

Lipani, L., B. G. R. Dupont, F. Doungmene, F. Marken, R. M. Tyrrell, R. H. Guy, and A. Ilie. 2018. Non-invasive, transdermal, path-selective and specific glucose monitoring via a graphene-based platform. *Nature Nanotechnology* 13(6):504-511. DOI: 10.1038/s41565-018-0112-4.

Litzler, E., and J. Young. 2012. Understanding the risk of attrition in undergraduate engineering: Results from the project to assess climate in engineering. *Journal of Engineering Education* 101(2):319-345. DOI: 10.1002/j.2168-9830.2012.tb00052.x.

Logar, N., L. D. Anadon, and V. Narayanamurti. 2014. Semiconductor research corporation: A case study in cooperative innovation partnerships. *Minerva* 52(2):237-261. DOI: 10.1007/s11024-014-9253-2.

Low, L. A., C. Mummery, B. R. Berridge, C. P. Austin, and D. A. Tagle. 2021. Organs-on-chips: Into the next decade. *Nature Reviews Drug Discovery* 20(5):345-361. DOI: 10.1038/s41573-020-0079-3.

Lu, R.-M., Y.-C. Hwang, I. J. Liu, C.-C. Lee, H.-Z. Tsai, H.-J. Li, and H.-C. Wu. 2020. Development of therapeutic antibodies for the treatment of diseases. *Journal of Biomedical Science* 27(1). DOI: 10.1186/s12929-019-0592-z.

Luo, Y., P. R. Westmoreland, D. Alkaya, R.V. Alves da Cruz, I.E. Grossmann, W.D. Provine, D.L. Silverstein, R.J. Steininger II, J.B. Talbot, A. Varma, T. McCreight, K. Chin, D. Schuster. 2015. *Chemical engineering academia-industry alignment: Expectations about new graduates*. American Institute of Chemical Engineers. https://www.aiche.org/sites/default/files/docs/conferences/2015che_academicindustryalignmentstudy.compressed.pdf.

Luo, X., B. Guo, J. Luo, F. Deng, S. Zhang, S. Luo, and J. Crittenden. 2015. Recovery of lithium from wastewater using development of li ion-imprinted polymers. *ACS Sustainable Chemistry & Engineering* 3(3):460-467. DOI: 10.1021/sc500659h.

Luo, S., T. Li, X. Wang, M. Faizan, and L. Zhang. 2020. High-throughput computational materials screening and discovery of optoelectronic semiconductors. *WIREs Computational Molecular Science* 11(1). DOI: 10.1002/wcms.1489.

Lutz, J.-F., M. Ouchi, D. R. Liu, and M. Sawamoto. 2013. Sequence-controlled polymers. *Science* 341(6146). DOI: 10.1126/science.1238149.

Lynd, L. R., J. H. Cushman, R. J. Nichols, and C. E. Wyman. 1991. Fuel ethanol from cellulosic biomass. *Science* 251(4999):1318-1323. DOI: 10.1126/science.251.4999.1318.

Ma, C. D., C. Wang, C. Acevedo-Velez, S. H. Gellman, and N. L. Abbott. 2015. Modulation of hydrophobic interactions by proximally immobilized ions. *Nature* 517(7534):347-350. DOI: 10.1038/nature14018.

Machado, R. M., and R. R. Broekhuis. 2003. Gas-liquid reaction process including ejector and monolith catalyst. U.S. Patent No. 6506361. Washington, DC: U.S. Patent and Trademark Office.

Machado, R. M., R. R. Broekhuis, A. F. Nordquist, B. P. Roy, and S. R. Carney. 2005. Applying monolith reactors for hydrogenations in the production of specialty chemicals—Process and economic considerations. *Catalysis Today* 105:305-317. DOI: 10.1016/j.cattod.2005.06.036.

Macrotrends. 2021a. Exxon research and development expenses 2006-2021. Retrieved December 3, 2021, from https://www.macrotrends.net/stocks/charts/XOM/exxon/research-development-expenses.

Macrotrends. 2021b. Corteva research and development expenses 2018-2021. Retrieved December 3, 2021, from https://www.macrotrends.net/stocks/charts/CTVA/corteva/research-development-expenses.

Maeurer, A., M. Schlummer, and O. Beck. 2012. Method for recycling plastic materials and use thereof. U.S. Patent No. 8138232. Washington, DC: U.S. Patent and Trademark Office.

Magnusdottir, S., A. Heinken, L. Kutt, D. A. Ravcheev, E. Bauer, A. Noronha, K. Greenhalgh, C. Jager, J. Baginska, P. Wilmes, R. M. Fleming, and I. Thiele. 2017. Generation of genome-scale metabolic reconstructions for 773 members of the human gut microbiota. *Nature Biotechnology* 35(1):81-89. DOI: 10.1038/nbt.3703.

Mahmood, A., M. Eqan, S. Pervez, H. A. Alghamdi, A. B. Tabinda, A. Yasar, K. Brindhadevi, and A. Pugazhendhi. 2020. COVID-19 and frequent use of hand sanitizers: Human health and environmental hazards by exposure pathways. *Science of the Total Environment* 742. DOI: 10.1016/j.scitotenv.2020.140561.

Makurvet, F. D. 2021. Biologics vs. small molecules: Drug costs and patient access. *Medicine in Drug Discovery* 9. DOI: 10.1016/j.medidd.2020.100075.

Malcom, S. M. 1996. Science and diversity: A compelling national interest. *Science.* 271(5257):1817-1819. http://www.jstor.org/stable/2889362.

Malcom, S., and L. Malcom-Piqueux. 2020. Institutional transformation: Supporting equity and excellence in STEMM. *Change: The Magazine of Higher Learning* 52(2):79-82. DOI: 10.1080/00091383.2020.1732792.

Mandal, J., Y. Fu, A. C. Overvig, M. Jia, K. Sun, N. N. Shi, H. Zhou, X. Xiao, N. Yu, and Y. Yang. 2018. Hierarchically porous polymer coatings for highly efficient passive daytime radiative cooling. *Science* 362(6412):315-319. DOI: 10.1126/science.aat9513.

Marbach, S., D. S. Dean, and L. Bocquet. 2018. Transport and dispersion across wiggling nanopores. *Nature Physics* 14(11):1108-1113. DOI: 10.1038/s41567-018-0239-0.

Marchetti, M. C., J. F. Joanny, S. Ramaswamy, T. B. Liverpool, J. Prost, M. Rao, and R. A. Simha. 2013. Hydrodynamics of soft active matter. *Reviews of Modern Physics* 85(3):1143-1189. DOI: 10.1103/RevModPhys.85.1143.

Marshall, J. 2007. Who needs oil? *New Scientist* 195(2611):28-31. DOI: https://doi.org/10.1016/S0262-4079(07)61712-6.

Martin, A. B., M. Hartman, D. Lassman, A. Catlin, and T. National Health Expenditure Accounts. 2021. National health care spending in 2019: Steady growth for the fourth consecutive year. *Health Affairs* 40(1):14-24. DOI: 10.1377/hlthaff.2020.02022.

Masias, A., J. Marcicki, and W. A. Paxton. 2021. Opportunities and challenges of lithium ion batteries in automotive applications. *ACS Energy Letters* 6(2):621-630. DOI: 10.1021/acsenergylett.0c02584.

Matthews, A. A., P. L. R. Ee, and R. Ge. 2020. Developing inhaled protein therapeutics for lung diseases. *Molecular Biomedicine* 1. DOI: 10.1186/s43556-020-00014-z.

Matthews, C. B., C. Wright, A. Kuo, N. Colant, M. Westoby, and J. C. Love. 2017. Reexamining opportunities for therapeutic protein production in eukaryotic microorganisms. *Biotechnology and Bioengineering* 114(11):2432-2444. DOI: 10.1002/bit.26378.

Maxson, A., and J. Phillips. 2011. Research and development for future coal generation. Retrieved August 16, 2021, from https://www.powermag.com/research-and-development-for-future-coal-generation.

Maziarka, Ł., A. Pocha, J. Kaczmarczyk, K. Rataj, T. Danel, and M. Warchoł. 2020. Mol-CycleGAN: A generative model for molecular optimization. *Journal of Cheminformatics* 12(1). DOI: 10.1186/s13321-019-0404-1.

Mbow, C., C. Rosenzweig, L.G. Barioni, T.G. Benton, M. Herrero, M. Krishnapillai, E. Liwenga, P. Pradhan, M.G. Rivera-Ferre,, and F. N. T. T. Sapkota, and Y. Xu. 2019. Food security. In *Climate change and land: An IPCC special report on climate change, desertification, land degradation, sustainable land management, food security, and greenhouse gas fluxes in terrestrial ecosystems*. P.R. Shukla, J. S., E. Calvo Buendia, V. Masson-Delmotte, H.-O. Pörtner, D.C. Roberts, P. Zhai, R. Slade, S. Connors, R. van Diemen, M. Ferrat, E. Haughey, S. Luz, S. Neogi, M. Pathak, J. Petzold, J. Portugal Pereira, P. Vyas, E. Huntley, K. Kissick, M. Belkacemi, J. Malley, eds. Geneva: Intergovernmental Panel on Climate Change.

McBride, J., and A. Chatzky. 2019. Is 'Made in China 2025' a threat to global trade? Retrieved September 13, 2021, from https://www.cfr.org/backgrounder/made-china-2025-threat-global-trade.

McCarty, N. S., and R. Ledesma-Amaro. 2019. Synthetic biology tools to engineer microbial communities for biotechnology. *Trends in Biotechnology* 37(2):181-197. DOI: 10.1016/j.tibtech.2018.11.002.

McCullough, M. B. A., and K. Williams. 2018. STEM researchers are needed to advance multi-level interventions for health disparities. *Journal of Public Health Policy and Planning* 2(1):71-73. https://www.alliedacademies.org/download.php?download=articles/stem-researchers-are-needed-to-advance-multilevel-interventions-for-health-disparities.pdf.

McGann, P. T., and C. Hoppe. 2017. The pressing need for point-of-care diagnostics for sickle cell disease: A review of current and future technologies. *Blood Cells, Molecules and Diseases* 67:104-113. DOI: 10.1016/j.bcmd.2017.08.010.

McGee, E. O. 2020. Interrogating structural racism in STEM higher education. *Educational Researcher* 49(9):633-644. DOI: 10.3102/0013189X20972718.

McKinsey Sustainability. 2016. The circular economy: Moving from theory to practice. Retrieved August 18, 2021, from https://www.mckinsey.com/business-functions/sustainability/our-insights/the-circular-economy-moving-from-theory-to-practice.

McMichael, S., P. Fernández-Ibáñez, and J. A. Byrne. 2021. A review of photoelectrocatalytic reactors for water and wastewater treatment. *Water* 13(9). DOI: 10.3390/w13091198.

McMurtrie, B. 2018. This is what Georgia Tech thinks college will look like in 2040. *The Chronicle of Higher Education*. https://www.chronicle.com/article/this-is-what-georgia-tech-thinks-college-will-look-like-in-2040.

McNeill, V. F. 2020. COVID-19 and the air we breathe. *ACS Earth and Space Chemistry* 4(5):674-675. DOI: 10.1021/acsearthspacechem.0c00093.

Mester, Z., and A. Z. Panagiotopoulos. 2015. Temperature-dependent solubilities and mean ionic activity coefficients of alkali halides in water from molecular dynamics simulations. *The Journal of Chemical Physics* 143(4). DOI: 10.1063/1.4926840.

Miandad, R., M. A. Barakat, A. S. Aburiazaiza, M. Rehan, and A. S. Nizami. 2016. Catalytic pyrolysis of plastic waste: A review. *Process Safety and Environmental Protection* 102:822-838. DOI: 10.1016/j.psep.2016.06.022.

Middleton, C. T., P. Marek, P. Cao, C. C. Chiu, S. Singh, A. M. Woys, J. J. de Pablo, D. P. Raleigh, and M. T. Zanni. 2012. Two-dimensional infrared spectroscopy reveals the complex behaviour of an amyloid fibril inhibitor. *Nature Chemistry* 4(5):355-360. DOI: 10.1038/nchem.1293.

Milbrandt, A., T. Seiple, D. Heimiller, R. Skaggs, and A. Coleman. 2018. Wet waste-to-energy resources in the United States. *Resources, Conservation and Recycling* 137:32-47. DOI: 10.1016/j.resconrec.2018.05.023.

Miller, K. K., and H. S. Alper. 2019. Yarrowia lipolytica: More than an oleaginous workhorse. *Applied Microbiology and Biotechnology* 103:9251-9262. DOI: 10.1007/s00253-019-10200-x.

MIT (Massachusetts Institute of Technology). 2021. Department of chemical engineering—History. Retrieved August 17, 2021, from https://cheme.mit.edu/about/history.

Mitchell, M. J., M. M. Billingsley, R. M. Haley, M. E. Wechsler, N. A. Peppas, and R. Langer. 2021. Engineering precision nanoparticles for drug delivery. *Nature Reviews Drug Discovery* 20(2):101-124. DOI: 10.1038/s41573-020-0090-8.

Monroe, J., M. Barry, A. DeStefano, P. Aydogan Gokturk, S. Jiao, D. Robinson-Brown, T. Webber, E. J. Crumlin, S. Han, and M. S. Shell. 2020. Water structure and properties at hydrophilic and hydrophobic surfaces. *Annual Review of Chemical and Biomolecular Engineering* 11:523-557. DOI: 10.1146/annurev-chembioeng-120919-114657.

Montoya, J. H., L. C. Seitz, P. Chakthranont, A. Vojvodic, T. F. Jaramillo, and J. K. Nørskov. 2017. Materials for solar fuels and chemicals. *Nature Materials* 16(1):70-81. DOI: 10.1038/nmat4778.

Morishita, M., and N. A. Peppas. 2006. Is the oral route possible for peptide and protein drug delivery? *Drug Discovery Today* 11:905-910. DOI: 10.1016/j.drudis.2006.08.005.

Morrow, W. R., J. Marano, A. Hasanbeigi, E. Masanet, and J. Sathaye. 2015. Efficiency improvement and CO2 emission reduction potentials in the United States petroleum refining industry. *Energy* 93:95-105. DOI: 10.1016/j.energy.2015.08.097.

Moss, B., O. Babacan, A. Kafizas, and A. Hankin. 2021. A review of inorganic photoelectrode developments and reactor scale-up challenges for solar hydrogen production. *Advanced Energy Materials* 11(13). DOI: 10.1002/aenm.202003286.

Moya, X., and N. D. Mathur. 2020. Caloric materials for cooling and heating. *Science* 370(6518):797-803. DOI: 10.1126/science.abb0973.

Mozur, P., and S. L. Myers. 2021. Xi's gambit: China plans for a world without American technology. *New York Times*. https://www.nytimes.com/2021/03/10/business/china-us-tech-rivalry.html.

Mullin, R. 2021. Cell and gene therapy: The next frontier in pharmaceutical services. *Chemical and Engineering News Digital Magazine* 99(14). https://cen.acs.org/business/outsourcing/Cell-and-gene-therapy-The-next-frontier-in-pharmaceutical-services/99/i14.

Muralikrishna, I. V., and V. Manickam. 2017. Chapter 5—Life cycle assessment. In *Environmental management*. Oxford: Butterworth-Heinemann.

Murphy, S. V., and A. Atala. 2014. 3D bioprinting of tissues and organs. *Nature Biotechnology* 32(8):773-785. DOI: 10.1038/nbt.2958.

NAE (National Academy of Engineering). 2008. *Changing the conversation: Messages for improving public understanding of engineering*. Washington, DC: The National Academies Press. DOI: 10.17226/12187.

NAE. 2018. *Understanding the educational and career pathways of engineers*. Washington, DC: The National Academies Press. DOI: 10.17226/25284.

Naidu, G., S. Ryu, R. Thiruvenkatachari, Y. Choi, S. Jeong, and S. Vigneswaran. 2019. A critical review on remediation, reuse, and resource recovery from acid mine drainage. *Environmental Pollution* 247:1110-1124. DOI: 10.1016/j.envpol.2019.01.085.

Napoli, M. T., E. Sciaky, D. J. Arya, and N. Balos. 2017. PIPELINES: Fostering university-community college partnerships and STEM professional success for underrepresented populations. Presented at *ASEE Annual Conference & Exposition*. Columbus, Ohio.

NASEM (National Academies of Sciences, Engineering, and Medicine). 2016. *Barriers and opportunities for 2-year and 4-year STEM degrees: Systemic change to support students' diverse pathways*. Washington, DC: The National Academies Press. DOI: 10.17226/21739.

NASEM. 2017. *Communities in action: Pathways to health equity*. Washington, DC: The National Academies Press. DOI: 10.17226/24624.

NASEM. 2018. *Sexual harassment of women: Climate, culture, and consequences in academic sciences, engineering, and medicine*. Washington, DC: The National Academies Press. DOI: 10.17226/24994.

NASEM. 2019a. *Gaseous carbon waste streams utilization: Status and research needs*. Washington, DC: The National Academies Press. DOI: 10.17226/25232.

NASEM. 2019b. *Negative emissions technologies and reliable sequestration: A research agenda*. Washington, DC: The National Academies Press. DOI: 10.17226/25259.

NASEM. 2019c. *Environmental engineering for the 21st century: Addressing grand challenges*. Washington, DC: The National Academies Press. DOI: 10.17226/25121.

NASEM. 2019d. *Frontiers of materials research: A decadal survey*. Washington, DC: The National Academies Press. DOI: 10.17226/25244.

NASEM. 2019e. *The science of effective mentorship in STEMM*. Washington, DC: The National Academies Press. DOI: 10.17226/25568.

NASEM. 2019f. *Deployment of deep decarbonization technologies: Proceedings of a workshop*. Washington, DC: The National Academies Press. DOI: 10.17226/25656.

NASEM. 2020a. *Promising practices for addressing the underrepresentation of women in science, engineering, and medicine: Opening doors*. Washington, DC: The National Academies Press. DOI: 10.17226/25585.

NASEM. 2020b. *The impacts of racism and bias on Black people pursuing careers in science, engineering, and medicine: Proceedings of a workshop*. Washington, DC: The National Academies Press. DOI: 10.17226/25849.

NASEM. 2021a. *Accelerating decarbonization of the U.S. energy system*. Washington, DC: The National Academies Press. DOI: 10.17226/25932.

NASEM. 2021b. *Innovations in pharmaceutical manufacturing on the horizon: Technical challenges, regulatory issues, and recommendations*. Washington, DC: The National Academies Press. DOI: 10.17226/26009.

NASEM. 2021c. *Diversity, equity, and inclusion in chemistry and chemical engineering: Proceedings of a workshop-in-brief*. Washington, DC: The National Academies Press. DOI: 10.17226/26334.

Nature. 2019. Human Microbiome Project, part 2. Retrieved August 16, 2021, from https://www.nature.com/collections/fiabfcjbfj.

NCES (National Center for Education Statistics). 2021. Expenditures. Retrieved March 18, 2021, 2021, from https://nces.ed.gov/fastfacts/display.asp?id=75.

NCSES (National Center for Science and Engineering Statistics). 2018. *Science and engineering degrees, by race and ethnicity of receipts: 2008-18*. Alexandria, VA: NCSES. https://ncsesdata.nsf.gov/sere/2018/index.html.

NCSES. 2020. *National patterns of R&D resources: 2017-18 data update*. Alexandria, VA: NCSES. https://ncses.nsf.gov/pubs/nsf20307/#general-notes.

NCSES. 2021. Explore data. Retreived 2021, from https://www.nsf.gov/statistics/data.cfm.

NIMS (National Institute for Materials Science). 2021. Polymer database (polyinfo). Retrieved August 27, 2021, from https://polymer.nims.go.jp/en.

NIST (National Institute of Standards and Technology). 2021. Thermophysical properties of fluid systems. Retrieved August 27, 2021, from https://webbook.nist.gov/chemistry/fluid.

NOAA (National Oceanic and Atmospheric Administration). 2020. Climate change: Atmospheric carbon dioxide. Retrieved September 11, 2021, from https://www.climate.gov/news-features/understanding-climate/climate-change-atmospheric-carbon-dioxide.

NOAA. 2021a. Deepwater Horizon. Retrieved August 16, 2021, from https://darrp.noaa.gov/oil-spills/deepwater-horizon.

NOAA. 2021b. A guide to plastic in the ocean. Retrieved August 16, 2021, from https://oceanservice.noaa.gov/hazards/marinedebris/plastics-in-the-ocean.html.

NREL (National Renewable Energy Laboratory). 2017. Concentrating solar power gen3 demonstration roadmap. Retrieved August 16, 2021, from https://www.nrel.gov/docs/fy17osti/67464.pdf.

NREL. 2020. NREL's top 2020 wind program accomplishments demonstrate a clear vision for wind energy advancement. Retrieved August 17, 2021, from https://www.nrel.gov/news/program/2020/2020-top-wind-accomplishments.html.

NRC (National Research Council). 1988. *Frontiers in chemical engineering: Research needs and opportunities*. Washington, DC: The National Academies Press. DOI: 10.17226/1095.

NRC. 2007. *International benchmarking of U.S. chemical engineering research competitiveness*. Washington, DC: The National Academies Press. DOI: 10.17226/11867.

NRC. 2012. *Discipline-based education research: Understanding and improving learning in undergraduate science and engineering*. Washington, DC: The National Academies Press. DOI: 10.17226/13362.

NRC. 2015. *Cost, effectiveness, and deployment of fuel economy technologies for light-duty vehicles*. Washington, DC: The National Academies Press. DOI: 10.17226/21744.

NSB (National Science Board). 2018. *Science and engineering indicators* NSB-2018-1. Alexandria, VA: National Science Foundation.

NSB. 2020. Merit review process fiscal year 2019 digest. Retrieved August 16, 2021, from https://www.nsf.gov/nsb/publications/2020/merit_review/FY-2019/nsb202038.pdf.

NSF (National Science Foundation). 2016. U.S. science and technology leadership increasingly challenged by advances in Asia. Retrieved 2021, from https://nsf.gov/news/news_summ.jsp?cntn_id=137394&org=NSF&from=news.

NSF. 2020. NSF & Congress. Retrieved August 16, 2021, from https://www.nsf.gov/about/congress/117/highlights/cu20.jsp.

Nelson, M. J., G. Nakhla, and J. Zhu. 2017. Fluidized-bed bioreactor applications for biological wastewater treatment: A review of research and developments. *Engineering* 3(3):330-342. DOI: 10.1016/j.Eng.2017.03.021.

Neri, G., P. M. Donaldson, and A. J. Cowan. 2017. The role of electrode-catalyst interactions in enabling efficient CO2 reduction with Mo(bpy)(CO)4 as revealed by vibrational sum-frequency generation spectroscopy. *Journal of the American Chemical Society* 139(39):13791-13797. DOI: 10.1021/jacs.7b06898.

Ng, A. 2016. What artificial intelligence can and can't do right now. *Harvard Business Review*. https://hbr.org/2016/11/what-artificial-intelligence-can-and-cant-do-right-now.

Ng, W. L., C. K. Chua, and Y.-F. Shen. 2019. Print me an organ! Why we are not there yet. *Progress in Polymer Science* 97. DOI: 10.1016/j.progpolymsci.2019.101145.

Nielsen. 2015. The sustainability imperative—New insights on consumer expectations. Retrieved August 18, 2021, from https://www.nielsen.com/wp-content/uploads/sites/3/2019/04/Global20Sustainability20Report_October202015.pdf.

Nietzel, M. T. 2021. Georgia Tech's online MS in computer science continues to thrive. Why that's important for the future of MOOCs. Retrieved September 8, 2021, from https://www.forbes.com/sites/michaeltnietzel/2021/07/01/georgia-techs-online-ms-in-co

mputer-science-continues-to-thrive-what-that-could-mean-for-the-future-of-moocs/?sh=750c608aa277.

Nikolau, B. J., M. A. D. N. Perera, L. Brachova, and B. Shanks. 2008. Platform biochemicals for a biorenewable chemical industry. *The Plant Journal* 54(4):536-545. DOI: 10.1111/j.1365-313X.2008.03484.x.

Nimpuno, N., and C. Scruggs. 2011. *Information on chemicals in electronic products: A study of needs, gaps, obstacles and solutions to provide and access information on chemicals in electronic products*. Copenhagen: Nordic Council of Ministers.

Nisbet, E. G., M. R. Manning, E. J. Dlugokencky, R. E. Fisher, D. Lowry, S. E. Michel, C. L. Myhre, S. M. Platt, G. Allen, P. Bousquet, R. Brownlow, M. Cain, J. L. France, O. Hermansen, R. Hossaini, A. E. Jones, I. Levin, A. C. Manning, G. Myhre, J. A. Pyle, B. H. Vaughn, N. J. Warwick, and J. W. C. White. 2019. Very strong atmospheric methane growth in the 4 years 2014–2017: Implications for the Paris Agreement. *Global Biogeochemical Cycles* 33(3):318-342. DOI: 10.1029/2018gb006009.

Nitopi, S., E. Bertheussen, S. B. Scott, X. Liu, A. K. Engstfeld, S. Horch, B. Seger, I. E. L. Stephens, K. Chan, C. Hahn, J. K. Norskov, T. F. Jaramillo, and I. Chorkendorff. 2019. Progress and perspectives of electrochemical CO2 Reduction on copper in aqueous electrolyte. *Chemical Reviews* 119(12):7610-7672. DOI: 10.1021/acs.chemrev.8b00705.

Nnodu, O., H. Isa, M. Nwegbu, C. Ohiaeri, S. Adegoke, R. Chianumba, N. Ugwu, B. Brown, J. Olaniyi, E. Okocha, J. Lawson, A. A. Hassan, I. Diaku-Akinwumi, A. Madu, O. Ezenwosu, Y. Tanko, U. Kangiwa, A. Girei, Y. Israel-Aina, A. Ladu, P. Egbuzu, U. Abjah, A. Okolo, N. Akbulut-Jeradi, M. Fernandez, F. B. Piel, and A. Adekile. 2019. HemoTypeSC, a low-cost point-of-care testing device for sickle cell disease: Promises and challenges. *Blood Cells, Molecules and Diseases* 78:22-28. DOI: 10.1016/j.bcmd.2019.01.007.

Noé, F., A. Tkatchenko, K. R. Müller, and C. Clementi. 2020. Machine learning for molecular simulation. *Annual Review of Physical Chemistry* 71:361-390. DOI: 10.1146/annurev-physchem-042018-052331.

Nordquist, A. F., F. C. Wilhelm, F. J. Waller, and R. M. Machado. 2002. Hydrogenation with monolith reactor under conditions of immiscible liquid phases. U.S. Patent No. 6479704. Washington, DC: U.S. Patent and Trademark Office.

Nuss, P. 2015. *Book review: Life cycle assessment handbook: A guide for environmentally sustainable products*. M. A. Curran, ed. Hoboken, NJ: PB - John Wiley & Sons, Inc., and Salem, MA: Scrivener Publishing LLC.

O'Brien, E. J., J. M. Monk, and B. O. Palsson. 2015. Using genome-scale models to predict biological capabilities. *Cell* 161(5):971-987. DOI: 10.1016/j.cell.2015.05.019.

Ochedi, F. O., D. Liu, J. Yu, A. Hussain, and Y. Liu. 2020. Photocatalytic, electrocatalytic and photoelectrocatalytic conversion of carbon dioxide: A review. *Environmental Chemistry Letters* 19(2):941-967. DOI: 10.1007/s10311-020-01131-5.

OECD (Organisation for Economic Co-operation and Development). 2012. OECD environmental outlook to 2050. *OECD Publishing*. DOI: 10.1787/9789264122246-en.

Ogden, J., L. Fulton, and D. Sperling. 2016. *Making the transition to light-duty electric-drive vehicles in the U.S.: Costs in perspective to 2035*. Davis: University of California, Davis. https://trid.trb.org/view/1441689.

Ogunnaike, B. A. 2019. 110th anniversary: Process and systems engineering perspectives on personalized medicine and the design of effective treatment of diseases. *Industrial & Engineering Chemistry Research* 58(44):20357-20369. DOI: 10.1021/acs.iecr.9b04228.

Ohio History Central. 2021. Standard oil company. Retrieved August 17, 2021, from https://ohiohistorycentral.org/w/Standard_Oil_Company.

Olivetti, E. A., J. M. Cole, E. Kim, O. Kononova, G. Ceder, T. Y.-J. Han, and A. M. Hiszpanski. 2020. Data-driven materials research enabled by natural language processing and information extraction. *Applied Physics Reviews* 7(4). DOI: 10.1063/5.0021106.

Olson, J., Y. Cao, J. Romero, P. Johnson, P.-L. Dallaire-Demers, N. Sawaya, P. Narang, I. Kivlichan, M. Wasielewski, and A. Aspuru-Guzik. 2016. Quantum information and computation for chemistry, report of an NSF workshop. *arXiv*. DOI: 1706.05413.

Oluwole, E. O., T. A. Adeyemo, G. E. Osanyin, O. O. Odukoya, P. J. Kanki, and B. B. Afolabi. 2020. Feasibility and acceptability of early infant screening for sickle cell disease in Lagos, Nigeria—A pilot study. *PLoS One* 15(12). DOI: 10.1371/journal.pone.0242861.

O'Neill, J., and J.-F. Zheng. 2019. The holistic approach to materials and processing for new and scaled devices. *Semiconductor Digest*. https://www.semiconductor-digest.com/the-holistic-approach-to-materials-and-processing-for-new-and-scaled-devices.

O'Regan, B., and M. Grätzel. 1991. A low-cost, high-efficiency solar cell based on dye-sensitized colloidal TiO2 films. *Nature* 353(6346):737-740. DOI: 10.1038/353737a0.

Our World in Data. 2019. Number of deaths by risk factor, world, 2017. Retrieved August 17, 2021 from https://ourworldindata.org/grapher/number-of-deaths-by-risk-factor.

Our World in Data. 2021a. Emissions of air pollutants, United States, 1970 to 2016. Retrieved August 17, 2021, from https://ourworldindata.org/grapher/emissions-of-air-pollutants?country=~USA.

Our World in Data. 2021b. Food: Greenhouse gas emissions across the supply chain. Retrieved August 17, 2021, from https://ourworldindata.org/grapher/food-emissions-supply-chain?country=Beef+%28beef+herd%29~Cheese~Poultry+Meat~Milk~Eggs~Rice~Pig+Meat~Peas~Bananas~Wheat+%26+Rye~Fish+%28farmed%29~Lamb+%26+Mutton~Beef+%28dairy+herd%29~Shrimps+%28farmed%29~Tofu~Maize.

Owoseni, O., E. Nyankson, Y. Zhang, S. J. Adams, J. He, G. L. McPherson, A. Bose, R. B. Gupta, and V. T. John. 2014. Release of surfactant cargo from interfacially-active halloysite clay nanotubes for oil spill remediation. *Langmuir* 30(45):13533-13541. DOI: 10.1021/la503687b.

Pacala, S., and R. Socolow. 2004. Stabilization wedges: Solving the climate problem for the next 50 years with current technologies. *Science* 305(5686):968-972. DOI: 10.1126/science.1100103.

Padervand, M., E. Lichtfouse, D. Robert, and C. Wang. 2020. Removal of microplastics from the environment. A review. *Environmental Chemistry Letters* 18(3):807-828. DOI: 10.1007/s10311-020-00983-1.

Pal, S., and K. A. Fichthorn. 1999. Accelerated molecular dynamics of infrequent events. *Chemical Engineering Journal* 74(1):77-83. DOI: 10.1016/S1385-8947(99)00055-8.

Palmer, J. C., P. H. Poole, F. Sciortino, and P. G. Debenedetti. 2018. Advances in computational studies of the liquid-liquid transition in water and water-like models. *Chemical Reviews* 118(18):9129-9151. DOI: 10.1021/acs.chemrev.8b00228.

Panagiotopoulos, A. Z. 1987. Direct determination of phase coexistence properties of fluids by Monte Carlo simulation in a new ensemble. *Molecular Physics* 61(4):813-826. DOI: 10.1080/00268978700101491.

Panuwatsuk, W., and N. A. Da Silva. 2003. Application of a gratuitous induction system in Kluyveromyces lactis for the expression of intracellular and secreted proteins during fed-batch culture. *Biotechnology and Bioengineering* 81(6):712-718. DOI: 10.1002/bit.10518.

Pappa, G., C. Boukouvalas, C. Giannaris, N. Ntaras, V. Zografos, K. Magoulas, A. Lygeros, and D. Tassios. 2001. The selective dissolution/precipitation technique for polymer recycling:

A pilot unit application. *Resources, Conservation and Recycling* 34(1):33-44. DOI: 10.1016/s0921-3449(01)00092-1.

Park, H., and K. Park. 1996. Biocompatibility issues of implantable drug delivery systems. *Pharmaceutical Research* 13(12):1770-1776. DOI: 10.1023/a:1016012520276.

Park, D. S., K. E. Joseph, M. Koehle, C. Krumm, L. Ren, J. N. Damen, M. H. Shete, H. S. Lee, X. Zuo, B. Lee, W. Fan, D. G. Vlachos, R. F. Lobo, M. Tsapatsis, and P. J. Dauenhauer. 2016. Tunable oleo-furan surfactants by acylation of renewable furans. *ACS Central Science* 2(11):820-824. DOI: 10.1021/acscentsci.6b00208.

Park, S.-Y., C.-H. Park, D.-H. Choi, J. K. Hong, and D.-Y. Lee. 2021. Bioprocess digital twins of mammalian cell culture for advanced biomanufacturing. *Current Opinion in Chemical Engineering* 33. DOI: 10.1016/j.coche.2021.100702.

Parsons, S., S. Raikova, and C. J. Chuck. 2020. The viability and desirability of replacing palm oil. *Nature Sustainability* 3(6):412-418. DOI: 10.1038/s41893-020-0487-8.

Paul, R., and R. Brennan. 2019. Discipline-based education research (DBER)—What is it, and why should engineering education research scholars be talking about it more? Presented at *Canadian Engineering Education Association Conference*. Ottawa, Ontario.

Peng, L., H. Dai, Y. Wu, Y. Peng, and X. Lu. 2018. A comprehensive review of the available media and approaches for phosphorus recovery from wastewater. *Water, Air, & Soil Pollution* 229(4). DOI: 10.1007/s11270-018-3706-4.

Pereao, O., C. Bode-Aluko, O. Fatoba, L. Petrik, and K. Laatikainen. 2018. Rare earth elements removal techniques from water/wastewater: A review. *Desalination and Water Treatment* 130:71-86. DOI: 10.5004/dwt.2018.22844.

Perlmutter, D. D. 1975. *Introduction to chemical process control*. Malabar, FL: Robert E. Krieger Publishing Company.

Peters, M., K. Timmerhaus, R. West, and M. Peters. 2002. *Plant design and economics for chemical engineers*. New York: McGraw-Hill Education.

Plastics Hall of Fame. 2021. Daniel Wayne Fox. Retrieved August 17, 2021, from https://www.plasticshof.org/members/daniel-wayne-fox.

PlasticsEurope. 2019. Plastics—The facts 2019: An analysis of European plastics production, demand and waste data. Retrieved 2021, from https://plasticseurope.org/wp-content/uploads/2021/10/2019-Plastics-the-facts.pdf.

Pozrikidis, C. 1997. Numerical studies of singularity formation at free surfaces and fluid interfaces in two-dimensional Stokes flow. *Journal of Fluid Mechanics* 331:145-167. DOI: 10.1017/S0022112096003813.

Prather, K. A., C. C. Wang, and R. T. Schooley. 2020. Reducing transmission of SARS-CoV-2. *Science* 368(6498):1422-1424. DOI: 10.1126/science.abc6197.

Prausnitz, M. R. 2017. Engineering microneedle patches for vaccination and drug delivery to skin. *Annual Review of Chemical and Biomolecular Engineering* 8:177-200. DOI: 10.1146/annurev-chembioeng-060816-101514.

Prausnitz, M. R., and R. Langer. 2008. Transdermal drug delivery. *Nature Biotechnology* 26(11):1261-1268. DOI: 10.1038/nbt.1504.

PRISMS Center (Center for Predictive Integrated Structural Materials Science). 2021. Materials common 2.0. Retrieved August 27, 2021, from https://materialscommons.org.

Ragauskas, A. J., G. T. Beckham, M. J. Biddy, R. Chandra, F. Chen, M. F. Davis, B. H. Davison, R. A. Dixon, P. Gilna, M. Keller, P. Langan, A. K. Naskar, J. N. Saddler, T. J. Tschaplinski, G. A. Tuskan, and C. E. Wyman. 2014. Lignin valorization: Improving lignin processing in the biorefinery. *Science* 344(6185). DOI: 10.1126/science.1246843.

Rahimi, A., and J. M. García. 2017. Chemical recycling of waste plastics for new materials production. *Nature Reviews Chemistry* 1(6). DOI: 10.1038/s41570-017-0046.

Raišienė, A. G., V. Rapuano, and K. Varkulevičiūtė. 2021. Sensitive men and hardy women: How do millennials, xennials and gen X manage to work from home? *Journal of Open Innovation: Technology, Market, and Complexity* 7(2). DOI: 10.3390/joitmc7020106.

Ramadan, Q., and M. Zourob. 2020. Organ-on-a-chip engineering: Toward bridging the gap between lab and industry. *Biomicrofluidics* 14(4). DOI: 10.1063/5.0011583.

Raman, A. P., M. A. Anoma, L. Zhu, E. Rephaeli, and S. Fan. 2014. Passive radiative cooling below ambient air temperature under direct sunlight. *Nature* 515(7528):540-544. DOI: 10.1038/nature13883.

Rao, D. P. 2015. The Story of "HIGEE". *Indian Chemical Engineer* 57(3-4):282-299. DOI: 10.1080/00194506.2015.1026946.

Ratnasari, D. K., M. A. Nahil, and P. T. Williams. 2017. Catalytic pyrolysis of waste plastics using staged catalysis for production of gasoline range hydrocarbon oils. *Journal of Analytical and Applied Pyrolysis* 124:631-637. DOI: 10.1016/j.jaap.2016.12.027.

Renzulli, K. A. 2019. Women reach leadership roles earlier than men do—But fewer make it to the top, according to LinkedIn. Retrieved September 8, 2021, from https://www.cnbc.com/2019/06/24/women-reach-leadership-roles-1point4-years-earlier-than-men-says-linkedin.html.

Rephaeli, E., A. Raman, and S. Fan. 2013. Ultrabroadband photonic structures to achieve high-performance daytime radiative cooling. *Nano Letters* 13(4):1457-1461. DOI: 10.1021/nl4004283.

Reynaert, S., C. F. Brooks, P. Moldenaers, J. Vermant, and G. G. Fuller. 2008. Analysis of the magnetic rod interfacial stress rheometer. *Journal of Rheology* 52(1):261-285. DOI: 10.1122/1.2798238.

Rezaiyan, J., and N. P. Cheremisinoff. 2005. *Gasification technologies: A primer for engineers and scientists*. Boca Raton, FL: CRC Press.

Richards, G. 1979. Third age of quantum chemistry. *Nature* 278(5704):507-507. DOI: 10.1038/278507a0.

Richardson, J., and J. Dantzler. 2002. Effect of a freshman engineering program on retention and academic performance. Presented at *32nd Annual Frontiers in Education*. Boston, MA..

Ritger, P. L., and N. A. Peppas. 1987. A simple equation for description of solute release I. Fickian and non-fickian release from non-swellable devices in the form of slabs, spheres, cylinders or discs. *Journal of Controlled Release* 5(1):23-36. DOI: 10.1016/0168-3659(87)90034-4.

Roh, S., A. H. Williams, R. S. Bang, S. D. Stoyanov, and O. D. Velev. 2019. Soft dendritic microparticles with unusual adhesion and structuring properties. *Nature Materials* 18(12):1315-1320. DOI: 10.1038/s41563-019-0508-z.

Rong, Y., Y. Hu, A. Mei, H. Tan, M. I. Saidaminov, S. I. Seok, M. D. McGehee, E. H. Sargent, and H. Han. 2018. Challenges for commercializing perovskite solar cells. *Science* 361(6408). DOI: 10.1126/science.aat8235.

Roque, B. M., M. Venegas, R. D. Kinley, R. de Nys, T. L. Duarte, X. Yang, and E. Kebreab. 2021. Red seaweed (Asparagopsis taxiformis) supplementation reduces enteric methane by over 80 percent in beef steers. *PLoS One* 16(3). DOI: 10.1371/journal.pone.0247820.

Rorrer, J. E., G. T. Beckham, and Y. Román-Leshkov. 2021. Conversion of polyolefin waste to liquid alkanes with Ru-based catalysts under mild conditions. *Journal of the American Chemical Society: Au* 1(1):8-12. DOI: 10.1021/jacsau.0c00041.

Rosales, A. M., and K. S. Anseth. 2016. The design of reversible hydrogels to capture extracellular matrix dynamics. *Nature Reviews Materials* 1(2). DOI: 10.1038/natrevmats.2015.12.

Rosales, A. M., R. A. Segalman, and R. N. Zuckermann. 2013. Polypeptoids: A model system to study the effect of monomer sequence on polymer properties and self-assembly. *Soft Matter* 9(35):8400-8414. DOI: 10.1039/C3SM51421H.

Ross, I., J. McDonough, J. Miles, P. Storch, P. Thelakkat Kochunarayanan, E. Kalve, J. Hurst, S. S. Dasgupta, and J. Burdick. 2018. A review of emerging technologies for remediation of PFASs. *Remediation Journal* 28(2):101-126. DOI: 10.1002/rem.21553.

Ruiz-Lopez, M. F., J. S. Francisco, M. T. C. Martins-Costa, and J. M. Anglada. 2020. Molecular reactions at aqueous interfaces. *Nature Reviews Chemistry* 4(9):459-475. DOI: 10.1038/s41570-020-0203-2.

Rzhetsky, A., J. G. Foster, I. T. Foster, and J. A. Evans. 2015. Choosing experiments to accelerate collective discovery. *Proceedings of the National Academy of Sciences* 112(47):14569-14574. DOI: 10.1073/pnas.1509757112.

Samet, J. M., K. Prather, G. Benjamin, S. Lakdawala, J. M. Lowe, A. Reingold, J. Volckens, and L. Marr. 2021. Airborne transmission of SARS-CoV-2: What we know. *Clinical Infectious Diseases* 73(10):1924-1926. DOI: 10.1093/cid/ciab039.

Samuel, M., D. Polson, D. Graham, W. Kordziel, T. Waite, G. Waters, P. S. Vinod, D. Fu, and R. Downey. 2000. Viscoelastic surfactant fracturing fluids: Applications in low permeability reservoirs. Presented at *SPE Rocky Mountain Regional/Low-Permeability Reservoirs Symposium and Exhibition*. Denver, CO.

Sánchez, A. 2019. The current role of chemical engineering in solving environmental problems. *Frontiers in Chemical Engineering* 1, Article 1. DOI: 10.3389/fceng.2019.00001.

Sanchez-Lengeling, B., and A. Aspuru-Guzik. 2018. Inverse molecular design using machine learning: Generative models for matter engineering. *Science* 361(6400):360-365. DOI: 10.1126/science.aat2663.

Sanderson, K. 2019. Automation: Chemistry shoots for the moon. *Nature* 568(7753):577-579. DOI: 10.1038/d41586-019-01246-y.

Santiesteban, J. G. and T. F. Degnan., Jr. 2021. Catalysis and the future of transportation fuels. *The Bridge 50th Anniversary Issue*. https://www.nae.edu/244855/Catalysis-and-the-Future-of-Transportation-Fuels.

Sargent & Lundy LLC Consulting Group. 2003. *Assessment of parabolic trough and power tower solar technology cost and performance forecasts* NREL/SR-550-34440. Chicago, IL: National Renewable Energy Laboratory (NREL).

Sarikurt, S., T. Kocabaş, and C. Sevik. 2020. High-throughput computational screening of 2D materials for thermoelectrics. *Journal of Materials Chemistry A* 8(37):19674-19683. DOI: 10.1039/d0ta04945j.

Sarsons, H. 2017. Gender differences in recognition for group work. https://scholar.harvard.edu/files/sarsons/files/full_v6.pdf.

Saygili, E., E. Yildiz-Ozturk, M. J. Green, A. M. Ghaemmaghami, and O. Yesil-Celiktas. 2021. Human lung-on-chips: Advanced systems for respiratory virus models and assessment of immune response. *Biomicrofluidics* 15(2). DOI: 10.1063/5.0038924.

Scanlon, B. R., B. L. Ruddell, P. M. Reed, R. I. Hook, C. Zheng, V. C. Tidwell, and S. Siebert. 2017. The food-energy-water nexus: Transforming science for society. *Water Resources Research* 53(5):3550-3556. DOI: 10.1002/2017wr020889.

Schepers, A., C. Li, A. Chhabra, B. T. Seney, and S. Bhatia. 2016. Engineering a perfusable 3D human liver platform from iPS cells. *Lab on a Chip* 16(14):2644-2653. DOI: 10.1039/c6lc00598e.

Schiffer, Z. J., and K. Manthiram. 2017. Electrification and decarbonization of the chemical industry. *Joule* 1(1):10-14. DOI: 10.1016/j.joule.2017.07.008.

Schöttker, B., K.-U. Saum, D. C. Muhlack, L. K. Hoppe, B. Holleczek, and H. Brenner. 2017. Polypharmacy and mortality: New insights from a large cohort of older adults by detection of effect modification by multi-morbidity and comprehensive correction of confounding by indication. *European Journal of Clinical Pharmacology* 73(8):1041-1048. DOI: 10.1007/s00228-017-2266-7.

Schowalter, W. R. 2003. The equations (of change) don't change: But the profession of engineering does. *Chemical Engineering Education* 37(4):242-247.

Schwaller, P., T. Laino, T. Gaudin, P. Bolgar, C. A. Hunter, C. Bekas, and A. A. Lee. 2019. Molecular Transformer: A model for uncertainty-calibrated chemical reaction prediction. *ACS Central Science* 5(9):1572-1583. DOI: 10.1021/acscentsci.9b00576.

Scriven, L. E. 1960. Dynamics of a fluid interface equation of motion for Newtonian surface fluids. *Chemical Engineering Science* 12(2):98-108. DOI: 10.1016/0009-2509(60)87003-0.

Scriven, L. E. 1991. On the emergence and evolution of chemical engineering. In *Advances in chemical engineering*. Colton, C. K., ed. San Diego, CA: Academic Press.

Seader, J. D., and E. J. Henley. 1998. *Separation process principles*. New York: Wiley.

Searchinger, T., R. Heimlich, R. A. Houghton, F. Dong, A. Elobeid, J. Fabiosa, S. Tokgoz, D. Hayes, and T.-H. Yu. 2008. Use of U.S. croplands for biofuels increases greenhouse gases through emissions from land-use change. *Science* 319(5867):1238-1240. DOI: 10.1126/science.1151861.

Seinfeld, J. H. 1991. Environmental chemical engineering. In *Advances in chemical engineering*. Colton, C. K., ed. San Diego, CA: Academic Press.

Seinfeld, J. H., and S. N. Pandis. 2016. *Atmospheric chemistry and physics: From air pollution to climate change*. New York: John Wiley & Sons, Inc.

Sharma, S. 2013. Ferrolectric nanofibers: Principle, processing and applications. *Advanced Materials Letters* 4(7):522-533. DOI: 10.5185/amlett.2012.9426.

Shashvatt, U., F. Amurrio, C. Portner, and L. Blaney. 2021. Phosphorus recovery by Donnan dialysis: Membrane selectivity, diffusion coefficients, and speciation effects. *Chemical Engineering Journal* 419. DOI: 10.1016/j.cej.2021.129626.

Shen, L., E. Worrell, and M. K. Patel. 2010. Open-loop recycling: A LCA case study of PET bottle-to-fibre recycling. *Resources, Conservation and Recycling* 55(1):34-52. DOI: 10.1016/j.resconrec.2010.06.014.

Sherman, Z. M., M. P. Howard, B. A. Lindquist, R. B. Jadrich, and T. M. Truskett. 2020. Inverse methods for design of soft materials. *The Journal of Chemical Physics* 152(14). DOI: 10.1063/1.5145177.

Shi, D., E. Dassau, and F. J. Doyle, 3rd. 2019. Multivariate learning framework for long-term adaptation in the artificial pancreas. *Bioengineering & Translational Medicine* 4(1):61-74. DOI: 10.1002/btm2.10119.

Shi, C., L. T. Reilly, V. S. Phani Kumar, M. W. Coile, S. R. Nicholson, L. J. Broadbelt, G. T. Beckham, and E. Y. X. Chen. 2021. Design principles for intrinsically circular polymers with tunable properties. *Chem* 7(11):2896-2912. DOI: 10.1016/j.chempr.2021.10.004.

Shim, S. H., R. Gupta, Y. L. Ling, D. B. Strasfeld, D. P. Raleigh, and M. T. Zanni. 2009. Two-dimensional IR spectroscopy and isotope labeling defines the pathway of amyloid formation with residue-specific resolution. *Proceedings of the National Academy of Sciences* 106(16):6614-6619. DOI: 10.1073/pnas.0805957106.

Sidorova, J., and M. Anisimova. 2014. NLP-inspired structural pattern recognition in chemical application. *Pattern Recognition Letters* 45:11-16. DOI: 10.1016/j.patrec.2014.02.012.

Silva, R. A., K. Hawboldt, and Y. Zhang. 2018. Application of resins with functional groups in the separation of metal ions/species—A review. *Mineral Processing and Extractive Metallurgy Review* 39(6):395-413. DOI: 10.1080/08827508.2018.1459619.

Simpson, G. B., and G. P. W. Jewitt. 2019. The development of the water-energy-food nexus as a framework for achieving resource security: A review. *Frontiers in Environmental Science* 7, Article 8. DOI: 10.3389/fenvs.2019.00008.

Sims, R. E. H., R. N. S., A. Adegbululgbe, J. Fenhann, I. Konstantinaviciute, W. Moomaw, H.B. Nimir, B. Schlamadinger, J. Torres-Martínez, C. Turner, Y. Uchiyama, S. J.V. Vuori, N. Wamukonya, X. Zhang 2007. Chapter 4: Energy supply. In *IPCC Fourth Assessment Report: Climate Change 2007: Contribution of Working Group III: Mitigation of Climate Change*. New York: Cambridge University Press.

Singh, A., N. A. Rorrer, S. R. Nicholson, E. Erickson, J. S. DesVeaux, A. F. T. Avelino, P. Lamers, A. Bhatt, Y. Zhang, G. Avery, L. Tao, A. R. Pickford, A. C. Carpenter, J. E. McGeehan, and G. T. Beckham. 2021. Techno-economic, life-cycle, and socioeconomic impact analysis of enzymatic recycling of poly(ethylene terephthalate). *Joule* 5(9):2479-2503. DOI: 10.1016/j.joule.2021.06.015.

Sinha, S., U. D. Irani, V. Manchaiah, and M. S. Bhamla. 2020. LoCHAid: An ultra-low-cost hearing aid for age-related hearing loss. *PLoS One* 15(9). DOI: 10.1371/journal.pone.0238922.

Siracusa, V., and I. Blanco. 2020. Bio-polyethylene (Bio-PE), bio-polypropylene (Bio-PP) and bio-poly (ethylene terephthalate) (Bio-PET): Recent developments in bio-based polymers analogous to petroleum-derived ones for packaging and engineering applications. *Polymers* 12(8). DOI: 10.3390/polym12081641.

Sivakumar, S., K. L. Wark, J. K. Gupta, N. L. Abbott, and F. Caruso. 2009. Liquid crystal emulsions as the basis of biological sensors for the optical detection of bacteria and viruses. *Advanced Functional Materials* 19(14):2260-2265. DOI: 10.1002/adfm.2009 00399.

Skoulidas, A. I., D. M. Ackerman, J. K. Johnson, and D. S. Sholl. 2002. Rapid transport of gases in carbon nanotubes. *Physical Review Letters* 89(18). DOI: 10.1103/PhysRevLett.89.185901.

Smith, R. C., R. K. Taggart, J. C. Hower, M. R. Wiesner, and H. Hsu-Kim. 2019. Selective recovery of rare earth elements from coal fly ash leachates using liquid membrane processes. *Environmental Science & Technology* 53(8):4490-4499. DOI: 10.1021/acs.est.9b00539.

Smith, C., A. K. Hill, and L. Torrente-Murciano. 2020. Current and future role of Haber–Bosch ammonia in a carbon-free energy landscape. *Energy & Environmental Science* 13(2):331-344. DOI: 10.1039/c9ee02873k.

Socorro, I. M., K. Taylor, and J. M. Goodman. 2005. ROBIA: A reaction prediction program. *Organic Letters* 7(16):3541-3544. DOI: 10.1021/ol0512738.

Song, X., Y. Guo, J. Zhang, N. Sun, G. Shen, X. Chang, W. Yu, Z. Tang, W. Chen, W. Wei, L. Wang, J. Zhou, X. Li, X. Li, J. Zhou, and Z. Xue. 2019. Fracturing with carbon dioxide: From microscopic mechanism to reservoir application. *Joule* 3(8):1913-1926. DOI: 10.1016/j.joule.2019.05.004.

Spielman, L. A. 1977. Particle capture from low-speed laminar flows. *Annual Review of Fluid Mechanics* 9(1):297-319. DOI: 10.1146/annurev.fl.09.010177.001501.

Spiker, M. L., H. A. B. Hiza, S. M. Siddiqi, and R. A. Neff. 2017. Wasted food, wasted nutrients: Nutrient loss from wasted food in the United States and comparison to gaps in dietary intake. *Journal of the Academy of Nutrition and Dietetics* 117(7):1031-1040.e22. DOI: 10.1016/j.jand.2017.03.015.

Squires, T. M., and T. G. Mason. 2009. Fluid mechanics of microrheology. *Annual Review of Fluid Mechanics* 42(1):413-438. DOI: 10.1146/annurev-fluid-121108-145608.

Stankiewicz, A. I., and J. A. Moulijn. 2000. Process intensification: Transforming chemical engineering. *CEP Magazine*. https://www.aiche.org/sites/default/files/docs/news/010022_cep_stankiewicz.pdf.

Statista. 2021. Forecast number of mobile devices worldwide from 2020 to 2025 (in billions). Retrieved September 11, 2021, from https://www.statista.com/statistics/245501/multiple-mobile-device-ownership-worldwide.

Staudinger, H. 1920. Über polymerisation. *Berichte der deutschen chemischen Gesellschaft (A and B Series)* 53(6):1073-1085. DOI: 10.1002/cber.19200530627.

Stebe, K. J., S. Y. Lin, and C. Maldarelli. 1991. Remobilizing surfactant retarded fluid particle interfaces. I. Stress-free conditions at the interfaces of micellar solutions of surfactants with fast sorption kinetics. *Physics of Fluids A: Fluid Dynamics* 3(1):3-20. DOI: 10.1063/1.857862.

Stone, H. A., and L. G. Leal. 1989. Relaxation and breakup of an initially extended drop in an otherwise quiescent fluid. *Journal of Fluid Mechanics* 198:399-427. DOI: 10.1017/S0022112089000194.

Striolo, A. 2006. The mechanism of water diffusion in narrow carbon nanotubes. *Nano Letters* 6(4):633-639. DOI: 10.1021/nl052254u.

Su, Z., J. D. Hostert, and J. N. Renner. 2020. Phosphate recovery by a surface-immobilized cerium affinity peptide. *ACS ES&T Water* 1(1):58-67. DOI: 10.1021/acsestwater.0c00001.

Sun, T., A. Dasgupta, Z. Zhao, M. Nurunnabi, and S. Mitragotri. 2020. Physical triggering strategies for drug delivery. *Advanced Drug Delivery Reviews* 158:36-62. DOI: 10.1016/j.addr.2020.06.010.

Swartz, J. R. 2012. Transforming biochemical engineering with cell-free biology. *American Institute of Chemical Engineers Journal* 58(1):5-13. DOI: 10.1002/aic.13701.

Takata, T., J. Jiang, Y. Sakata, M. Nakabayashi, N. Shibata, V. Nandal, K. Seki, T. Hisatomi, and K. Domen. 2020. Photocatalytic water splitting with a quantum efficiency of almost unity. *Nature* 581(7809):411-414. DOI: 10.1038/s41586-020-2278-9.

Takatori, S. C., and J. F. Brady. 2015. Towards a thermodynamics of active matter. *Physical Review E* 91(3). DOI: 10.1103/PhysRevE.91.032117.

Takeuchi, I., J. Lauterbach, and M. J. Fasolka. 2005. Combinatorial materials synthesis. *Materials Today* 8(10):18-26. DOI: 10.1016/S1369-7021(05)71121-4.

Talmadge, M. S., R. M. Baldwin, M. J. Biddy, R. L. McCormick, G. T. Beckham, G. A. Ferguson, S. Czernik, K. A. Magrini-Bair, T. D. Foust, P. D. Metelski, C. Hetrick, and M. R. Nimlos. 2014. A perspective on oxygenated species in the refinery integration of pyrolysis oil. *Green Chemistry* 16(2):407-453. DOI: 10.1039/c3gc41951g.

Talvitie, J., A. Mikola, A. Koistinen, and O. Setala. 2017. Solutions to microplastic pollution—Removal of microplastics from wastewater effluent with advanced wastewater treatment technologies. *Water Research* 123:401-407. DOI: 10.1016/j.watres.2017.07.005.

Tan, D. H. S., A. Banerjee, Z. Chen, and Y. S. Meng. 2020a. From nanoscale interface characterization to sustainable energy storage using all-solid-state batteries. *Nature Nanotechnology* 15(3):170-180. DOI: 10.1038/s41565-020-0657-x.

Tan, Y., J. Shen, T. Si, C. L. Ho, Y. Li, and L. Dai. 2020b. Engineered live biotherapeutics: Progress and challenges. *Biotechnology Journal* 15(10). DOI: 10.1002/biot.202000155.

TARSC (Training and Research Support Centre). 2015. Innovations for health: Use of appropriate technologies in primary health care in Zimbabwe—Report of an assessment. Retrieved August 18, 2021, from https://www.tarsc.org/publications/documents/AppTech%20PHC%20Zim%20rep%20April2015.pdf.

Taylor, G. I. 1934. The formation of emulsions in definable fields of flow. *Proceedings of the Royal Society of London. Series A, Containing Papers of a Mathematical and Physical Character* 146(858):501-523. DOI: 10.1098/rspa.1934.0169.

Taylor, R., and R. Krishna. 2000. Modelling reactive distillation. *Chemical Engineering Science* 55(22):5183-5229. DOI: 10.1016/s0009-2509(00)00120-2.

Teesalu, T., K. N. Sugahara, and E. Ruoslahti. 2012. Mapping of vascular ZIP codes by phage display. *Methods in Enzymology* 503:35-56. DOI: 10.1016/B978-0-12-396962-0.00002-1.

Tesar, J. E. 1996. *The Macmillan visual almanac*. B. S. Glassman, ed. London: Macmillan General Reference.

Teuten, E. L., J. M. Saquing, D. R. U. Knappe, M. A. Barlaz, S. Jonsson, A. Björn, S. J. Rowland, R. C. Thompson, T. S. Galloway, R. Yamashita, D. Ochi, Y. Watanuki, C. Moore, P. H. Viet, T. S. Tana, M. Prudente, R. Boonyatumanond, M. P. Zakaria, K. Akkhavong, Y. Ogata, H. Hirai, S. Iwasa, K. Mizukawa, Y. Hagino, A. Imamura, M. Saha, and H. Takada. 2009. Transport and release of chemicals from plastics to the environment and to wildlife. *Philosophical Transactions of the Royal Society B: Biological Sciences* 364(1526):2027-2045. DOI: doi:10.1098/rstb.2008.0284.

The Pew Charitable Trusts. 2019. Two decades of change in federal and state higher education funding. Retrieved September 8, 2021, from https://www.pewtrusts.org/en/research-and-analysis/issue-briefs/2019/10/two-decades-of-change-in-federal-and-state-higher-education-funding.

The Royal Society. 2018a. Options for producing low-carbon hydrogen at scale. Retrieved August 17, 2021, from https://royalsociety.org/~/media/policy/projects/hydrogen-production/energy-briefing-green-hydrogen.pdf.

The Royal Society. 2018b. Greenhouse gas removal. Retrieved August 18, 2021, from https://royalsociety.org/-/media/policy/projects/greenhouse-gas-removal/royal-society-greenhouse-gas-removal-report-2018.pdf.

Thollander, P., M. Karlsson, P. Rohdin, W. Johan, and J. Rosenqvist. 2020. *Introduction to industrial energy efficiency*. Cambridge, MA: Academic Press.

Thoman, D. B., E. R. Brown, A. Z. Mason, A. G. Harmsen, and J. L. Smith. 2015. The role of altruistic values in motivating underrepresented minority students for biomedicine. *BioScience* 65(2):183-188. DOI: 10.1093/biosci/biu199.

Thomassen, G., M. Van Dael, S. Van Passel, and F. You. 2019. How to assess the potential of emerging green technologies? Towards a prospective environmental and techno-economic assessment framework. *Green Chemistry* 21(18):4868-4886. DOI: 10.1039/C9GC02223F.

Tian, Y., S. E. Demirel, M. M. F. Hasan, and E. N. Pistikopoulos. 2018. An overview of process systems engineering approaches for process intensification: State of the art. *Chemical Engineering and Processing—Process Intensification* 133:160-210. DOI: 10.1016/j.cep.2018.07.014.

Tian, X., S. D. Stranks, and F. You. 2020a. Life cycle energy use and environmental implications of high-performance perovskite tandem solar cells. *Science Advances* 6(31). DOI: 10.1126/sciadv.abb0055.

Tian, Y., J. R. Lhermitte, L. Bai, T. Vo, H. L. Xin, H. Li, R. Li, M. Fukuto, K. G. Yager, J. S. Kahn, Y. Xiong, B. Minevich, S. K. Kumar, and O. Gang. 2020b. Ordered three-dimensional nanomaterials using DNA-prescribed and valence-controlled material voxels. *Nature Materials* 19(7):789-796. DOI: 10.1038/s41563-019-0550-x.

Tilman, D., C. Balzer, J. Hill, and B. L. Befort. 2011. Global food demand and the sustainable intensification of agriculture. *Proceedings of the National Academy of Sciences* 108(50):20260-20264. DOI: 10.1073/pnas.1116437108.

Tolle, K. M., D. S. W. Tansley, and A. J. G. Hey. 2011. The fourth paradigm: Data-intensive scientific discovery. *Proceedings of the IEEE* 99(8):1334-1337. DOI: 10.1109/JPROC.2011.2155130.

Tong, Y., I. Y. Zhang, and R. K. Campen. 2018. Experimentally quantifying anion polarizability at the air/water interface. *Nature Communications* 9. DOI: 10.1038/s41467-018-03598-x.

Tremblay, J.-F. 2018. Golden times for electronic materials suppliers. *Chemical & Engineering News* 96(28). https://cen.acs.org/materials/electronic-materials/Golden-times-electronic-materials-suppliers/96/i28.

Tsurushita, N., P. R. Hinton, and S. Kumar. 2005. Design of humanized antibodies: From anti-Tac to Zenapax. *Methods* 36(1):69-83. DOI: 10.1016/j.ymeth.2005.01.007.

Tuck, C. O., E. Perez, I. T. Horvath, R. A. Sheldon, and M. Poliakoff. 2012. Valorization of biomass: Deriving more value from waste. *Science* 337(6095):695-699. DOI: 10.1126/science.1218930.

Tumbleston, J. R., D. Shirvanyants, N. Ermoshkin, R. Janusziewicz, A. R. Johnson, D. Kelly, K. Chen, R. Pinschmidt, J. P. Rolland, A. Ermoshkin, E. T. Samulski, and J. M. DeSimone. 2015. Additive manufacturing: Continuous liquid interface production of 3D objects. *Science* 347(6228):1349-1352. DOI: 10.1126/science.aaa2397.

Tunuguntla, R. H., R. Y. Henley, Y. C. Yao, T. A. Pham, M. Wanunu, and A. Noy. 2017. Enhanced water permeability and tunable ion selectivity in subnanometer carbon nanotube porins. *Science* 357(6353):792-796. DOI: 10.1126/science.aan2438.

Tursi, A., T. F. Mastropietro, R. Bruno, M. Baratta, J. Ferrando-Soria, A. I. Mashin, F. P. Nicoletta, E. Pardo, G. De Filpo, and D. Armentano. 2021. Synthesis and enhanced capture properties of a new BioMOF@SWCNT-BP: Recovery of the endangered rare-earth elements from aqueous systems. *Advanced Materials Interfaces* 8(16). DOI: 10.1002/admi.202100730.

Turton, R., R. C. Bailie, W. B. Whiting, J. A. Shaeiwitz, and D. Bhattacharyya. 2012. *Analysis, synthesis, and design of chemical processes*. Upper Saddle River, NJ: Prentice Hall.

Uekert, T., M. A. Bajada, T. Schubert, C. M. Pichler, and E. Reisner. 2020. Scalable photocatalyst panels for photoreforming of plastic, biomass and mixed waste in flow. *ChemSusChem* 14(19):4190-4197. DOI: 10.1002/cssc.202002580.

UN (United Nations). 2017. World population projected to reach 9.8 billion in 2050, and 11.2 billion in 2100. Retrieved August 16, 2021, from https://www.un.org/development/desa/en/news/population/world-population-prospects-2017.html.

UNEP (United Nations Environment Programme). 2020. *Handbook for the Montreal protocol on substances that deplete the ozone: 14th edition.* Nairobi, Kenya: Ozone Secretariat.

United Nations Environment Programme. 2021. Our planet is drowning in plastic pollution—It's time for change! https://www.unep.org/interactive/beat-plastic-pollution.

UNFCC (United Nations Framework Convention on Climate Change). 2015. Adoption of the Paris Agreement. 21st Conference of the Parties. Paris, United Nations. https://unfccc.int/sites/default/files/english_paris_agreement.pdf.

University of Hertfordshire. 2021. PPDB: Pesticide properties database. Retrieved August 27, 2021, from http://sitem.herts.ac.uk/aeru/ppdb.

University of Minnesota. 2003. Professor L. E. (Skip) Scriven. Retrieved August 17, 2021, from http://www.chemeng.ntua.gr/dep/boudouvis/U%20of%20M%20CEMS-scriven.htm.

U.S. Census Bureau. 2018. An aging nation: Projected number of children and older adults. Retrieved March 18, 2021, from https://www.census.gov/library/visualizations/2018/comm/historic-first.html.

U.S. Census Bureau. 2019a. Percent of population 25 Years and over by detailed attainment level: 2000-2019. Retrieved March 18, 2021, 2021, from https://www.census.gov/content/dam/Census/library/visualizations/time-series/demo/fig11.png.

U.S. Census Bureau. 2019b. Percent of population age 25 and over by educational attainment. Retrieved March 18, 2021, 2021, from https://www.census.gov/content/dam/Census/library/visualizations/time-series/demo/fig2.png.

USPTO (U.S. Patent and Trademark Office) 2021. U.S. colleges and universities—Utility patent grants 1969-2012. Retrieved September 8, 2021, from https://www.uspto.gov/web/offices/ac/ido/oeip/taf/univ/doc/doc_info_2012.htm.

van Anders, G., D. Klotsa, A. S. Karas, P. M. Dodd, and S. C. Glotzer. 2015. Digital alchemy for materials design: Colloids and beyond. *ACS Nano* 9(10):9542-9553. DOI: 10.1021/acsnano.5b04181.

Vargason, A. M., A. C. Anselmo, and S. Mitragotri. 2021. The evolution of commercial drug delivery technologies. *Nature Biomedical Engineering* 5:951-967. DOI: 10.1038/s41551-021-00698-w.

Vaucher, A. C., F. Zipoli, J. Geluykens, V. H. Nair, P. Schwaller, and T. Laino. 2020. Automated extraction of chemical synthesis actions from experimental procedures. *Nature Communications* 11(1):3601. DOI: 10.1038/s41467-020-17266-6.

Vazquez, M. 2018. Engaging biomedical engineering in health disparities challenges. *Journal of Community Medicine and Health Education* 8(2). DOI: 10.4172/2161-0711.1000595.

Veers, P., K. Dykes, E. Lantz, S. Barth, C. L. Bottasso, O. Carlson, A. Clifton, J. Green, P. Green, H. Holttinen, D. Laird, V. Lehtomäki, J. K. Lundquist, J. Manwell, M. Marquis, C. Meneveau, P. Moriarty, X. Munduate, M. Muskulus, J. Naughton, L. Pao, J. Paquette, J. Peinke, A. Robertson, J. Sanz Rodrigo, A. M. Sempreviva, J. C. Smith, A. Tuohy, and R. Wiser. 2019. Grand challenges in the science of wind energy. *Science* 366(6464). DOI: 10.1126/science.aau2027.

Venkatasubramanian, V. 2019. The promise of artificial intelligence in chemical engineering: Is it here, finally? *AIChE Journal* 65(2):466-478. DOI: 10.1002/aic.16489.

Verhougstraete, M. P., J. K. Gerald, C. P. Gerba, and K. A. Reynolds. 2019. Cost-benefit of point-of-use devices for lead reduction. *Environmental Research* 171:260-265. DOI: 10.1016/j.envres.2019.01.016.

Villagomez-Salas, S., P. Manikandan, S. F. Acuna Guzman, and V. G. Pol. 2018. Amorphous carbon chips li-ion battery anodes produced through polyethylene waste upcycling. *ACS Omega* 3(12):17520-17527. DOI: 10.1021/acsomega.8b02290.

Voiland, A., and R. Simmon. 2010. Aerosols: Tiny particles, big impact. Retrieved August 16, 2021, from https://earthobservatory.nasa.gov/features/Aerosols.

Wahman, D. G., M. D. Pinelli, M. R. Schock, and D. A. Lytle. 2021. Theoretical equilibrium lead(II) solubility revisited: Open source code and practical relationships. *AWWA Water Science* 3(5). DOI: 10.1002/aws2.1250.

Walker, S., and R. Rothman. 2020. Life cycle assessment of bio-based and fossil-based plastic: A review. *Journal of Cleaner Production* 261. DOI: 10.1016/j.jclepro.2020.121158.

Walker, T. W., N. Frelka, Z. Shen, A. K. Chew, J. Banick, S. Grey, M. S. Kim, J. A. Dumesic, R. C. Van Lehn, and G. W. Huber. 2020. Recycling of multilayer plastic packaging materials by solvent-targeted recovery and precipitation. *Science Advances* 6(47). DOI: 10.1126/sciadv.aba7599.

Wang, R. E., S. A. Wu, J. A. Evans, J. B. Tenenbaum, D. C. Parkes, and M. Kleiman-Weiner. 2020. Too many cooks: Bayesian inference for coordinating multi-agent collaboration. *arXiv*. DOI: arXiv:2003.11778.

Wang, L., D. Rehman, P.-F. Sun, A. Deshmukh, L. Zhang, Q. Han, Z. Yang, Z. Wang, H.-D. Park, J. H. Lienhard, and C. Y. Tang. 2021a. Novel positively charged metal-coordinated nanofiltration membrane for lithium recovery. *ACS Applied Materials & Interfaces* 13(14):16906-16915. DOI: 10.1021/acsami.1c02252.

Wang, L. L., M. E. Janes, N. Kumbhojkar, N. Kapate, J. R. Clegg, S. Prakash, M. K. Heavey, Z. Zhao, A. C. Anselmo, and S. Mitragotri. 2021b. Cell therapies in the clinic. *Bioengineering & Translational Medicine* 6(2). DOI: 10.1002/btm2.10214.

Wang, T., Y. Wu, L. Shi, X. Hu, M. Chen, and L. Wu. 2021c. A structural polymer for highly efficient all-day passive radiative cooling. *Nature Communications* 12(1). DOI: 10.1038/s41467-020-20646-7.

Waring, M. J., J. Arrowsmith, A. R. Leach, P. D. Leeson, S. Mandrell, R. M. Owen, G. Pairaudeau, W. D. Pennie, S. D. Pickett, J. Wang, O. Wallace, and A. Weir. 2015. An analysis of the attrition of drug candidates from four major pharmaceutical companies. *Nature Reviews Drug Discovery* 14(7):475-486. DOI: 10.1038/nrd4609.

Warnock, S. J., R. Sujanani, E. S. Zofchak, S. Zhao, T. J. Dilenschneider, K. G. Hanson, S. Mukherjee, V. Ganesan, B. D. Freeman, M. M. Abu-Omar, and C. M. Bates. 2021. Engineering Li/Na selectivity in 12-Crown-4–functionalized polymer membranes. *Proceedings of the National Academy of Sciences* 118(37). DOI: 10.1073/pnas.2022197118.

Washington State Department of Ecology. 2021. Monitoring Hanford's groundwater and protecting the Columbia River. Retrieved August 17, 2021, from https://ecology.wa.gov/Waste-Toxics/Nuclear-waste/Hanford-cleanup/Protecting-air-water/Groundwater-monitoring.

Wei, J. 1991. Centennial symposium of chemical engineering: Opening remarks. In *Advances in chemical engineering*. Colton, C. K., ed. San Diego, CA: Academic Press.

Weisbrod, A., A. Bjork, D. McLaughlin, T. Federle, K. McDonough, J. Malcolm, and R. Cina. 2016. Framework for evaluating sustainably sourced renewable materials. *Supply Chain Forum: An International Journal* 17(4):259-272. DOI: 10.1080/16258312.2016.1258895.

Wells, E., and A. S. Robinson. 2017. Cellular engineering for therapeutic protein production: Product quality, host modification, and process improvement. *Biotechnology Journal* 12(1). DOI: 10.1002/biot.201600105.

Welp, K., A. Cartolano, D. Parrillo, R. Boehme, R. Machado, and S. Caram. 2006. Monolith catalytic reactor coupled to static mixer. U.S. Patent No. 7109378. Washington, DC: U.S. Patent and Trademark Office.

Welp, K., A. Cartolano, D. Parrillo, R. Boehme, R. Machado, and S. Caram 2009. Monolith catalytic reactor coupled to static mixer. U.S. Patent No. 7595029. Washington, DC: U.S. Patent and Trademark Office.

Westmoreland, P. R., and C. McCabe. 2018. Revisiting the future of chemical engineering. *CEP Magazine*. https://www.aiche.org/resources/publications/cep/2018/october/revisiting-future-chemical-engineering.

Whitehead, T. A., S. Banta, W. E. Bentley, M. J. Betenbaugh, C. Chan, D. S. Clark, C. A. Hoesli, M. C. Jewett, B. Junker, M. Koffas, R. Kshirsagar, A. Lewis, C. T. Li, C. Maranas, E. Terry Papoutsakis, K. L. J. Prather, S. Schaffer, L. Segatori, and I. Wheeldon. 2020. The importance and future of biochemical engineering. *Biotechnology and Bioengineering* 117(8):2305-2318. DOI: 10.1002/bit.27364.

Whiteside, A. 2019. Language-based software's accurate predictions translate to benefits for chemists. *Chemistry World News*. https://www.chemistryworld.com/news/language-based-softwares-accurate-predictions-translate-to-benefits-for-chemists/4010437.article.

WHO (World Health Organization). 2016. *Ambient air pollution: A global assessment of exposure and burden of disease*. Geneva: WHO. https://apps.who.int/iris/rest/bitstreams/1061179/retrieve.

WHO. 2018. Ambient (outdoor) air pollution. Geneva: WHO. https://www.who.int/news-room/fact-sheets/detail/ambient-(outdoor)-air-quality-and-health.

Wilkerson, C. G., S. D. Mansfield, F. Lu, S. Withers, J.-Y. Park, S. D. Karlen, E. Gonzales-Vigil, D. Padmakshan, F. Unda, J. Rencoret, and J. Ralph. 2014. Monolignol ferulate transferase introduces chemically labile linkages into the lignin backbone. *Science* 344(6179):90-93. DOI: 10.1126/science.1250161.

Wilkes, J. 2002. *A century of chemical engineering at the University of Michigan*. Ann Arbor, MI: University of Michigan.

Willson, V. L., T. Monogue, and C. Malave. 1995. First year comparative evaluation of the Texas A&M freshman integrated engineering program. *Proceedings Frontiers in Education 1995 25th Annual Conference. Engineering Education for the 21st Century*. DOI: 10.1109/FIE.1995.483114.

Winter, E., M. C. Clark, and C. Burns. 2021. Team-based learning brings academic rigor, collaboration, and community to online learning. In *Resilient pedagogy: Practical teaching strategies to overcome distance, disruption, and distraction*. Thurston, T. N., K. Lundstrom and C. Gonzalesz, eds. Logan: Utah State University.

Wischmeyer, P. E., D. McDonald, and R. Knight. 2016. Role of the microbiome, probiotics, and "dysbiosis therapy" in critical illness. *Current Opinion in Critical Care* 22(4):347-353. DOI: 10.1097/MCC.0000000000000321.

Wittrup, K. D. 2001. Protein engineering by cell-surface display. *Current Opinion in Biotechnology* 12(4):395-399. DOI: 10.1016/s0958-1669(00)00233-0.

WNA (World Nuclear Association). 2021. Nuclear power in the USA. Retrieved August 17, 2021, from https://www.world-nuclear.org/information-library/country-profiles/countries-t-z/usa-nuclear-power.aspx.

Wolkowicz, K. L., E. M. Aiello, E. Vargas, H. Teymourian, F. Tehrani, J. Wang, J. E. Pinsker, F. J. Doyle, 3rd, M. E. Patti, L. M. Laffel, and E. Dassau. 2021. A review of biomarkers in the context of type 1 diabetes: Biological sensing for enhanced glucose control. *Bioengineering & Translational Medicine* 6(2). DOI: 10.1002/btm2.10201.

World Economic Forum. 2011. *Water security: The water-food-energy-climate nexus*. Washington, DC: Island Press.

Wu, F., J. Maier, and Y. Yu. 2020. Guidelines and trends for next-generation rechargeable lithium and lithium-ion batteries. *Chemical Society Reviews* 49(5):1569-1614. DOI: 10.1039/c7cs00863e.

Wu, L., D. Wang, and J. A. Evans. 2019. Large teams develop and small teams disrupt science and technology. *Nature* 566(7744):378-382. DOI: 10.1038/s41586-019-0941-9.

Xin Yu, J., V. M. Hubbard-Lucey, and J. Tang. 2019. Immuno-oncology drug development goes global. *Nature Reviews Drug Discovery* 18(12):899-900. DOI: 10.1038/d41573-019-00167-9.

Xu, Z., P. Lei, R. Zhai, Z. Wen, and M. Jin. 2019. Recent advances in lignin valorization with bacterial cultures: microorganisms, metabolic pathways, and bio-products. *Biotechnology for Biofuels* 12. DOI: 10.1186/s13068-019-1376-0.

Yan, Q., and J. J. de Pablo. 1999. Hyper-parallel tempering Monte Carlo: Application to the Lennard-Jones fluid and the restricted primitive model. *The Journal of Chemical Physics* 111(21):9509-9516. DOI: 10.1063/1.480282.

Yang, H. C., Y. Xie, J. Hou, A. K. Cheetham, V. Chen, and S. B. Darling. 2018. Janus membranes: Creating asymmetry for energy efficiency. *Advanced Materials* 30(43). DOI: 10.1002/adma.201801495.

Yang, M., N. R. Baral, B. A. Simmons, J. C. Mortimer, P. M. Shih, and C. D. Scown. 2020. Accumulation of high-value bioproducts in planta can improve the economics of advanced biofuels. *Proceedings of the National Academy of Sciences* 117(15):8639-8648. DOI: 10.1073/pnas.2000053117.

Ye, R.-P., J. Ding, W. Gong, M. D. Argyle, Q. Zhong, Y. Wang, C. K. Russell, Z. Xu, A. G. Russell, Q. Li, M. Fan, and Y.-G. Yao. 2019. CO2 hydrogenation to high-value products via heterogeneous catalysis. *Nature Communications* 10(1). DOI: 10.1038/s41467-019-13638-9.

Yates, F. E.; Urquhart, J.1962. Control of plasma concentrations of adrenocortical hormones. *Physiological Reviews* 42, 359–443. DOI: 10.1152/physrev.1962.42.3.359.

Yeh, Y.-C., T.-H. Huang, S.-C. Yang, C.-C. Chen, and J.-Y. Fang. 2020. Nano-based drug delivery or targeting to eradicate bacteria for infection mitigation: A review of recent advances. *Frontiers in Chemistry* 8(286). DOI: 10.3389/fchem.2020.00286.

Yin, Y., Y. Wang, J. A. Evans, and D. Wang. 2019. Quantifying the dynamics of failure across science, startups and security. *Nature* 575(7781):190-194. DOI: 10.1038/s41586-019-1725-y.

Zakzeski, J., P. C. Bruijnincx, A. L. Jongerius, and B. M. Weckhuysen. 2010. The catalytic valorization of lignin for the production of renewable chemicals. *Chemical Reviews* 110(6):3552-3599. DOI: 10.1021/cr900354u.

Zero Waste Scotland. 2013. Plastics to oil products—Final report. Retrieved August 18, 2021, from https://www.zerowastescotland.org.uk/research-evidence/plastic-oil-report.

Zhai, Y., Y. Ma, S. N. David, D. Zhao, R. Lou, G. Tan, R. Yang, and X. Yin. 2017. Scalable-manufactured randomized glass-polymer hybrid metamaterial for daytime radiative cooling. *Science* 355(6329):1062-1066. DOI: 10.1126/science.aai7899.

Zhang, Z., X. Sui, P. Li, G. Xie, X. Y. Kong, K. Xiao, L. Gao, L. Wen, and L. Jiang. 2017. Ultrathin and ion-selective Janus membranes for high-performance osmotic energy conversion. *Journal of the American Chemical Society* 139(26):8905-8914. DOI: 10.1021/jacs.7b02794.

Zhang, M., P. T. Corona, N. Ruocco, D. Alvarez, P. Malo de Molina, S. Mitragotri, and M. E. Helgeson. 2018a. Controlling complex nanoemulsion morphology using asymmetric cosurfactants for the preparation of polymer nanocapsules. *Langmuir* 34(3):978-990. DOI: 10.1021/acs.langmuir.7b02843.

Zhang, X., M. Fevre, G. O. Jones, and R. M. Waymouth. 2018b. Catalysis as an enabling science for sustainable polymers. *Chemical Reviews* 118(2):839-885. DOI: 10.1021/acs.chemrev.7b00329.

Zhang, C., D. Hu, R. Yang, and Z. Liu. 2020a. Effect of sodium alginate on phosphorus recovery by vivianite precipitation. *Journal of Environmental Sciences (China)* 93:164-169. DOI: 10.1016/j.jes.2020.04.007.

Zhang, F., M. Zeng, R. D. Yappert, J. Sun, Y.-H. Lee, A. M. LaPointe, B. Peters, M. M. Abu-Omar, and S. L. Scott. 2020b. Polyethylene upcycling to long-chain alkylaromatics by tandem hydrogenolysis/aromatization. *Science* 370(6515):437-441. DOI: 10.1126/science.abc5441.

Zhao, Y. B., X. D. Lv, and H. G. Ni. 2018. Solvent-based separation and recycling of waste plastics: A review. *Chemosphere* 209:707-720. DOI: 10.1016/j.chemosphere.2018.06.095.

Zhu, X., J. Hao, B. Bao, Y. Zhou, H. Zhang, J. Pang, Z. Jiang, and L. Jiang. 2018. Unique ion rectification in hypersaline environment: A high-performance and sustainable power generator system. *Science Advances* 4(10). DOI: 10.1126/sciadv.aau1665.

Zimmermann, A. W., and R. Schomäcker. 2017. Assessing early-stage CO2 utilization technologies—Comparing apples and oranges? *Energy Technology* 5(6):850-860. DOI: 10.1002/ente.201600805.

Appendix A
List of Acronyms

AA	acrylic acid
ACS	American Chemical Society
AGORA	assembly of gut organisms through reconstruction and analysis
AI	artificial intelligence
AIChE	American Institute of Chemical Engineers
AM	additive manufacturing
API	active pharmaceutical ingredient
API	American Petroleum Institute
ASEE	American Society for Engineering Education
ASSB	all-solid-state batteries
AWEA	American Wind Energy Association
BAU	business as usual
BEV	battery electric vehicle
BF	blast furnace
BIO	Biotechnology Innovation Organization
BIPOC	Black, Indigenous, and People of Color
BLS	Bureau of Labor Statistics
BTX	benzene, toluene, xylene
CAR	chimeric antigen receptor
CCUS	carbon capture, use, and storage (or carbon capture, utilization, and sequestration)
CFC	chlorofluorocarbon
CHO	Chinese hamster ovary
CHP	combined heat and power
CMS	U.S. Centers for Medicare & Medicaid Services
CNN	convolutional neural network
COBRA	constraints-based reconstruction and analysis
CPU	central processing unit
CRISPR	clustered regularly interspaced short palindromic repeats
DAC	direct air capture
DBER	discipline-based education research
DMF	N,N-dimethylformamide
DNA	deoxyribonucleic acid
DNN	deep neural networks
DOE	U.S. Department of Energy

DRI	direct reduction of iron
ECM	extracellular matrix
ED	electrodialysis
EG	ethylene glycol
EMF	Energy Modeling Forum
EOR	enhanced oil recovery
EPA	U.S. Environmental Protection Agency
EV	electric vehicle
Fab	fragment antigen-binding
FAO	Food and Agriculture Organization of the United Nations
Fc	fragment crystallizable
FCC	fluid catalytic cracking
FCEV	fuel cell electric vehicle
FDA	U.S. Food and Drug Administration
FDCA	furan-2,5-dicarboxylic acid
FMT	fecal matter transplant
FORTRAN	Formula Translation
FOSSI	Future of STEM Scholars Initiative
GAN	generative adversarial networks
GDP	gross domestic product
GE	General Electric
GEM	genome-scale metabolic model
GGE	gallon gasoline equivalent
GHG	greenhouse gas
GI	gastrointestinal
GM	genetically modified
GPU	graphics processing unit
HCFC	hydrochlorofluorocarbon
HDA	high-density amorphous
HIV	human immunodeficiency viruses
HMP	Human Microbiome Project
HT	high temperature
IAM	integrated assessment management
IBD	inflammatory bowel disease
IC	integrated circuit
ICE	internal combustion engine
IDM	independent device manufacturer
IEA	International Energy Agency
ILI	Institute for Learning and Innovation

IoT	Internet of Things
IR	infrared
ISO	International Standards Organization
IUPAC	International Union of Pure and Applied Chemistry
LAB	linear alkyl benzene
LCA	life-cycle assessment
LDA	low-density amorphous
LNG	liquefied natural gas
LPG	liquefied propane gas
LPN	liquid nanoparticle
mAb	monoclonal antibody
MD	molecular dynamics
MEA	membrane electrode assembly
MEG	monoethylene glycol
MF	microfiltration
MIT	Massachusetts Institute of Technology
ML	machine learning
MPI	multiple principal investigators
mRNA	messenger ribonucleic acid
MS	mass spectrometry
MSW	municipal solid waste
MW	molecular weight
NAE	National Academy of Engineering
NASEM	National Academies of Sciences, Engineering, and Medicine
NET	negative emissions technologies
NF	nanofiltration
NG	natural gas
NGL	natural gas liquids
NHL	N-acyl-L-homoserine lactones
NIH	National Institutes of Health
NMR	nuclear magnetic resonance
NN	neural networks
NOAA	U.S. National Oceanic and Atmospheric Administration
NOM	natural organic matter
NRC	National Research Council
NREL	National Renewable Energy Laboratory
NSF	National Science Foundation
OC	organic carbon
OCM	oxidative coupling of methane
ORNL	Oak Ridge National Laboratory

PAH	polycyclic aromatic hydrocarbon
PAM	polyacrylamide
PBAT	polybutyrate adipate terephthalate
PBS	polybutylene succinate
PCB	polychlorinated biphenyl
PCL	polycaprolactone
PCR	polymerase chain reaction
PDRC	passive daytime radiative cooling
PE	polyethylene
PEC	photoelectrochemical cells
PEF	polyethylene furanoate
PEG	polyethylene glycol
PEM	proton exchange membrane
PET	polyethylene terephthalate
PGMEA	propylene glycol methyl ether acetate
PHA	polyhydroxy alkanoate
PI	process intensification
PLA	polylactic acid
PM	particulate matter
PMMA	polymethylmethacrylate
PNAS	Proceedings of the National Academy of Sciences
PP	polypropylene
PS	polystyrene
PSC	perovskite solar cells
PTA	purified terephthalic acid
PTT	polytrimethylene terephthalate
PUR	polyurethane
PV	photovoltaic
PVC	polyvinyl chloride
PVOH	polyvinyl alcohol
QSAR	quantitative structure–property relationships
QSPR	quantitative structure–activity relationships
R&D	research and development
RD&D	research, development, and demonstration
REU	Research Experience for Undergraduates
RNA	ribonucleic acid
RNN	recurrent neural network
RO	reverse osmosis
scFv	single-chain variable fragment
SCM	supplementary cementitious materials

SCR	selective catalytic reduction
SEM	strategic energy management
SFG	sum frequency generation
siRNA	small interfering RNA
SOEC	solid oxide electrolysis cell
SMARTS	SMILES arbitrary target specification
SMILES	simplified molecular-input line-entry system
SPI	single principal investigator
STEM	science, technology, engineering, and math
TEA	technoeconomic assessment
THF	tetrahydrofuran
TRISO	TRi-structural ISOtropic
TRL	technology readiness level
UF	ultrafiltration
UN	United Nations
USD	United States dollar
WEF	water–energy–food
WHO	World Health Organization
WHP	waste heat to power
WtE	waste-to-energy

Appendix B
Journals Used in International Benchmarking

The following list of journals was compiled for the analysis of publications by chemical engineers discussed in Chapter 10. The original list, minus journals that have gone out of print, was taken from Appendix 3B of the National Academies report *International Benchmarking of U.S. Chemical Engineering Research Competitiveness* (NRC, 2007). Journals that began publication or that have risen in prominence since 2007 were added by the committee and are italicized in the list below.

ACS Applied Energy Materials
ACS Catalysis
ACS Central Science
ACS Omega
ACS Sensors
ACS Sustainable Chemistry & Engineering
Acta Materialia
Advanced Drug Delivery Reviews
Advanced Functional Materials
Advanced Materials
Advanced Science
Advanced Structural and Chemical Imaging
Advances in Nano Research
Aerosol Science and Technology
AIChE Journal
Analytical Methods
Angewandte Chemie
Annals of Biomedical Engineering
Annals of Operations Research
Annual Review of Biophysics
Annual Review of Chemical and Biomolecular Engineering
Applied and Environmental Microbiology
Applied Bionics and Biomechanics
Applied Catalysis A: General
Applied Catalysis B: Environmental
Applied Sciences (Switzerland)
Arabian Journal of Chemistry
Atmospheric Chemistry and Physics
Atmospheric Environment
Automatica
Avicenna Journal of Medical Biotechnology
Biocatalysis and Agricultural Biotechnology
Biochip Journal
Bioengineered
Bioengineering
Biofabrication
Biofuel Research Journal
Biofuels, Bioproducts and Biorefining
Bioinformatics
Biomacromolecules
Biomaterials
Biomicrofluidics
Biomimetics
BioNanoScience
Bioprocess and Biosystems Engineering
BioResources
Biotechnology and Bioengineering
Biotechnology and Genetic Engineering Reviews
Biotechnology Progress
Canadian Journal of Chemical Engineering
Carbon
Carbon Letters
Case Studies in Thermal Engineering
Catalysis Science and Technology
Catalysis Today
Catalysis, Structure and Reactivity
Catalysts
Chem
ChemCatChem
ChemElectroChem
Chemical Engineering and Processing - Process Intensification
Chemical Engineering and Technology
Chemical Engineering Research and Design
Chemical Engineering Science
Chemie-Ingenieur-Technik
Chemistry of Materials
Chemosphere
ChemSusChem
Cogent Engineering
Colloids and Interface Science Communications
Colloids and Surfaces A: Physicochemical and Engineering Aspects
Colloids and Surfaces B: Biointerfaces
Combustion and Flame
Combustion Science and Technology

Composite Structures
Composites Science and Technology
Computational Optimization and Applications
Computational Particle Mechanics
Computers and Chemical Engineering
Cosmetics
Crystals
Current Opinion in Biomedical Engineering
Current Opinion in Green and Sustainable Chemistry
CYTA - Journal of Food
Dose-Response
Ecological Economics
Egyptian Journal of Petroleum
Electrochimica Acta
Energy & Fuels
Engineering
Engineering in Agriculture, Environment and Food
Engineering Science and Technology, an International Journal
Environmental Progress and Sustainable Energy
Environmental Science & Technology
Environmental Toxicology and Chemistry
Enzyme and Microbial Technology
European Journal of Pharmaceutical Sciences
Express Polymer Letters
Extreme Mechanics Letters
Fluid Phase Equilibria
Food and Bioprocess Technology
Food Structure
Frontiers in Bioengineering and Biotechnology
Frontiers of Chemical Science and Engineering
Fuel
Fuel Processing Technology
Granular Matter
Green Processing and Synthesis
Ground Water
Indonesian Journal of Science and Technology
Industrial & Engineering Chemistry Research
INFORMS Journal on Computing
Inorganic Chemistry
Inorganic Materials
Interface Focus
International Journal of Air-Conditioning and Refrigeration
International Journal of Chemical Kinetics
International Journal of Corrosion
International Journal of Industrial Chemistry
International Journal of Multiphase Flow
International Review of Aerospace Engineering
International Review on Modelling and Simulations
Iranian Journal of Catalysis
Johnson Matthey Technology Review
Journal of Aerosol Science
Journal of Agricultural Engineering
Journal of Analytical Methods in Chemistry
Journal of Applied Biomaterials and Functional Materials
Journal of Applied Electrochemistry
Journal of Biomaterials Science, Polymer Edition
Journal of Biomedical Materials Research
Journal of Biomedical Nanotechnology
Journal of Catalysis
Journal of Chemical Physics
Journal of Chemical Thermodynamics
Journal of CO_2 Utilization
Journal of Colloid and Interface Science
Journal of Combustion
Journal of Contaminant Hydrology

Journal of Controlled Release
Journal of Diabetes Science and Technology
Journal of Engineering
Journal of Environmental Chemical Engineering
Journal of Flow Chemistry
Journal of Fluid Mechanics
Journal of Food Measurement and Characterization
Journal of Geophysical Research
Journal of Global Optimization
Journal of Hazardous, Toxic, and Radioactive Waste
Journal of Materials Research
Journal of Materials Science
Journal of Materials Science: Materials in Medicine
Journal of Membrane Science
Journal of Membrane Science and Research
Journal of Nanofluids
Journal of Nanoparticle Research
Journal of Non-Newtonian Fluid Mechanics
Journal of Optimization Theory and Applications
Journal of Orthopaedic Research
Journal of Physical Chemistry B
Journal of Polymer Engineering
Journal of Polymer Science, Part A: Polymer Chemistry
Journal of Polymer Science, Part B: Polymer Physics
Journal of Power Sources
Journal of Process Control
Journal of Rheology
Journal of the Air and Waste Management Association
Journal of the American Ceramic Society
Journal of the American Chemical Society
Journal of the Electrochemical Society
Journal of the European Ceramic Society
Journal of the Taiwan Institute of Chemical Engineers
Journal of Thermal Science and Engineering Applications
Journal of Vacuum Science and Technology B: Nanotechnology and Microelectronics
Journal of Water Process Engineering
Journal of Water Reuse and Desalination
KONA Powder and Particle Journal
Langmuir
Macromolecular Reaction Engineering
Macromolecules
Materials Horizons
Materials Research Bulletin
Materials Today Chemistry
Mathematical Programming, Series B
Membranes
Metabolic Engineering
Microbial Biotechnology
Molecular Catalysis
Molecular Simulation
Molecular Systems Design and Engineering
Nano Futures
Nano Letters
Nanomaterials
Nanotechnology for Environmental Engineering
Nanotechnology Reviews
Nanotechnology, Science and Applications
Nature
Nature Biomedical Engineering
Nature Biotechnology
Nature Catalysis
Nature Chemistry
Nature Materials
Nature Reviews Chemistry
New Biotechnology
Nonlinear Engineering

Optimization and Engineering
Particuology
Pharmaceutical Research
Physical Review Fluids
Physical Review Letters
Physicochemical Problems of Mineral Processing
Polymer
Polymer Chemistry
Polymer Composites
Polymer Engineering and Science
Polymer-Plastics Technology and Materials
Powder Technology
Proceedings of the Combustion Institute
Proceedings of the National Academy of Sciences of the United States of America
Process Biochemistry
Process Safety Progress
Progress in Energy and Combustion Science
Progress in Polymer Science
Propulsion and Power Research
Protein Science
Reaction Chemistry and Engineering
Reaction Kinetics, Mechanisms and Catalysis
Rheologica Acta
RSC Advances
Safety and Health at Work
Science
Science and Technology for the Built Environment
Separation and Purification Reviews
Separation and Purification Technology
Separation Science and Technology
Separations
SIAM Journal of Scientific Computing
SIAM Journal on Optimization
Solar Energy Materials and Solar Cells
Solid State Ionics
South African Journal of Chemical Engineering
SPE Journal
Studies in Surface Science and Catalysis
Surface and Interface Analysis
Surface Innovations
Surface Topography: Metrology and Properties
Synthetic Biology
Tellus, Series B: Chemical and Physical Meteorology
Thermal Science and Engineering Progress
Tissue Engineering
Tissue Engineering - Part A
Tissue Engineering - Part B: Reviews
Tissue Engineering - Part C: Methods
Toxics
Vodohospodarsky Casopis/Journal of Hydrology & Hydromechanics
Water Research
Water Resources Research
Wiley Interdisciplinary Reviews: Nanomedicine and Nanobiotechnology

Appendix C
Summary of Results of the Chemical Engineering Community Questionnaire

The committee originally planned to hold workshops and other meetings in conjunction with the annual meetings of relevant professional societies to solicit input for this study from the broader chemical engineering community. Unfortunately, the COVID-19 pandemic precluded such in-person gatherings. As an alternative, in spring 2021 the committee distributed a questionnaire to members of the chemical engineering community to gather broad input on challenges and opportunities for the discipline, as well as key needs in education and training.

The web link to the online questionnaire was distributed via email to subscribers to the mailing list for the National Academies' Board on Chemical Sciences and Technology (BCST). The questionnaire was open to anyone who wished to provide input. Additionally, members of the committee shared the invitation with their own professional networks. There were 43 complete responses and 249 partial responses. All questions were optional, including basic demographic information, which was collected only to help ensure an appropriate range of perspectives.

This appendix summarizes the input provided. Some responses have been edited or condensed for clarity. Note that the ideas and suggestions summarized here are those of the anonymous respondents to the online questionnaire and do not represent the views of the committee or the National Academies.

1. Across all of chemical engineering, what three fields of chemical engineering will be the most intellectually exciting/promising in the next decade?

- Energy and sustainable/renewable energy – 15
- Process automation/control /design/safety/data analytics – 8
- Environmental sustainability/engineering – 7
- Pharmaceuticals, therapeutics, personalized medicine, health care – 7
- Data science/AI – 6
- Biomedical engineering applications – 4
- Decarbonization (energy, transportation, etc.) – 4
- Electrochemistry, energy storage, batteries – 4
- Materials (engineering/processing/computational design) – 4
- Agriculture (artificial photosynthesis, safe food) – 3
- Bioprocess and process engineering – 3
- Biotechnology – 3
- Catalysis – 3
- Food process engineering – 3
- Green chemistry – 3
- Hydrogen (synthesis, use, storage) – 3

- Bio/Biochemical engineering – 2
- Biomolecular engineering – 2
- Environmentally friendly materials (i.e., biodegradable end products) – 2
- Modeling/Simulations – 2
- Multiscale modeling in biological systems – 2
- Nanotechnology – 2
- Plastics alternatives – 2
- Carbon, capture, utilization, storage – 1
- Chemical technology – 1
- Cosmetics – 1
- Educational technology, training – 1
- Entropy vs enthalpy-controlled systems – 1
- Global engineering – 1
- Green materials – 1
- Manufacturing of complex, non-Newtonian fluids - 1
- Membrane science – 1
- Reaction engineering - 1
- Recycling plastics and critical materials – 1
- Research and development – 1
- Water resources management – 1

2. Thinking about the next 10 to 30 years, what three fields of chemical engineering will have the greatest impact on emerging technologies, national needs, and/or the wider science and engineering enterprise?

- Energy/Renewable energy (its availability & processes) – 11
- Green/environmental engineering & sustainability – 8
- Catalysis – 6
- Electrochemistry/Batteries/ Electronic materials – 5
- Materials science/processing/discovery/ engineering – 5
- Biotechnology/Biopharmaceuticals – 4
- Circular economy/LCA (recycling plastics/products) – 4
- Decarbonization – 4
- Pharmaceuticals (i.e., additive manufacturing, biopharma) – 4
- Biomolecular/Bioengineering – 3
- Food security/waste – 3
- Process control/data analytics/ development and scale-up – 3
- Transport phenomena – 3
- Biomedical engineering – 2
- Clean water resources – 2
- Computing & AI (in all facets) – 2
- Health care/testing for public health – 2
- Rheology studies – 2
- Automation – 1
- Biochemistry/Biochemical engineering – 1
- Bio-defense – 1
- Bioprocess engineering – 1
- Carbon capture – 1
- Chemical technology – 1
- Cost-effective biodegradable end products – 1
- Food waste/loss – 1
- Hydrogen economy – 1
- Manufacturing – 1
- Multiscale simulation – 1
- Nanotechnology – 1
- Novel sensing – 1
- Nuclear – 1
- Plastics alternatives – 1
- Process energy optimization – 1
- Research and development – 1
- Separations – 1
- Systems integration – 1

3. Within your personal field of research or professional focus, what are the major goals for the next 10 to 30 years, and what are the major barriers to getting there?

Goal(s)	Barrier(s)
Decarbonization & Reducing GHG Emissions	
Decarbonization of global economy by 2050	Lack of political consensus—because of this, other countries will develop winning technologies at scale
Massive reductions in GHG emissions with politically acceptable impacts of standard of living	Adoption of alternative technology
In the last decades, chemical processes were based on petrochemistry. Not only fuels, but also most of the chemical commodities. To replace this fossil carbon by renewable carbon (biomass, CO_2) is still a major challenge. Although in the last 15–20 years a big effort was paid to these processes, they are far from being competitive with petrochemical. Similar situation applies to hydrogen energy.	
50% reduction in greenhouse gases by 2030, net zero by 2050	
Low cost and scalable technology for carbon capture and storage, point source and direct air; Novel tech for low emission fuels for hard to decarbonize sectors, biofuels and hydrogen; Scalable and affordable negative emission biomass based technologies	
Sustainability, carbon capture, circular economy	
Circular Economy	
We are targeting to engage in circular economy.	The major barrier is the know-how to engage in it.
Sustainability and next-gen manufacturing tech	Lack of public understanding/appreciation for what goes into developing products
Sustainability	Adoption of alternative technology

Sustainability, carbon capture, circular economy Plastics alternatives or recycling	Cost effectiveness, matching performance/purity of recycled and materials
Improve agricultural practices	
Development of energy storage systems to make renewable energy more economically and practically feasible	Fragmented research across many disciplines that don't speak the same language
Engineering of molecules that can interrogate and modulate physiological environments with greater precision and sensitivity than current diagnostics and therapeutics.	Proper mechanistic understanding of the system and the ability to design molecules to achieve the precise goal without off-target effects.
Biopharmaceutical Processes Move biopharmaceutical process development from being largely trial-and-error experimentation to becoming a systematic technology based on mechanistic models, data analytics, and process control Biopharmaceutical manufacturing processes that have the potential to replace the current processes while having major increases in quality, development time, and/or cost	Individuals who control the funding are trial-and-error experimentalists and have a vested interest in maximizing the research funding to their own approach and activities Powerful vested interests want to continue to have the research funding go overwhelmingly into the existing established processes rather than to competing processes. Another barrier is a strong resistance to any new technology
Connecting industry to academia in a robust, respectful, and collaborative manner; connecting the various engineering disciplines to act on common problems that require consensus and convergence thinking	Things that have kept the groups apart over the years—including elitism, skeptical colleagues, and seemingly separate goals.
Robust engineering of cells for therapeutic delivery and tissue repair/regeneration	Understanding/defining principles for engineering cells for robust transgene expression and control
(i) Mathematical modeling of phase changes in flowing soft-matter systems (ii) Mathematical modeling of particle-laden interfaces	Major barriers for both are the proper problem formulation and computational resources
Digitization and automation of industry	
Industrial waste water recycling	Cost effectiveness and disposal options of reject streams

Breaking silos and creating partnerships to address grand challenges	
Lost-cost biodegradable products	
Materials de novo synthesis, analysis, then delivery to systems-level in context of specific applications	Shareholder short-term optimization in industry, see-sawing values for federal research grants
Fundamental understanding of physicochemical properties of catalysts and electrocatalysts relationship with activity and selectivity	Absence of adequate tools for atomic-level images of catalyst structure and composition under working conditions, multiscale simulation of performance of electrochemical systems
Sustaining and expanding manufacturing capability in the United States	Competing with off-shore sites that do not have the same labor and environmental requirements and more government subsidies and indirect government involvement
Innovative discoveries representing future directions of technology	Movement away from curiosity-driven research and tightening of funding with a focus on "deliverables"
Make oil and gas space more attractive to consumers	Make it greener with suitable applications for CO_2 emissions
Development based on green chemistry new processes that account their environmental impact	
Good teaching schools for young chemical engineers	

4. Again, thinking about your personal field of research or professional focus, what is the societal relevance of your work, and what are the barriers to translation and/or scale-up?

Societal relevance	**Barrier**
Decarbonization, GHG Emission Reduction, Reducing Climate Change Decarbonization of the global economy by 2050 Zero-carbon technology to sustain humanity; energy storage Meeting the growing energy needs for the world as economies prosper while mitigating environmental impact including emissions Mitigating the impacts of climate change	Lack of political consensus Huge fossil fuel infrastructure makes it hard to scale renewable energy practically and economically Science-based, technology-neutral policy that incentivizes all relevant technologies to meet the dual challenge; ecosystems that promote partnerships and collaborations; skills and competencies for novel process development and scale-up Suitable regulatory policies
Improve Human Health & Affordable Health Care Human health (antibiotic-resistant infectious disease, cancer, cardiovascular disease, etc.) Well-controlled cell-based therapeutics could reshape how we treat disease Improved biopharmaceutical manufacturing would result in higher quality products at lower cost and shorter time to market – more promising drugs make it to market Cancer drugs, drugs for COVID, COVID vaccine	Physiological complexity Need to understand and develop cheap, simple quality control to ensure quality production of cells that are specific to patients to make it scalable Acquisition of funding. Biopharma doesn't want to spend money to translate a technology that will help competitors, new tech draws resources from existing tech, incentivizing deemphasis on new tech. Academic reviewers also have vested interest in preventing translation of tech of other researchers Time for the heavily regulated industry to accept new manufacturing methods that will meet regulatory scrutiny and general risk aversion; continuous manufacturing is being implemented in biologics drug substance manufacturing

Sustainability/Circular Economy It will create employment and make our planet cleaner (circular economy) Environmental sustainability	Availability of funds Cost effectiveness, performance of the alternatives or recycled options and trade-off with energy
Connecting diverse groups of STEM learners and practitioners to utilize all parts of our communities	
	Political partisanship. Technoeconomic challenges can be solved, politics cannot
Materials processing is necessary to turn materials into useful products	Chemical engineering research has shifted away from process-related questions toward chemical-/molecular-level details. Need fundamental research in materials processing, such as fluid mechanics and heat/mass transport. This needs industrial interaction, should be encouraged through establishment and generous support of programs like NSF GOALI
Catalytic upgrading of bioplatform molecules	Processes are not competitive, but regulations are pulling to phase out use of oil and even natural gas
Food engineering and sustainable technologies—innovated food technologies geared toward maternal and child health for local and global development supporting UNSDGs 2&17	
Reducing water use across industries	Unit ops and processes in water are 100 years or older with little innovation. Need to rethink priorities at national education level to revitalize critical thinking and research in water use and unit operations/processes
Ensure a better future	Changing long established practices and adopting new approaches and technologies
Safe, efficient, environmentally sound petrochemicals production	

Mitigate the human damage to the ecosystems	Free market capitalism and engineering triumphalism
Electrocatalysts will play an increasingly important role for fuel cells and electrolyzers for water-splitting to generate hydrogen and reducing CO_2. Developing means for the efficient capture of CO_2 and its conversion to chemicals holds immense opportunity	
Manufacturing provides more GDP impact and better salaries overall compared to service and other industries. Necessary for national security and not rely on outside sources excessively for strategic materials and capabilities	Need government to actively maintain a level playing field and use tariffs, etc., to enforce it, rather than lowering standards
Economic, social, food, and water systems are going to be disrupted by climate change if goals are not met	Barriers are political
	Isolation of industrial scientist and engineers reduces the connection between academic knowledge and advances from industry knowledge and practice; industrial scientists are not able to participate in meaningful research projects and meetings, creating among other things, echo chamber for academics
Plastics manufacturing, creating low-cost products that will be applied to food preservation and health security	
Oil and gas space is getting a lot of negative publicity, but it provides a cheap and reliable source of energy	More consumer friendly public relation pointers so that society as a whole realizes the importance of oil and gas energy space. Also challenge to mitigate CO_2 emissions and find suitable applications/conversion technologies for CO_2
To improve the processes and making them more friendly toward the environment.	New mentality

5. What are some of the key ideas/principles/drivers that have rotated out over the last 10 years in your field of research or professional focus? In other words, as new capabilities emerge in the field, which areas are making way for the new concepts?

Theme	Individual Responses
AI/Machine learning/Data/ Computational tools	Traditional experimental phase equilibrium and property measurement is nearly nonexistent in academia or industry. New AI and machine learning tools need data for training—where are students going to learn it? Developing biopharmaceutical processes based only on trial-and-error experimentation is being rotated out in the last 10 years in industry, as new capabilities emerge in process data analytics and machine learning, mechanistical modeling, and process control. Academia with rare exceptions have been slow to make way for these approaches which are not new in all of chemical engineering but have become increasingly important in biopharmaceutical process development as practiced in the leading biopharma companies and equipment vendors. The changes are happening in industry, whether trial-and-error experimentalists like it or not. Either academia can lead and contribute to these developments and contribute to society, translation, and scale-up, or they can keep gripping tighter and tighter on the past while harming U.S. competitiveness directly and indirectly by not producing the trained people needed by these changes in the industry. Computational tools are revolutionizing our understanding of gene regulation Manufacturing process, introduction of technology in industry
Circular economy	Material balance, recycling, the chemistry for producing certain materials Reliance on recycling is replacing only use of raw materials The production of commodity fuels from hydrocarbon sources is fading and will become, at best, like paper production today. Sustainability and life-cycle metrics will drive innovation, and transportation fuels from HC fails these tests. Similarly with current plastics production. Circular polymers with minimal waste is the new concept Recycling waste material (lithium batteries, plastic, municipal raw material, agricultural waste material)

Traditional processes/ChemE fundamentals are being lost—Losing depth in the field	Traditional process engineering has suffered in the quest for federal R&D funding at the "leading edge." It has materially harmed our national capability to produce competent engineers at the B.S. level for industries that exist today. None. The fundamentals (applied mathematics, fluid mechanics, heat and mass transport, thermodynamics, and reaction engineering) are all still very relevant. I studied polymers in graduate school and ended up in biotechnology. Our graduate programs are teaching people how to become hyper experts in a field. We live in an interdisciplinary world, and we have to build skillsets in diverse areas. Traditional chemical engineering core competencies must be strengthened, and certainly not abandoned. However, instead of forcing undergrads to take two semesters of organic chemistry, how about encouraging them to take a course in biochemistry and another course in biochemical engineering instead? Teach chemical engineers the science underlying carbon capture and clean energy technologies. The chip shortage is a reminder that there is room for innovation and beefing up the supply chain. The field has shifted away from fundamentals and analysis and more toward empirical "gee-whiz" results. We are losing depth.
Bioengineering is rotating out; technology is taking over	Biochemical engineering, biomedical engineering are two key areas that have rotated out of the way. The field of nanotechnology, green-chemistry, and bio-molecular engineering would be some of them. Some of the more industry-oriented technologies, such as the divided-wall-column, do show green technology potential in the near future.
Other	Benefits and limitations of remote education and remote teamwork. No need for all the irrelevant/redundant chemistry classes that are in the canonical curriculum. Computational/informatic/theoretical approaches are being hybridized with experimental approaches, both high-throughput/midfidelity and low-throughput/high fidelity.

	In my opinion, one of the key problems observed nowadays is the very different speeds of scientific development (very fast) and the aspects related to the scaling up and process engineering (traditionally a core of chemical engineering). We observe the fast development of new and very active catalysts for different processes (OER, HER, hydrogenation, photocatalysts, single-atom catalysts, MOFs), but there are very few attempts of scaling up. UN Sustainable Development Goals Hidden Hunger (micronutrient deficiencies); public–private partnership to address community needs. In cosmetics where I am working currently, trend is toward skin-friendly organic products. Six Sigma and Lean are over-blown concepts that have become a distraction and their own industry. The idea that climate science is questionable is rotating out, as climate change becomes more obvious. Heavy oil conversion process and catalyst development; hydroprocessing and hydroconversion catalyst and materials for conventional fuels; conventional petrochemical process R&D.

6. What are the five most important areas of technical knowledge for chemical engineers to learn during an undergraduate degree?

- Thermodynamics – 16
- Computing, Machine Learning, Statistics, Data Science – 12
- Mass Transfer/Mass and Energy Balances – 9
- Transport Phenomena – 9
- Kinetics – 8
- Reaction Engineering – 8
- Applied Mathematics – 7
- Economics (Manufacturing and Scale-up) – 7
- Process Design/Engineering/ Simulation – 7
- Process Control – 6
- Fluid Mechanics/Dynamics – 5
- Unit Operation Principles – 5
- Biology/Earth Sciences/Geology – 4
- Circular Economy/Sustainability – 4
- Heat Transfer – 4
- Systems Thinking/Engineering (i.e. Connecting engineering – techno-economic-social-cultural-geographical models for process flow) – 4
- Chemistry – 3
- Fact-based Analysis – 3
- Science of next generation clean energy generation and technologies – 3
- Catalysis – 2
- Chemical Reactor Engineering – 2
- Green Chemistry/Technologies – 2
- Heat & Material Balances – 2
- Nanotechnology – 2
- Biochemistry & Biochemical Engineering – 1
- Entrepreneurship – 1
- Food & Nutrition Security – 1
- Humanities – 1
- Leadership – 1
- Manufacturing – 1
- Momentum Transfer – 1
- R&D – 1
- Reach Methodology – 1
- Physics – 1
- Separations – 1
- Synthesis of new materials) – 1
- Vaccine Development – 1

7. What are the five most important areas of technical knowledge for chemical engineers to learn during a graduate degree?

- Numerical Methods/Statistics/ Mathematical Modeling – 12
- Thermodynamics – 7
- Advanced Transport/Transport Phenomena – 6
- Modeling & Simulation – 5
- Circular Economy & Sustainability (Hydrogen Economy, Decarbonization, etc.) – 4
- Reaction Engineering – 4
- Heat & Mass Transfer – 3
- Kinetics – 3
- Process Engineering/Systems/ Design – 3
- Research Methodology & Experimental Design – 3
- Specialized areas relevant to Thesis – 3
- Advanced Reactors/Reactor Design – 2
- AI/Computer Science – 2
- Applied Mathematics – 2
- Biology – 2
- Biomolecular Engineering/Bioengineering – 2
- Bio and Biomedical Applications (Vaccine Development) – 2
- Catalysis – 2
- Data Science/Big Data Analytics – 2
- Economics/Cost Estimation – 2
- Good Writing Skills/Technical Writing – 2
- Green Processes/Technologies – 2
- Nanotechnology – 2
- Renewable Energy – 2
- Self-Awareness, Emotional Intelligence, Conflict Resolution – 2
- Systems Thinking (i.e. connecting engineering-techno-economic-social-cultural-geographical models for process flow) – 2
- Time Management & Prioritization/ Project Management – 2
- Water Resources Management/Treatment – 2
- Active Materials and Low Energy Separations – 1
- Advanced Manufacturing – 1
- Advanced Chemical Synthesis – 1
- Biodegradable Science – 1
- Chemistry – 1
- Communicating difficult concepts to a wide audience – 1
- Convergence Technologies – 1
- Digitalization – 1
- Drug Development – 1
- Environmental Impact Assessment – 1
- Fluid Mechanics – 1
- Food and Nutrition Security – 1
- Heat and Material Balances – 1
- Industry Standards – 1
- Instrumentation, Control, Digital Signals and Control – 1
- Leadership – 1
- Materials Science – 1
- Model Discrimination & Parameter Estimation – 1
- Momentum Transfer – 1
- Operations Research – 1
- Problem Solving – 1

- Process Innovation and Scale-up – 1
- Process Intensification – 1
- Physics – 1
- Rheology (non-Newtonian) – 1
- Thermofluids – 1
- Unit Operations – 1
- Unit Operations in Mars – 1

8. Based on your own experience or those of your recent hires, what skills are missing from chemical engineering undergraduate and/or graduate education (please include skills that are important, but you/your employees have needed to learn outside of a standard chemical engineering curriculum)?

- Computing, Data Science and Analytics, Statistics – 9
- Effective Writing & Communication Skills – 9
- Humanities & Social Sciences (all students benefit from these courses, benefits to interdisciplinary learning) – 3
- Creativity – 2
- Economics – 2
- Holistic Process Development (Need students to be good at lab, experimentation/simulation, and modeling) – 2
- Understanding phenomena and their relationship to complex processes – 2
- Advanced Math (Linear Algebra) – 1
- Analytical Thinking – 1
- Basic General Knowledge of Experimental Work – 1
- Biology from an engineering perspective – 1
- Climate Change, Sustainability & Circularity – 1
- Entrepreneurship – 1
- Fundamental Chemistry – 1
- Green Chemistry – 1
- Hands-on Experience – 1
- Independence – 1
- Industrial Exposure/Industry Standards – 1
- Leadership – 1
- Operation Research Skills – 1
- Overall Familiarity with Instrumentation and Control Systems – 1
- Particle Technology and Solids Handling – 1
- Practical Experience – 1
- Process Control/Process Dynamics/Process Simulation – 1
- Project Flowsheet – 1
- Societal Impacts of Technology – 1
- Transport Phenomena – 1

9. In what way(s) might interdisciplinary or emerging topics (for example biology, sustainability, data science, polymers, nanomaterials, etc.) be better integrated into chemical engineering undergraduate and/or graduate education?

- Circular Economy/Sustainability – 8
- Data Science, AI, Robotics – 6
- Incorporate broader array of real-world problems within the ChE core – 5
- Community Engagement and Practical Experience (Project, Internships, Field Trips) – 3
- Interdisciplinary topics are useful as a way of synthesizing knowledge from core classes, but should not be a substitute for acquiring fundamental numerical skills that are only learned in college – 3
- Polymer Chemistry – 3
- Biology/Biological Engineering – 2
- Collaborating with and Exposure to Industry – 2
- Integrate emerging topics into core curriculum rather than creating specialized ones – 2
- Nanotechnology – 2
- New Requirement: ChEs to take classes outside of the ChE Department – 2
- Applied ChE – 1
- Applied Statistics – 1
- Establish Forums at Universities to Encourage Entrepreneurship Around Innovative Technologies – 1
- Focused seminars/courses (grad) – 1
- Geosciences – 1
- Guest lectures (undergrad) – 1
- Project-based work with other Engineers – 1
- Quantum Computing – 1
- Separate modules (grad) – 1
- Transversal issues can be introduced in classical undergrad modules (undergrad) – 1

10. What are the biggest barriers to increasing the diversity of the chemical engineering workforce?

- Current pool of undergrads is not very diverse; Diversity in STEM needs to start at a younger age (i.e., high school) – 5
- Perception/Marketing issue—positive aspects (financial benefits, sustainability) of ChE not emphasized enough and negative perceptions (too hard or boring, not forward looking enough) of ChE need to be squashed – 5
- Lack of resources to attract and retain high-quality people (including scholarships) – 3
- Negative American political ethos (anti-immigration policy, prejudice, bigotry, discrimination, sexism, homophobia, racism) – 3
- Not enough mentors or role models to seek guidance – 3
- Develop STEM programs that support URM throughout college – 2
- Better involvement of women in STEM – 1
- Declining higher education enrollments – 1
- Increase opportunities for a broader community – 1
- Lack of diversity within faculty/educational institutions are the barrier – 1
- Lack of incentives for encouraging diversity – 1
- Lack of outreach to URM students – 1
- Lack of time to invest in this activity – 1
- Limited job and growth opportunities in conventional ChE fields – 1
- Myth of Meritocracy – 1
- Narrow-minded thinking – 1
- Other fields such as health care, medicine, bioengineering, computer and data science, finance, etc., are deemed to be more exciting than conventional ChE – 1
- URM aren't attracted to ChE in part because of its lack of diversity – 1

11. How does the U.S. remain competitive and at the cutting edge in chemical engineering? And how do we establish and maintain important international collaborations?

- Attract more foreign students to study in the U.S. (and reforming U.S. immigration policy) – 5
- Modify ChE education (i.e., need more innovation in undergrad curriculum (ABET should be radically modified); we need to reconnect academia with the actual *practice* of engineering; need to increase STEM education at a younger age) – 5
- International collaborations should be with countries that are committed to our same values of democracy, intellectual property, and openness – 3
- AIChE growing a global network of student chapters – 2
- ChE not prioritized in the U.S.; funding agencies do not provide incentives for U.S. faculty to work on unresolved ChE problems – 2
- Diversity of opinion and looking at the problems from multiple perspectives are key to innovative solutions – 2
- Identify the wide breadth of opportunities where chemical engineers can be a driving force; expose the community (at all career stages) to these opportunities – 2
- Systems-thinking approach coupled with a quest for innovation; make ChE more interdisciplinary – 2
- To establish and maintain important international collaborations, parties should ensure mutual benefit by protecting intellectual property – 2
- U.S. universities would need to increase its hiring/teaching in chemical engineering topical areas of importance to growing high-tech industries (topical areas include particle technology, biopharmaceutical manufacturing, advanced industrial polymers, energy challenges and lead in discovery, development and deployment of new breakthrough technologies for low-carbon future, etc.) – 2
- By engaging with different entities in our professional societies, research endeavors and transparent interactions in the international/global realm – 1
- Establish ecosystems to further partnership and collaboration with academia, government, and industry to accelerate advancement of new technologies – 1
- Incentivize domestic students to pursue ChE – 1
- International collaborations are less important, build expertise *and* facilities here in the U.S. – 1
- Maintenance of an interactive academia and industry focus; along with support of industrial development into the future – 1
- More global collaboration to share best practices and ideas – 1
- Processing of advanced materials has not received nearly the amount of focus it should relative to its importance; this problem could be addressed by providing more research support for the fundamentals that play a key role in materials processing, such as fluid mechanics and heat/mass transport – 1

- U.S. universities should be encouraged to set up their campuses in other countries, and offer the same enriched engineering curriculum experience and quality of education as it would on their main campuses in the mainland USA – 1

12. Please use this box for any additional input related to the committee's charge or the questions above.

- The focus needs to capture nonresearch elements.
- Chemical engineering should include basic general knowledge like who is the president/prime minister of their country and US, UK, China, etc. Should have their field training during their course of work. Sound grasp on computer tools like Excel, Word, etc. And also chemical engineering or any other course should have the capacity to generate research skills in students.
- Focus on non-hyped, fact-based analysis.
- The hallmark of chemical engineers is their ability to blend in a seamless manner knowledge from the fields of chemistry, biology, physics, and mathematics.
- Chemical engineering input is needed to assess proposed solutions to climate change and other modern challenges, so that solutions that do not make sense can be ruled out early, rather than wasting time and money on them. An example is government funding of research to use coal in green hydrogen processes.
- We focus so much on "identifying a set of ... new chemical engineering areas," but what we miss is the ability to discover truly new areas or solutions to grand challenges. The investments the U.S. is making in science and engineering lag far behind especially Asia (China, Singapore, Korea) and the investments made by industry are even worse. How do we reverse this trend? How do we recommit U.S. enterprises to innovation in the chemical and biological sciences? Perhaps the richest area today for chemical engineers in terms of new science and technology is pharmaceutical manufacturing, especially biologics, and "new" technologies (or finally recognized technologies) like mRNA vaccines.
- I'd like to see AIChE spend more time teaching/mentoring younger students.
- At present, the chemical engineering community is losing its potential chemical engineers of the future to other branches of science, and engineering, for the simple fact that these students who are in the high school age group are not exposed to the wonders of chemical engineering, and they do not have role models to look up to. For instance, news such as self-driving cars, or autonomous drones, really gets students excited; however, the same is not made visible to these students by the chemical engineering field, and industry experts. Right now, quite a number of students identify the field of chemical engineering with global warming challenges and CO_2 emissions, which is not a fair assessment.
- Chemical engineering, especially in the leading academic institutions, has increasingly turned to the microscopic. Process technology has been a technologically mature field, with incremental advances, and mostly an industry with a sedate pace of new construction and retrofits. But handling large volumes of materials cannot remain at the micro scale outside the lab. It requires the competent design, construction, and operation of process operations.

Appendix D
Acknowledgments

The committee would like to acknowledge the intellectual contributions of the following individuals:

Noubar Afeyan, Flagship Pioneering and Moderna Therapeutics
Alina Alexeenko, Purdue University
Kristi Anseth, University of Colorado Boulder
Frances Arnold, California Institute of Technology
Alán Aspuru-Guzik, University of Toronto
Norman Augustine, Lockheed Martin (retired)
David Awschalom, University of Chicago
William Banholzer, University of Wisconsin–Madison
Zhenan Bao, Stanford University
Carlos Barroso, CJB and Associates
Saad Bhamla, Georgia Institute of Technology
Donna Blackmond, Scripps Research
Santanu Chaudhuri, Argonne National Laboratory
Shannon Ciston, Lawrence Berkeley National Laboratory, Molecular Foundry
Ismaila Dabo, The Pennsylvania State University
Cathy Davidson, City University of New York
Pablo Debenedetti, Princeton University
Joseph DeSimone, Stanford University
Francis (Frank) Doyle, Harvard University
Lee Ellen Drechsler, The Procter & Gamble Company
Allessandro Faldi, ExxonMobil Research and Engineering Company
Glenn Frederickson, University of California, Santa Barbara
Benny Freeman, The University of Texas at Austin
Shishir Gadam, Bristol Myers Squibb
Salvador Garcia Muñoz, Carnegie Mellon University
Dario Gil, IBM
Rajamani Gounder, Purdue University
Michael Graetzel, École Polytechnique Fédérale de Lausanne
Ignacio Grossman, Carnegie Mellon University
Supratik Guha, Argonne National Laboratory
Frank Gupton, Virginia Commonwealth University
Eric Hagemeister, The Procter & Gamble Company
Nick Halla, Impossible Foods
William Hammack, University of Illinois Urbana-Champaign
Evelynn Hammonds, Harvard University
Phillip Hustad, The 3M Company
Ah-Hyung (Alissa) Park, Columbia University
Robert Johnson, ExxonMobil Research and Engineering Company
Christopher Jones, Georgia Institute of Technology

Cherie Kagan, University of Pennsylvania
Lynn Katz, The University of Texas at Austin
Jay Keasling, University of California, Berkeley
Ermias Kebreab, University of California, Davis
Konstantin Konstantinov, Codiak Biosciences
Christine Lambert, Ford Research & Advanced Engineering
Dan Lambert, Savannah River National Laboratory
Michael Lawson, National Renewable Energy Laboratory
John Layman, The Procter & Gamble Company
Kelvin Lee, The National Institute for Innovation in Manufacturing Biopharmaceuticals
Thomas Lograsso, Ames Laboratory, Critical Materials Institute
Lee Lynd, Dartmouth College
Hang Lu, Georgia Institute of Technology
Julius Lucks, Northwestern University
Gargi Maheshwari, Bristol Myers Squibb
Benjamin Maurer, National Renewable Energy Laboratory
Paul McKenzie, CSL Behring
Faye McNeill, Columbia University
Mark Meili, The Procter & Gamble Company
Carl Mesters, Shell (retired)
Eric Miller, U.S. Department of Energy, Office of Energy Efficiency and Renewable Energy
Ahmad Moini, BASF Corporation
Raul Miranda, U.S. Department of Energy, Office of Science
Lynn Orr, Stanford University
Jim Pfaendtner, University of Washington
Bryan Pivovar, National Renewable Energy Laboratory
Katie Randolph, U.S. Department of Energy, Office of Energy Efficiency and Renewable Energy
Jeffrey Reimer, University of California, Berkeley
Gintaris (Rex) Reklaitis, Purdue University
William Ristenpart, University of California, Davis
James Rogers, Apeel Sciences
Don Roe, The Procter & Gamble Company
Kirsten Sinclair Rosselot, Process Profiles
Tony Ryan, University of Sheffield
Aaron Sarafinas, Sarafinas Process & Mixing Consulting, LLC
David Sedlak, University of California, Berkeley
Jeffrey Selingo, Arizona State University
Avi Shultz, U.S. Department of Energy, Office of Energy Efficiency and Renewable Energy
Justin G. Sink, ExxonMobil Research and Engineering Company
Mark Sivik, The Procter & Gamble Company
Henry Snaith, University of Oxford
Scott Stanley, The Procter & Gamble Company
George Stephanopoulos, Arizona State University and Massachusetts Institute of Technology
Vijay Swarup, ExxonMobil Research and Engineering Company
Kazuhiro Takanabe, University of Tokyo
Robert Thresher, National Renewable Energy Laboratory

Jean Tom, Bristol Myers Squibb
Annabelle Watts, The 3M Company
Phillip Westmoreland, North Carolina State University
Dane Wittrup, Massachusetts Institute of Technology
Omar Yaghi, University of California, Berkeley
Yushan Yan, University of Delaware
Aleksey Yezerets, Cummins Inc.
Joe Zasadzinski, University of Minnesota
Stacey Zones, Chevron Energy Technology Company

The committee would like to acknowledge the financial contributions of the following organizations:

The American Chemical Society
The American Institute of Chemical Engineers
Colorado School of Mines
Georgia Institute of Technology
The Johns Hopkins University
Louisiana State University
Massachusetts Institute of Technology
North Carolina State University
Northwestern University
The Pennsylvania State University
Princeton University
Purdue University
Rice University
Texas A&M University
University at Buffalo
University of Arkansas
University of California, Berkeley
University of California, Davis
University of California, Los Angeles
University of California, Merced
University of Delaware
University of Florida
University of Houston
University of Maryland, Baltimore County
University of Michigan
University of Minnesota
University of Notre Dame
The University of Texas at Austin
University of Virginia
University of Wisconsin
West Virginia University
The American Chemistry Council
Arkema
Bristol Myers Squibb Company
The Dow Chemical Company
DuPont de Nemours, Inc.
Eastman Chemical Company
Evonik Industries
ExxonMobil Corporation
Honeywell International, Inc.
PPG Industries, Inc.
The Procter & Gamble Company
Shell Global

Appendix E
Committee Member and Staff Biographical Sketches

Eric W. Kaler (Chair), NAE, is president of Case Western Reserve University. Previously, he was president emeritus and professor of chemical engineering and materials science at the University of Minnesota and served on the faculty of the Department of Chemical Engineering at the University of Washington. Dr. Kaler also served as associate professor and dean of the College of Engineering at the University of Delaware. Dr. Kaler's research interests are in complex fluids containing surfactants, polymers, proteins, or colloidal particles, either separately or in mixtures. Additionally, he studies statistical mechanics and thermodynamics. Dr. Kaler received one of the first Presidential Young Investigator Awards from the National Science Foundation in 1984. He has received numerous other awards for his research and is a fellow of several scientific societies. Dr. Kaler has authored or coauthored more than 200 peer-reviewed papers and holds 10 U.S. patents. He was elected to the National Academy of Engineering in 2010 and named a fellow of the American Academy of Arts and Sciences in 2014. Dr. Kaler earned a PhD in chemical engineering from the University of Minnesota in 1982.

Monty M. Alger, NAE, is professor of chemical engineering at The Pennsylvania State University. His experience in the chemical and energy industries includes positions as vice president and chief technology officer at Air Products and Chemicals Inc., and as senior vice president of research at Myriant. Dr. Alger spent 23 years at General Electric (GE), where he led technology development at the Global Research Center of GE Plastics and was general manager of technology for the advanced materials business. Prior to GE, he was director of the Massachusetts Institute of Technology (MIT) Chemical Engineering Practice School Station at GE Plastics. Dr. Alger has served on advisory boards for several universities and organizations, including the Shenhua National Institute of Clean and Low-Carbon Energy and PTT Global Chemical (Thailand). He is a fellow and past president of the American Institute of Chemical Engineers. Dr. Alger has an SB and SM in chemical engineering from MIT and a PhD in chemical engineering from the University of Illinois at Urbana-Champaign.

Gilda A. Barabino, NAE, NAM, is president of Olin College of Engineering and professor of biomedical and chemical engineering. She previously served as Daniel and Frances Berg professor and dean at the City College of New York (CCNY) Grove School of Engineering. Prior to joining CCNY, Dr. Barabino was associate chair for graduate studies and professor in the Wallace H. Coulter Department of Biomedical Engineering at the Georgia Institute of Technology and Emory University. At Georgia Tech, she also served as inaugural vice provost for academic diversity. Dr. Barabino is a noted investigator in the areas of sickle cell disease; cellular and tissue engineering; and the role of race, ethnicity, and gender in science and engineering. She is president-elect of the

American Association for the Advancement of Science, the world's largest interdisciplinary scientific society. Dr. Barabino is also an active member of the National Academy of Engineering and the National Academy of Medicine and serves on numerous committees of the National Academies, including the Roundtable on Black Men and Black Women in Science, Engineering, and Medicine; the Health and Medicine Division Committee; and the Committee on Women in Science Engineering and Medicine, which she chairs. She consults nationally and internationally on STEM (science, technology, engineering, and mathematics) education and research, diversity in higher education, policy, and faculty and workforce development. Dr. Barabino received a PhD in chemical engineering from Rice University.

Gregg T. Beckham is a senior research fellow and group leader at the National Renewable Energy Laboratory (NREL), where he leads an interdisciplinary team of biologists, chemists, engineers, and material scientists developing green processes and products from lignocellulosic biomass and waste plastics. He has published more than 200 peer-reviewed articles. Dr. Beckham was awarded the American Chemical Society (ACS) OpenEye Outstanding Junior Faculty Award, the American Institute of Chemical Engineers (AIChE) Computational Science and Engineering Forum Young Investigator Award, an inaugural *ACS Sustainable Chemistry and Engineering* Lectureship, the Society for Industrial Microbiology and Biotechnology (SIMB) Young Investigator Award, the Royal Society of Chemistry Beilby Medal and Prize, the SIMB Charles D. Scott Award, the BioEnvironmental Polymer Society Outstanding Young Scientist Award, and the NREL Innovator of the Year Award. He is also founding cochair of both the Lignin Gordon Research Conference (2018) and the Plastics Upcycling and Recycling Gordon Research Conference (2022). He testified before the U.S. House of Representatives Committee on Science, Space, and Technology's Subcommittee on Research and Technology on "Closing the Loop: Emerging Technologies in Plastics Recycling" and coorganized the National Academies of Sciences, Engineering, and Medicine Symposium on "Closing the Loop on the Plastics Dilemma," both in 2019. At NREL, Dr. Beckham cofounded and leads the U.S. Department of Energy (DOE)–funded BOTTLE Consortium, which develops advanced plastics recycling and redesign strategies. He also was a founding member of the Agile BioFoundry and several other DOE-funded consortia. Before NREL, Dr. Beckham served as director of a Massachusetts Institute of Technology (MIT) Practice School Station in Singapore. He received an MSCEP in 2004 and PhD in chemical engineering in 2007 at MIT.

Dimitris I. Collias is research fellow in the Corporate Research and Development Department of the Procter & Gamble (P&G) Company, leading the development of technologies in the bio- and circular economy space, such as bioacrylic acid, biosurfactants alcohols, recycled polyolefins, and recycled superabsorbent polymers, which are currently in various stages of commercialization. His experience in industrial materials processing, properties, and delivery systems spans 29 years. Dr. Collias has experience in technology development and management; has worked extensively with outside industrial partners, start up companies, and universities; and has seen many of his technical developments commercialized in various P&G products. He coauthored the book *Polymer Processing—Principles and Design*, is coauthoring and coediting a book on *Circular Economy of Polymers—Topics in Recycling Technologies*, has published more than 40 articles in journals and conference proceedings, and holds more than 100 granted U.S. patents and numerous patent applications. Dr. Collias has earned many awards, such as the 2020 American Chemical Society's (ACS's) Affordable Green Chemistry Award, P&G's CTO Pathfinder Awards in 2019 and 2010, P&G's IP Strategy Award in 2018, Los Alamos National Laboratory's Recognition and Commendation in 2009, and P&G's Cost Innovation Award in 2007. He is a member of ACS, the American Institute of Chemical Engineers, the Society of Rheology, the Society of Plastics Engineers, and the American Oil Chemists' Society. Dr. Collias earned a diploma in chemical engineering from the National Technical University of Athens, Greece, and a PhD in chemical engineering from Princeton University.

Juan J. de Pablo, NAE, is Liew Family professor and executive vice president for national laboratories, science strategy, innovation, and global initiatives at the University of Chicago. He is also a senior scientist at Argonne National Laboratory. Dr. de Pablo is a leader in developing models and simulations of molecular and large-scale phenomena, including advanced molecular simulation methods and artificial-intelligence–based algorithms; he also conducts supercomputer simulations to design and find applications for new materials, including protein optimization and aggregation, DNA folding and hybridization, glassy materials, block copolymers, liquid crystals, and active molecular systems. He holds more than 20 patents and has authored or coauthored approximately 650 publications. Dr. de Pablo received the DuPont Medal for Excellence in Nutrition and Health Sciences, the Intel Patterning Science Award, the Charles Stine Award from the American Institute of Chemical Engineers (AIChE), and the Polymer Physics Prize from the American Physical Society (APS). He is also a member of AIChE and has served as chair of the awards selection subcommittee. Dr. de Pablo is founding editor of *Molecular Systems Designing and Engineering* and deputy editor of *Science Advances*. He has served as chair of the Mathematical and Physical Sciences Advisory Committee of the National Science Foundation and the Committee on Condensed Matter and Materials Research at the National Academies of Sciences, Engineering, and Medicine, and he is a fellow of the American Academy of Arts and Sciences and of the APS. Dr. de Pablo was elected as foreign correspondent member of the Mexican Academy of Sciences in 2014, and he was elected into the U.S. National Academy of Engineering in 2016. He earned a PhD in chemical engineering at the University of California, Berkeley.

Sharon C. Glotzer, NAS, NAE, is Anthony C. Lembke Department chair of chemical engineering and John W. Cahn distinguished university professor at the University of Michigan. Her research on computational assembly science and engineering aims toward predictive materials design of colloidal and soft matter. Dr. Glotzer is a fellow of the American Physical Society (APS), the American Association for the Advancement of Science, the American Institute of Chemical Engineers (AIChE), and the Materials Research Society (MRS). She has received numerous awards, including the Nanoscale

Science & Engineering Forum Award, the Alpha Chi Sigma Award, and the Charles M.A. Stine Award from AIChE; the Kavli Lectureship from the MRS and the MRS Medal; the Aneesur Rahman Prize in Computational Physics from the APS; and the Presidential Early Career Award for Scientists and Engineers. Dr. Glotzer has participated in many activities of the National Academies of Sciences, Engineering, and Medicine, and she serves currently on the Division on Engineering and Physical Sciences Committee and served previously on the Board on Chemical Sciences and Technology (2015–2020). She is also a member of the American Academy of Arts and Sciences. Dr. Glotzer earned a PhD in physics from Boston University.

Paula Hammond, NAS, NAE, NAM, is instititute professor and department head of the Chemical Engineering Department at the Massachusetts Institute of Technology (MIT); she is also a member of MIT's Koch Institute for Integrative Cancer Research and the MIT Energy Initiative, and she is a founding member of the MIT Institute for Soldier Nanotechnologies. She previously served as executive officer (associate chair) of the Chemical Engineering Department and is associate editor for the journal *ACS Nano*. Dr. Hammond's research is focused on the self-assembly of polymeric nanomaterials, particularly the use of electrostatics and other complementary interactions to generate functional materials with highly controlled architectures, including the development of new biomaterials and electrochemical energy devices. She served on the Board on Chemical Sciences and Technology of the National Academies of Sciences, Engineering, and Medicine from 2006 to 2009, and was chair of the American Institute of Chemical Engineers (AIChE) Materials Science and Engineering Division. Dr. Hammond has been involved for several years in developing Polymer programming, and in the past has served as faculty advisor for the AIChE MIT student chapter. She received the William Grimes Award and the Distinguished Scientist Award from Harvard University, and she is a fellow of the American Physical Society, the Polymer Division of the American Chemical Society, and the American Institute of Biological and Medical Engineers. Dr. Hammond earned a PhD in chemical engineering from MIT.

Enrique Iglesia, NAE, is Theodore Vermeulen chair in chemical engineering at the University of California, Berkeley, and laboratory fellow at the Pacific Northwest National Laboratory. His research addresses the synthesis and structural and mechanistic characterization of porous inorganic catalysts useful in energy conversion, chemical synthesis, and environmental control. He is a fellow of the American Chemical Society (ACS) and the American Institute of Chemical Engineers (AIChE), and an honorary fellow of the Chinese Chemical Society. Dr. Iglesia received a Senior Scientist Award from the Alexander von Humboldt Foundation and doctors honoris causa degrees from the Universidad Politecnica de Valencia and the Technical University of Munich. He is a member of the American Academy of Arts and Sciences, the National Institute of Inventors, and the Real Academia de Ciencias (Spain). Dr. Iglesia received the Olah, Somorjai, and Murphree awards from the ACS; the Wilhelm, Alpha Chi Sigma, and Walker awards from the AIChE; the Emmett, Burwell, Gault, Boudart, and Distinguished Service awards from the European and North American Catalysis Societies; and the Cross

Canada Lectureship from the Chemical Institute of Canada. He is former editor-in-chief of *The Journal of Catalysis* and has served as president of the North American Catalysis Society and the International Association of Catalysis Societies. His dedication to teaching has been recognized with several campus awards, most notably the Noyce Prize, the most prestigious teaching award in the physical sciences at Berkeley. Dr. Iglesia received a B.S. from Princeton University and a PhD from Stanford University, both in chemical engineering.

Sangtae Kim, NAE, is distinguished professor of chemical engineering at Purdue University, where he was also inaugural Donald W. Feddersen distinguished professor of mechanical engineering. He served previously as executive director, Morgridge Institute for Research; inaugural division director, National Science Foundation Cyberinfrastructure Division; and vice president of research and development in information technology at Eli Lilly and Warner Lambert. Dr. Kim started his career as a faculty member in chemical engineering at the University of Wisconsin, where he developed mathematical and computational methods for microhydrodynamics and coauthored a book on this topic, published in 1991. He is a fellow of the American Institute of Medical and Biological Engineers and the American Institute of Chemical Engineers (AIChE), and a trustee of the AIChE Foundation. Dr. Kim received the 2013 Ho-Am Prize in Engineering, AICHE's George Lappin and Colburn awards, and the 1992 Award for Initiatives in Research from the National Academies of Sciences, Engineering, and Medicine. He received a PhD in chemical engineering from Princeton University.

Samir Mitragotri, NAE, NAM, is Hiller professor of bioengineering and Hansjorg Wyss professor of biologically inspired engineering at Harvard University, John A. Paulson School of Engineering and Applied Sciences. His research is focused on transdermal, oral, and targeted drug delivery systems. Dr. Mitragotri has made groundbreaking contributions to the field of biological barriers and drug delivery, advancing fundamental understanding of biological barriers and enabling the development of new materials and technologies for diagnosis and treatment of various ailments, including diabetes and cardiovascular, skin, and infectious diseases. He is an elected fellow of the National Academy of Inventors, the American Association for the Advancement of Science, the Biomedical Engineering Society, the American Institute for Medical and Biological Engineering, and the American Association of Pharmaceutical Scientists. Dr. Mitragotri is an author of more than 350 publications; an inventor on more than 200 patents or patent applications; and a recipient of the American Institute of Chemical Engineers's Colburn and Acrivos Professional Progress awards, as well as the Society for Biomaterials's Clemson award. He received a PhD in chemical engineering from the Massachusetts Institute of Technology.

Babatunde A. Ogunnaike, NAE, was the William L. Friend Chaired professor of chemical engineering at the University of Delaware, where he began serving after a 13-year research career with DuPont. His research focused on process control, modeling and simulation, systems biology, and applied statistics. Dr. Ogunnaike made notable contributions to the modeling and control of industrial polymer reactors and to understanding biological control systems. He was the author or coauthor of four books, including the widely used textbook *Process Dynamics, Modeling and Control* and *Random Phenomena: Fundamentals of Probability and Statistics for Engineers*. Dr. Ogunnaike received the American Institute of Chemical Engineers' (AIChE's) CAST Computing Practice Award, the University of Delaware's College of Engineering Excellence in Teaching Award, the International Society of Automation's Eckman Award, and the American Automation and Control Council's Control Engineering Practice Award. He was a fellow of the AIChE, the American Association for the Advancement of Science, the International Federation of Automatic Control, and the Nigerian Academy of Engineering. Dr. Ogunnaike received a PhD in chemical engineering from the University of Wisconsin–Madison.

Anne Robinson is trustee professor and department head of chemical engineering at Carnegie Mellon University. She holds several patents and has authored more than 100 publications in the areas of protein (re)folding and aggregation, protein biophysics, and protein expression of therapeutically relevant protein molecules. From 2015 to 2017, Dr. Robinson served on the board of directors of the American Institute of Chemical Engineers (AIChE); she is on the advisory board of *Biotechnology and Bioengineering* and the editorial board of *Biotechnology Journal*, and has been an ad hoc reviewer for many National Institutes of Health and National Science Foundation study sections. She is also a member of the Advisory Committee for Pharmaceutical Sciences of the Food and Drug Administration. Dr. Robinson received a DuPont Young Professor Award and a National Science Foundation Presidential Early Career Award for Science and Engineering, and she is a fellow of the American Institute for Medical and Biological Engineering and of the AIChE. Dr. Robinson received a PhD in chemical engineering from the University of Illinois Urbana-Champaign.

José G. Santiesteban, NAE, is recently retired from ExxonMobil, where he served for more than 30 years in a number of technical leadership and management roles, including, mostly recently, strategy manager for ExxonMobil Research and Engineering Company. In this role, he led a team for developing strategic technology direction, providing research guidance, and ensuring robustness of the research and development portfolio. His expertise is in heterogeneous catalysis, including design, synthesis, physical chemical characterization of novel catalytic materials, and reaction mechanisms and kinetics. Dr. Santiesteban is inventor or coinventor on more than 85 U.S. patents, editor of two special catalysis journals, and coauthor of more than 20 referenced publications. He has led and made significant technical contributions to the discovery, development, and commercialization of more than 20 novel catalyst technologies for the production of high-performing lubricants, clean fuels, and petrochemicals. Dr. Santiesteban was elected a member of The Academy of Medicine, Engineering and Science of Texas in 2018. He received the Society of Hispanic Professional Engineers 2018 Innovator Award and the "Key to the City" of Parral from Chihuahua in 2016; he also received multiple technical and leadership awards within ExxonMobil Research and Engineering Company and Mobil Research and Development Company. Dr. Sanitesteban is a board member of the Board on Energy and Environmental Systems of the National Academies of Sciences, Engineering, and Medicine and a senior member of the American Institute of Chemical Engineers and the North American Catalysis Society. He has served on the advisory board of various academic and research institutions around the world. Dr. Santiesteban received a PhD in physical chemistry from Lehigh University.

Rachel A. Segalman, NAE, is Schlinger department chair of chemical engineering and Edward Noble Kramer professor of materials at the University of California, Santa Barbara. Previously, she served as acting director at Lawrence Berkeley Laboratories in the Materials Science Division and as professor of chemical engineering at the University of California, Berkeley. Dr. Segalman's research is focused on molecular structure control over soft matter on molecular through nanoscopic length scales to optimize properties for applications ranging from energy to biomaterials. She is particularly interested in materials for energy applications, such as batteries, photovoltaics, fuel cells, and thermoelectrics. Dr. Segalman received the Dillon Medal from the American Physical Society (APS), the Presidential Early Career Award in Science and Engineering, and the Innovation Award from the *Journal of Polymer Science*. She has been elected to the American Academy of Arts and Sciences and as a fellow of APS, and is an elected senior member of the American Institute of Chemical Engineers. Dr. Segalman received a PhD in chemical engineering from the University of California, Santa Barbara.

David Sholl is director of the Transformational Decarbonization Initiative at the Oak Ridge National Laboratory. From 2013 to 2021, he was school chair of chemical and biomolecular engineering at the Georgia Institute of Technology. His research uses computational materials modeling to accelerate development of new materials for energy-related applications, including generation and storage of gaseous and liquid fuels and CO_2 mitigation. Before his appointment at Georgia Tech, Dr. Sholl served on the faculty of Carnegie Mellon University for 10 years. He has published more than 360 papers, which have been cited more than 21,000 times. He has also written a textbook on density functional theory, a quantum-chemistry method applied widely through the physical sciences and engineering. Dr. Sholl served as a member on the National Academies of Sciences, Engineering, and Medicine Committee on a Research Agenda for a New Era in Separations Science. He was senior editor of the American Chemical Society journal *Langmuir* for 10 years, and he was instrumental in the development of the Rapid Advancement in Process Intensification Deployment (RAPID) Institute, a $70 million, U.S. Department of Energy–funded manufacturing institute focused on process intensification run by the American Institute of Chemical Engineers. Dr. Sholl received a PhD in applied mathematics from the University of Colorado Boulder.

Kathleen J. Stebe, NAE, is Goodwin professor in the School of Engineering and Applied Sciences at the University of Pennsylvania. Her research is focused on directed assembly in soft matter and at fluid interfaces. After training at the Levich Institute under the guidance of Charles Maldarelli, she spent a postdoctoral year in Compiegne, France, under the guidance of Dominique Barthes-Biesel. Thereafter, Dr. Stebe joined the faculty of the Department of Chemical Engineering at The Johns Hopkins University, and later joined the Department of Chemical and Biomolecular Engineering at the University of Pennsylvania. She has been recognized by the American Academy of Arts and Sciences, by the Johns Hopkins Society of Scholars, and as a fellow of the American Physical Society and of the Radcliffe Institute. Dr. Stebe received a PhD in chemical engineering at the City College of New York.

STAFF

Maggie L. Walser is associate executive director of the Division on Earth and Life Studies and has been with the National Academies of Sciences, Engineering, and Medicine since 2010. She previously served as senior program officer with the Board on Chemical Sciences and Technology and as director of education and capacity building for the Gulf Research Program, where she contributed to strategic planning for the program and oversaw education and training activities and fellowship programs that support early-career scientists. From 2010 to 2014, Dr. Walser was a program officer with the Board on Atmospheric Sciences and Climate and worked on such topics as climate science, weather research and policy, climate change and water security, and Arctic research priorities. Before joining the staff of the National Academies, she was congressional science fellow with the American Geophysical Union (AGU)/American Association for the Advancement of Science, and worked on water and energy policy and legislation with the U.S. Senate Committee on Energy and Natural Resources. Dr. Walser is past president of the AGU Science and Society Section. She holds bachelor's degrees in chemistry and chemical engineering and a PhD in chemistry from the University of California, Irvine.

Brittany P. Bishop was a Christine Mirzayan science and technology policy fellow in 2020 with the Board on Chemical Sciences and Technology of the National Academies of Sciences, Engineering, and Medicine. Her research interests include clean energy and renewable technologies. Dr. Bishop earned a B.S.E. in chemical engineering from Case Western Reserve University, where she researched continuous flow nanocrystal synthesis techniques, and a PhD in chemical engineering and nanotechnology and molecular engineering from the University of Washington.

Kesiah Clement was research associate with the Board on Chemical Sciences and Technology of the National Academies of Sciences, Engineering, and Medicine, which she joined as a program assistant in July 2019. She graduated from Georgetown University's School of Foreign Service with a B.S. in science, technology, and international affairs focusing on global environmental health. In 2021, Ms. Clement left the National Academies to pursue a JD at the University of Colorado Boulder Law School.

Liana Vaccari is program officer with the Board on Chemical Sciences and Technology of the National Academies of Sciences, Engineering, and Medicine. Prior to this role, she acted as the Resiliency Working Group lead at the New Jersey Department of Transportation as a science fellow of the Rutgers University Eagleton Institute of Politics, where she coordinated internal efforts to incorporate projected risks from climate change into project prioritization, asset management, and other agency processes. Previously, Dr. Vaccari worked at the Consortium for Ocean Leadership, managing community engagement for the Ocean Observatories Initiative, a federally funded resource for ocean scientists. She was a Mirzayan science and technology policy graduate fellow in 2018 with the Ocean Studies Board at the National Academies, where she contributed to studies examining the use of dispersants in oil spills and coral reef resilience. Dr. Vaccari earned a B.E. in chemical engineering from the Stevens Institute of Technology, a B.S. in chemistry from New York University, and a PhD in chemical engineering from the University of Pennsylvania.

Jessica Wolfman is research assistant for the Board on Chemical Sciences and Technology (BCST). She joined the National Academies of Sciences, Engineering, and Medicine in 2017 as a senior program assistant with both BCST and the Board on Environmental Studies and Toxicology. Ms. Wolfman has worked on a variety of activities at the National Academies, including studies examining the states of chemical separations, the future of chemical engineering, and the chemical economy. She currently leads a webinar series for the Chemical Sciences Roundtable, a standing body within the BCST. Ms. Wolfman graduated Phi Beta Kappa from Dickinson College with a B.S. in earth sciences and a minor in mathematics.

Elise Zaidi is communications and media associate for the Division on Earth and Life Studies of the National Academies of Sciences, Engineering, and Medicine. Her primary responsibilities include promoting report releases, creating derivative products for National Academies projects and publications, and formulating targeted outreach campaigns for committee and study activities. Prior to starting her work with the National Academies in July 2019, Ms. Zaidi held positions with the Council on Foreign Relations and the Pan American Health Organization. She graduated from The George Washington University with a B.A. in international affairs with a concentration in global public health and a minor in journalism and mass communication.